Springer Tracts in Mechanical Engineering

Springer Tracts in Mechanical Engineering (STME) publishes the latest developments in Mechanical Engineering - quickly, informally and with high quality. The intent is to cover all the main branches of mechanical engineering, both theoretical and applied, including:

- Engineering Design
- Machinery and Machine Elements
- Mechanical Structures and Stress Analysis
- Automotive Engineering
- Engine Technology
- Aerospace Technology and Astronautics
- Nanotechnology and Microengineering
- Control, Robotics, Mechatronics
- MEMS
- Theoretical and Applied Mechanics
- Dynamical Systems, Control
- Fluids Mechanics
- Engineering Thermodynamics, Heat and Mass Transfer
- Manufacturing
- Precision Engineering, Instrumentation, Measurement
- Materials Engineering
- Tribology and Surface Technology

Within the scope of the series are monographs, professional books or graduate textbooks, edited volumes as well as outstanding PhD theses and books purposely devoted to support education in mechanical engineering at graduate and post-graduate levels.

Indexed by SCOPUS, zbMATH, SCImago.

Please check our Lecture Notes in Mechanical Engineering at https://link.springer.com/bookseries/11236 if you are interested in conference proceedings.

To submit a proposal or for further inquiries, please contact the Springer Editor **in your region**:

Ms. Ella Zhang (China)
Email: ella.zhang@springernature.com
Swati Meherishi (India)
Email: swati.meherishi@springer.com
Dr. Leontina Di Cecco (All other countries)
Email: leontina.dicecco@springer.com

All books published in the series are submitted for consideration in Web of Science.

Ramesh Narayanaswamy · Malay Kumar Das ·
K. Muralidhar
Editors

Sustainable Water Harvesting

Principles and Technologies

Springer

Editors
Ramesh Narayanaswamy
School of Civil and Mechanical Engineering
Curtin University
Perth, WA, Australia

Malay Kumar Das
Department of Mechanical Engineering
Indian Institute of Technology Kanpur
Kanpur, Uttar Pradesh, India

K. Muralidhar
Department of Mechanical Engineering
Indian Institute of Technology Kanpur
Kanpur, Uttar Pradesh, India

ISSN 2195-9862 ISSN 2195-9870 (electronic)
Springer Tracts in Mechanical Engineering
ISBN 978-981-95-3372-5 ISBN 978-981-95-3373-2 (eBook)
https://doi.org/10.1007/978-981-95-3373-2

This Springer imprint is published by the registered company Springer Nature Singapore Pte Ltd.
The registered company address is: 152 Beach Road, #21-01/04 Gateway East, Singapore 189721, Singapore

Preface

Water is essential to sustain human life, plants and the animal kingdom. As for human life, water is increasingly becoming a precious commodity for a variety of reasons. The scarcity of fresh water currently affects human life much more than anyone could ever have imagined. A report by the United Nations indicates that water scarcity currently affects over 40% of the human population in the entire world, with an estimated 783 million people having no access to clean water. Alarming results of a study published in leading journals indicate over 4 billion people live with severe water scarcity for at least one month every year; some of the severely affected regions of the world include parts of Africa, Middle east, India and Australia.

Technologies for obtaining water from conventional water bodies such as lakes, rivers and the oceans exist in the modern world and are often supplemented by processes such as desalination and the use of recycled water. In spite of the availability of these technologies, due to various limitations, water distribution systems in the world are incapable of providing fresh water to satisfy the needs of the rising human population. The changing global rainfall patterns, excessive drought, climate changes, man-made wastages of water due to poor planning and implementation schemes woefully contribute toward water scarcity. Water as a commodity is thus increasingly becoming a strategic resource, and its scarcity motivates the need to identify and develop technologies for alternate sources of water to feed and sustain life on this planet. Against this background, the droplet condensation of atmospheric water vapor and trapping of water agglomerates, known as fog-harvesting, is significant and provide a source for generating water. Companion technologies that use solar energy to increase surface temperature leading to pumping of moisture into air are equally significant.

The earth's atmosphere is the storehouse of about 13,000 cubic km of water volume present in the form of water vapor and droplets and is a good resource to tap into, for creating fresh water. The harvesting of atmospheric water vapor is an inherent life mechanism that exists in certain varieties of plants and insects, having bestowed with unique characteristics in their structural anatomy configured by nature, to generate water by making use of the atmospheric water vapor. What bio-inspired examples have led scientists to the emergence of fog capturing screens and engineered surfaces for droplet condensation in a variety of applications such as steam condenser design, cooling of high heat flux electronic components and energy-efficient building designs.

Though the potential for obtaining water by fog capture and from the condensation capture of vapor in the atmosphere using engineered surfaces has been shown in the literature, the fundamental scientific understanding of the multi-phase phenomena leading to the development of technologically capable systems is embryonic and largely unknown. Accordingly, this manuscript addresses the subjects of solar-driven evaporation, moisture condensation and fog harvesting from the atmosphere to generate water. The perspective adopted includes a historical overview, the present-day water crisis, fundamentals of phase change processes, laboratory-scale experiments, technology demonstrators, commercialization, and environmental and policy measures. The intended audience for the book includes engineers involved in building alternative technologies for water purification and researchers who wish to analyze and improve existing ones, while conceiving newer configurations. It is hoped that the approach followed herein will attract a broad audience working toward an understanding of sustainable water harvesting principles and technologies to create a sustainable lifestyle.

This monograph is dedicated to late Prof. Sameer Khandekar, Professor of Mechanical Engineering, Indian Institute of Technology Kanpur, India. His inspiring ideas and commitment to this monograph at the embryonic stage have stood with us despite his absence, enabling the present authors to complete this work.

The editors gratefully acknowledge financial support from the Ministry of Education, Government of India, under the SPARC scheme (Ref. SPARC/2019-2020/P2379/SL dated 28.07.2023).

Perth, Australia — Ramesh Narayanaswamy
Kanpur, India — Malay Kumar Das
Kanpur, India — K. Muralidhar

Contents

Editors and Contributors

About the Editors

Ramesh Narayanaswamy is currently a Professor of Mechanical Engineering in the School of Civil and Mechanical Engineering at Curtin University, Australia. His teaching and research in fluid mechanics and heat transfer include natural convection and radiation, jet impingement, droplet evaporation, and biofluid mechanics. He has expertise in both computational and experimental methodologies of research, including optical methods in fluid flow and heat transfer. He has supervised several Ph.D. students supported through research grants and scholarships. His current research focus on sustainable water generation methods supports the UN Sustainable Development Goals, while building research partnerships to support life on this planet.

Malay Kumar Das began his academic career in 2008 and is currently a Professor in the Department of Mechanical Engineering at the Indian Institute of Technology (IIT) Kanpur. He has undertaken multiple research projects in the fields of gas hydrate, transport via porous media, and electrochemical energy conversion. He has also mentored a number of graduate students. He co-authored a textbook on thermal system design and edited a monograph on transport in porous media. His current research includes data-driven thermofluid dynamics, marine hydrates, and carbon dioxide sequestration.

K. Muralidhar is Professor Emeritus in the Department of Mechanical Engineering at the Indian Institute of Technology (IIT) Kanpur, where he started his career in 1987. He has guided quite a few doctoral students and has completed funded research projects in the areas of wake dynamics, transport in porous media, crystal growth, biomedical flows, and clean water technologies. He has coordinated a national initiative on solar hydrogen generation for DST, a multi-institutional project on CFD code development on unstructured grids for DAE, and an initiative on futuristic mechanics with IGCAR Kalpakkam. With multiple monographs on transport in porous media, droplet dynamics and dropwise condensation, and schlieren and shadowgraph imaging, he also serves as the editor-in-chief of the Journal of Flow Visualization and Image Processing. His present work relates to moisture condensation over mesh-like surfaces and blood rheology. He is also involved in an institute initiative to develop a blood pump for the human heart. He is an elected Fellow of the American Society of Thermal and Fluid Engineers, the Indian National Academy of Engineering, and the National Academy of Sciences, India.

Contributors

Manish Bhendura IIT, Roorkee, India

Debartha Chatterjee Indian Institute of Technology Kanpur, Kanpur, Uttar Pradesh, India

Malay K. Das Department of Mechanical Engineering, Indian Institute of Technology Kanpur, Kanpur, Uttar Pradesh, India

Sohom Goswami IIT Kanpur, Kanpur, India

Tarun Kulshrestha Department of Engineering, School of Computing, Engineering and Mathematical Sciences, La Trobe University, Bendigo, VIC, Australia;
Department of Mechanical Engineering, Indian Institute of Technology Kanpur, Kanpur, Uttar Pradesh, India

K. Muralidhar Department of Mechanical Engineering, Indian Institute of Technology Kanpur, Kanpur, Uttar Pradesh, India

Ramesh Narayanaswamy School of Civil and Mechanical Engineering, Curtin University, Bentley, WA, Australia

Basant Singh Sikarwar Amity University, Noida, India

Introduction

Ramesh Narayanaswamy

1 Background and Significance

Water is a naturally available high-value resource and undoubtedly the source of all life forms on earth. It is an essential and powerful ingredient that is required for the existence of this planet in the solar system, besides helping the functioning and regulation of metabolism of the human body and supporting a myriad of plant and animal life forms. Besides supporting life forms, water is an essential ingredient used in several stages of all industrial manufacturing processes, electrical power generation, irrigation for agricultural produce, animal industry, manufacturing of varieties of food, the textile industry, and several other applications. The list is perhaps endless, and beyond anything else, the presence of water bodies on earth serve to cause seasonal rainfalls and affect climate in several ways supporting the maintenance of a balanced ecosystem while also nourishing flora and fauna. In short, water is indispensable for the survival of our planet earth and for all its forms of life. Every piece of water body on this plant is closely associated with the origin of human civilisation, and hence very closely linked with traditional and cultural sensitivities of human population and the civil society.

About 70% of the earth contains water in one or other form. Visible forms of water containment in its liquid form on the earth consists of water in the oceans, rivers, lakes, and ponds. Data based on Shiklomanov (1993) indicates that the oceans contain about 97% of available water on the globe, and that is all saline, which means that fresh water availability on the globe is limited to about 3%. Of the 3% of freshwater in the globe, 69% is found in glaciers and ice caps, about 30% as ground water, and about 1% as surface freshwater. Out of this 1% surface freshwater sources, we have ground ice and permafrost accounting for about 70%, lakes consisting of about 21%, the atmosphere containing about 3%, and the rest of the water in several forms on the

R. Narayanaswamy (✉)
Curtin University, Bentley, WA, Australia
e-mail: r.narayanaswamy@curtin.edu.au

R. Narayanaswamy et al. (eds.), *Sustainable Water Harvesting*, Springer Tracts in Mechanical Engineering, https://doi.org/10.1007/978-981-95-3373-2_1

surface (soil, wetlands, rivers, and lakes) accounting for about 6%. We must realise that this data is very significant and, subject to environmental and climate changes, water use; that this data is based on estimates available in 1993, according to the cited reference. The most recent ones would be the more accurate; nonetheless, the above data provides us with an overall estimate of the percentages of water found in different forms on the earth, and in that sense, the data is very important and useful.

The availability of global water, of course varies with seasonal changes and comes with a price—a fact echoed by the ever-present struggles for daily water among human populations worldwide, in the midst of repercussions arising from conflict issues due to geographic boundaries of countries not always coinciding with the availability or sharing of the boundaries of the source of water. This often has led to protracted relationship struggles between countries propelling the escalated status of water as a strategic commodity.

From a deeper point of view, water is very closely associated with the culture and traditions of any community and venerated by the custodians or traditional owners of the land in every part of the globe since the dawn of human civilisation. Historical evidence points to the fact that nearly all civilisations owed its existence to the source of water, as has been the case with the origin of civilisations such as the Indus Valley Civilisation (Indus River, near the Himalayas), the Egyptian Civilisation (Nile River), and the Mayan Civilisation (Usumacinta River Valley, near Mexico). The importance of the relationship between water and traditional culture is beautifully conveyed through some examples such as the River Ganges in India (Eck Diana 1983); the Fitzroy River in Australia (Alessandro et al. 2021); water from wells located on the grounds of Nashinoki Shrine in Kyoto, Japan (Niponica 2015).

Indeed, beyond its ability as a natural and safe thirst-quencher, water is a magical scientific liquid that exists in the solid, liquid, and gaseous states for a range of temperatures and pressures suitable for several commercial and industrial applications. From every point of view, water is thus an indispensable element on earth so much so, that the absence of it affects all forms of life on this planet, and threatens the very existence of this plant (Fig. 1).

An assessment of global water demands provides insight into the state of available water worldwide, its use, and its vulnerability. These are introduced below, through an evaluation of the various sources of freshwater and water usage by various sectors.

2 Water Scarcity: Need for Alternatives

Water scarcity is a persistent issue in the semi-arid and arid regions of the world. Water scarcity arises due to several reasons: absence of freshwater from its natural sources; management of water distribution network disrupting supply and water wastage; non-uniform distribution of natural freshwater sources over the globe; inadequate industrial, commercial, residential wastewater reclamation; unsustainable water use by communities; and, population density and excessive water consumption. In addition to this, certain parts of the globe for example, the outback regions of Australia

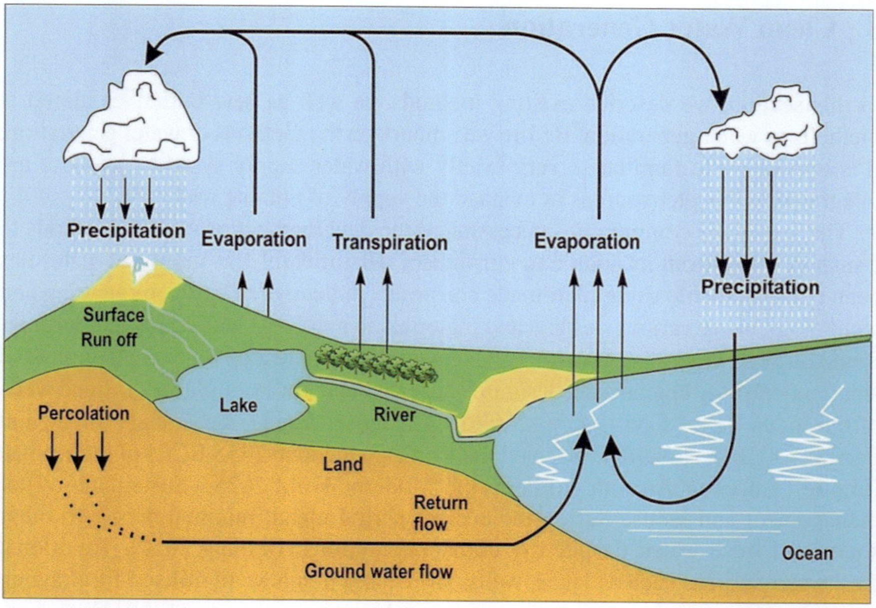

Fig. 1 Hydrological cycle links various sources of water. The cycle is a large-scale distillation process occurring through various natural media. This cycle is under threat owing to climate change, leading to maldistribution in space and time as well as inconvenient interruptions. These interruptions cause water scarcity from a water creation viewpoint

(Australia: State of the Environment 2021) and some freshwater sources in Canada (Derry et al. 2023), have excessive salinisation due to man-made causes and landscape changes. Freshwater sources are also likely to have trace contaminants due to its proximity with mineral and mining sites contributed mainly through leaching into water sources from mineral deposits. In certain parts of the globe, leaching from sewage and septic tanks, as well as drainage of water from abandoned mines (Abandoned Mine Drainage 2024) is often a very major health hazard. In those cases, however, it is not the availability of freshwater, but the quality of freshwater to be used as potable water is a major concern. On an average, half of the population anywhere in the world is currently undergoing water scarcity during some parts of the year, and the lack of freshwater has often contributed to quality of life of the population. Evidence-based research indicates (Mekonnen and Hoekstra 2016) that about two thirds of the global population experience low to high levels of water scarcity each year monthly, and almost all areas of the globe are affected by water stress or scarcity, regardless of whether the countries are developed, developing, or under-developed. Recognition of the importance of water and its availability has triggered the United Nations (https://www.un.org/sustainabledevelopment/) to categorise water as one of its sustainable development goals—United Nations Sustainable Development Goals: 6.

3 Clean Water Generation

In this section, we describe existing methods, as well as new initiatives aimed at methods of water generation. Before we embark on the methods of water generation, it is imperative to familiarise very briefly with water supply systems, as these are integrated with water sources to manage the supply of potable water.

The growth of communities necessitated the development of various methods to transport water from its source to consumers. Historically, the supply of water has been made available using man-made constructs indicative of early engineering and technology using various designs and materials to suit local landscape and purpose. Some of these ingenious constructs have withstood the test of time, challenging the smart modern day engineering designs. These included for example, brick-lined wells of the Indus Valley Civilisation (2500 BCE) (Jansen 2011), the 'qanats' in ancient Persia (700 BCE) (Wulff 1968), and the Roman aqueducts (455 BCE) of the Roman Empire (Traces of Ancient Rome in the Modern World 2025). Sustainable water management and its use during the ancient period meant minimal access to other sources of fresh water, despite dry climate conditions (Deming 2020). Traditional freshwater sources such as lakes, wells, rivers, and ponds were utilised for drawing water for agricultural as well as domestic needs.

Water consumption escalated with the advent of steam engine and the Industrial Revolution, provoking industrial growth as well as population growth, and so did water earn the title of being a strategic commodity over centuries due also to countries being at war, independence of colonies, and from droughts and water stresses due to climate change. Moreover, communities started migrating to inner city suburbs and population density hence increased, and so did the demands on freshwater in these locations. The availability of liveable habitats in and around cities started dwindling, and this triggered the growth of urban landscape to stretch beyond the comfort zones of cities, to regions far away from water supply and existing pipeline distribution networks. Thus, modern day water scarcity has now more or less become synonymous with urban growth, unless alternate methods of water generation is in place to meet the ever-increasing demands for water consumption to feed far out communities. Over centuries, agricultural growth had to be stepped-up to meet the consumption requirements of a growing population, and hence water use by the agricultural sector also increased with population growth, to enhance year-round yields of produce, while battling the vagaries of an unpredictable rhythms of a hydrological cycle, generating rainfall and unforeseen seasonal changes that affected varieties of crops and its production.

4 Sources of Freshwater: Surface Water and Ground Water

Naturally occurring sources of surface freshwater include rivers, lakes, and ponds. These containments get filled due to rainfall and from surface run-off water. Due to heavy dependence on rainfall, seasonal changes due to climate change issues affect rainfall in catchment areas, and hence the availability of freshwater from these sources is not always guaranteed. Some of the great freshwater lakes include Lake Superior in North America, Wular Lake in Jammu and Kashmir in India, and Lake Victoria in Africa.

A closer look at the history of water availability from these freshwater lakes indicate strong seasonal changes affecting the quantity of water for use. Often most countries resort to ground water made available from man-made aquifers to augment the demands of freshwater supplied from lakes and rivers for use by the community. Besides, wells and aquifers depend heavily on the porosity of the soil in which they are constructed.

Modern living requirements often lead to excessive consumption of water both in the residential/commercial and the industrial sector, that creates shortage of water obtainable from traditional sources, and hence recourse is often made to supplement the water distribution system with technologies such as rainwater harvesting, desalination, solar distillation, and water harvesting using atmospheric fog capture. These passive and active additional methods are choices influenced more by economic considerations, need, and suitability of purpose aided by geographical considerations of the region. For example, fog capture has shown tremendous potential in geographical locations such as Morocco, Nepal, Peru, where fog availability is perennial. On the other hand, desalination is the preferred choice in regions such as the United Arab Emirates, Saudi Arabia, Israel, USA, and Australia, due to the easy access to oceans and abundant solar energy to offset some of the energy needs required to run desalination plants. Tropical regions of the globe benefit from rainwater harvesting due to predictable precise monsoon arrivals each year.

5 Global Water: Agricultural and Urban Outlook

The agriculture industry consumes about 70% of the world's freshwater for irrigation (UN Water Development Report 2024). Based on 2020 estimates by the Food and Agricultural Organisation (2022), global agricultural land was estimated to be 4.7 billion hectares, which is approximately one-third of the total land area. Out of this, cropland accounted for one third of global agricultural land (1.6 billion hectares, 12% of global land area), with 22% of cropland equipped for irrigation (350 million hectares). India was identified as the country with the largest area of cropland (169 million hectares), followed by the United States of America (160 million hectares), and China (136 million hectares). China and India had the largest areas for cropland irrigation, both with 70 million hectares, while the United States of America had 27 million hectares.

Global population has been predicted to grow to 9.2 billion people by 2050, with the majority of growth predicted to take place in developing countries. In consonant with this growth, by 2050, agricultural production will require a 50% increase on the 2012 values (Mekonnen and Gerbens-Leenes 2020) with 90% of the required increase in crop production forecast to come as a result of increased yields and increased cropping intensity (FAO 2009), which will obviously consume more water. Production in developing countries will need to double to accommodate the rising population and food demand, and globally, the area of land equipped for irrigation will increase.

Over the next 25 years, urban water demand is expected to increase by 80% as a result of rapid urbanisation (Florke et al. 2018), while climate change is predicted to exacerbate these water challenges, by triggering changes to spatiotemporal dynamics of both temperature and precipitation. The global urban deficit of surface water is predicted to reach 6764 million cubic metres by 2050, which will constrain the available groundwater and lead to competition between urban and agricultural water demands. To cope with the rising water and food demands that come because of population growth and urbanisation, new methods of water generation are required so that all these technologies co-exist without depleting the global water resources. Post 2020, as the world has witnessed climate change to significantly alter global temperatures and rainfall (examples such as higher average summer temperatures (India), flooding (Spain, the UK), and wildfires (California—USA), evaporation rates from natural water bodies are predicted to increase in the not-so-distant future. This evaporative loss is irreparable and can affect residual stock of existing water bodies causing scarcity of water.

6 Global Water: Industrial Outlook

Non-agricultural industrial water consumption relates to the use of water for various industrial applications such as electric power generation, cooling, process water, wastewater treatment, cleaning, steam production, manufacturing, processed and packaged foods and dairy. About 20% of the global freshwater is consumed by the industries. Although efforts by industries for captive restoration of water from wastewater are underway, there appears to be more work to be done in this area of work. The overall water consumption requirements of these plants are very high, for example thermal power plants in India consumed about 88% of freshwater (Benchmarking Industrial Water Use Efficiency in India 2024), while thermal power plants in the USA consumed about 48 trillion gallons of water (https://www.eia.gov/todayinenergy/detail.php?id=56820).

7 Passive and Active Methods for Potable Water Generation

Generating water using scientifically robust methods came into existence as a means to augment the excessive demand from natural freshwater resources. These are classified broadly into active and passive methods. The active methods include desalination, drawing water from man-made aquifers, active devices that capture fog from the atmosphere and force them to condense using thermodynamically sound principles. These are known as active methods as they consume electrical energy for its operation to produce water. In some cases, part of the electrical energy requirement is met from renewable sources such as solar collectors or wind farms. The passive methods, on the other hand do not consume any electrical energy to generate water. Passive fog harvesting using wire meshes (Klemm et al. 2012) is a classic example of this category. Advanced study on selectively coated radiation shield (Haechler et al. 2021) for water harvesting has demonstrated the availability of uninterrupted water to yields close to the scientific limit.

8 Objectives of the Monograph

Climate change is poised to affect the quantity and quality of available surface and ground water resources for freshwater. In turn, this is likely to create competing requirements from residential, agricultural, industrial, and other water uses—cumulatively creating water stress as well as scarcity in several parts of the globe. Water security is characterised by indices such as water availability, demand, and quality, and evidences (Climate Change 2014) already point to the effect of climate change on all of these indices. If the already stressed freshwater ecosystems are not to be disturbed, while we are compelled to meet modern demands on water for various uses, it is imperative that sustainable water harvesting methods must be conceptualised, evolved, and implemented very well into our way of life, to offset the dire effects of competing water use demands and climate change on water availability on this planet. This monograph is intended to offer an overview of existing as well as potentially viable developmental systems for sustainable water harvesting. Accordingly, this text addresses passive methods of water harvesting that includes solar-driven techniques using adsorption, evaporation, moisture condensation, and fog harvesting from the atmosphere to generate water. Applicable theories and research findings are used as supporting evidence to provide credence and viability of scientifically proven alternate, non-traditional water generation methods.

References

Abandoned Mine Drainage (2024) United State environmental protection agency. https://www.epa.gov/nps/abandoned-mine-drainage. Accessed 22 April 2025

Alessandro P, Erin O, Ann P (2021) Pursuit: Australia's rivers are ancestral beings. The University of Melbourne

Australia: State of the Environment (2021) Commonwealth of Australia report. https://soe.dcceew.gov.au/inland-water/. Accessed 22 April 2025

Benchmarking Industrial Water Use Efficiency in India (2024) Policy brief. The Energy Resources Institute

Climate Change (2014) Impacts, adaptation, and vulnerability. Part A: global and sectoral aspects. Contribution of Working Group II to the Fifth Assessment Report of the Intergovernmental Panel on Climate Change

Deming D (2020) The aqueducts and water supply of ancient Rome. Groundwater 58(1):152–161

Derry A, Cañedo-Argüelles M, Melles SJ (2023) Freshwater ecosystems are becoming increasingly salty. Here's why this is a concern. The Conversation. Accessed 22 April 2025

Eck Diana L (1983) Banaras. Penguin Books, New York

Flörke M, Schneider C, McDonald RI (2018) Water competition between cities and agriculture driven by climate change and urban growth. Nat Sustain 1(1):51–58. https://doi.org/10.1038/s41893-017-0006-8

Food and Agricultural Organization (2009) Global agriculture towards 2050. In: How to feed the world in 2050. Food and Agriculture Organisation of the United Nations, Rome

Food and Agricultural Organization (2022) Land use statistics and indicators. In: Global, regional, and country trends 2000–2020 FAOSTAT analytical briefs, p 15. https://doi.org/10.4060/cc0963en. Accessed 22 April 2025

Haechler I, Park H, Schnoering G, Gulich T, Rohner M, Tripathy A, Milionis A, Schutzius TM, Poulikakos D (2021) Exploiting radiative cooling for uninterrupted 24-h water harvesting from the atmosphere. Sci Adv 7:eabf397823

https://www.un.org/sustainabledevelopment/. Accessed 22 April 2025

Jansen M (2011) Mohenjo-Daro, Indus Valley civilization: water supply and water use in one of the largest bronze age cities of the third millennium BC. In: Geo: a new world of knowledge. ICOMOS Open Archive. https://openarchive.icomos.org/. Accessed 22 April 2025

Japan—Land of Water (2015) In Niponica, No. 15 ministry of foreign affairs of Japan Kasumigaseki 2-2-1, Chiyoda-ku, Tokyo 100–8919, Japan

Klemm O, Schemenauer RS, Lummerich A, Cereceda P, Marzol V, Corell D, van Heerden J, Reinhard D, Gherezghiher T, Olivier J, Osses P, Sarsour J, Frost E, Estrela MJ, Valiente JA, Fessehaye GM (2012) Fog as a fresh-water resource: overview and perspectives. Ambio 41:221–234. https://doi.org/10.1007/s13280-012-0247-8

Mekonnen MM, Gerbens-Leenes W (2020) The water footprint of global food production. Water 12(10):696. https://doi.org/10.3390/w12102696

Mekonnen MM, Hoekstra AY (2016) Four billion people facing severe water scarcity. Sci Adv 2:e1500323

Shiklomanov IA (1993) World freshwater resources. In: Gleick PH (ed) Water in crisis: a guide to the world's fresh water resources. Oxford University Press, Oxford

The United Nations Water Development Report (2024) Water for prosperity and peace. UNESCO World Water Assessment Programme

Traces of Ancient Rome in the Modern World—National Geographic. Robert Harding Library. Accessed 22 April 2025

Wulff HE (1968) The Qanats of Iran. Sci Am 218:94–105

Sustainable Water Harvesting

Ramesh Narayanaswamy

1 Historical Background

One of the popular methods of sustainable water harvesting is rainwater harvesting. There are many forms by which rainwater harvesting takes place. Most simple ones essentially have channels on rooftops of buildings that collect water instead of the water being allowed to runoff freely and is directed to be stored in cisterns or percolation ponds near the building. The cisterns and percolation ponds store the rainwater, which is then pumped out for crop irrigation or cleaning purposes. Cost-effectiveness of this approach has proved to be manageable by most communities, and hence rainwater harvesting has been adopted by many people all around the world.

Despite its popularity, rainwater harvesting has one of its main downsides—the dependency on rainfall, and hence, this method is very popular only among the rainfall-affluent zones of the planet. The pumping out of water from the tank requires electrical energy, that in some cases are supported by solar energy from photovoltaic cells or central solar towers. It is therefore a mix of "green" and "non-green" approach, depending on the installation. The advantages of rainwater harvesting are that this method reduces the chances of flooding in and around homes and buildings where this is installed and reduces the load on the stormwater drainage system of the water supply and management network of the locality.

Artificially constructed traditional rainwater collection tanks for large scale application deserves mention. These include the classical *Johad* (water pond) found in many parts of Northern India, and the stepwells (Wipperman 2020) found in ancient temples or palaces that allow rainwater to be stored for later use.

R. Narayanaswamy (✉)
Curtin University, Perth, WA, Australia
e-mail: r.narayanaswamy@curtin.edu.au

R. Narayanaswamy et al. (eds.), *Sustainable Water Harvesting*, Springer Tracts in Mechanical Engineering, https://doi.org/10.1007/978-981-95-3373-2_2

2 Atmospheric Water Generators

One of the earliest fog-capturing method can be found from the Rain Tree of Hierro of the Canary Islands (Hutchinson 1919). The early inhabitants of the island found that this tree could supply additional water (apart from rainfall) for the entire township, by drawing fog from the atmosphere. Since then, several methods have been invented in the past for converting atmospheric humidity into dew and are outlined in the methods for recovery of atmospheric humidity (Nelson 2003). A clear survey of literature points out that some of these methods were tried out, and subsequently not followed-up, while some others existed as patents only and not prototyped.

Over time, consistent efforts to obtain water from sources in addition to rainwater harvesting have been explored, although the use of rainwater harvesting still continues. Broadly, these efforts make use of either atmospheric water vapour (fog), using atmospheric water generators employing active or passive techniques, or solar energy using solar distillation systems.

2.1 Atmospheric Water Vapour as a Source of Water

Earth's atmosphere contains a mixture of gases, water in solid and liquid forms as well as fine suspended solid particles. The atmosphere consists of permanent as well as variable gases (Jacobson 2005). Permanent gases include Nitrogen (78% volume), Oxygen (21% volume), Argon (1% volume), and low-volume percent gases such as Argon, Neon, Helium, Krypton, and Xenon. The major variable gas component is water vapour, occupying about 0.25% volume. Other variable gases (in smaller amounts) include Carbon dioxide, Methane, and Ozone. Water vapour as a gas in invisible; it becomes visible when droplets start to form due to condensation triggered by nucleation sites constituted by aerosol particles.

Water vapour is the most abundant variable gas in the atmosphere, although its presence in the atmosphere varies according to the hydrological cycle and locations on the globe. Hence, the tropical regions have the benefit of a higher water vapour content in the atmosphere. The polar regions have almost zero water vapour content in the atmosphere above the regions. The distribution of water vapour and its movement (made possible by winds) affects water generation using atmospheric water generators, and this is principally the reason why only certain areas of the globe are effective in enabling significant yields through fog harvesting.

The water vapour in the atmosphere is mainly due to evaporation of surface water from the earth. The mixing ratio of water vapour to air in the atmosphere varies from region to region, with the equatorial regions contributing to about 4% of the atmospheric water vapour due to higher evaporation rates, concomitant with higher average ambient temperatures. As we mentioned earlier, fog is a collection of water vapour and that moves around with a gentle wind, it is prudent to assume that atmospheric water vapour-based terrestrial fog collectors will work best in fog-prone areas. A pre-requisite for water vapour condensation in the atmosphere is the need for hygroscopic nuclei released from terrestrial earth (volcanoes, oceans, combustion products, dust particles, etc.) by various means.

When we say fog is a collection of water vapour, we must realise that upon condensation phenomena in the atmosphere, the invisible water vapour gets transformed into visible liquid droplets and this is what constitutes fog. One can expect liquid droplets of varying sizes to constitute the fog. Typical sizes of water droplets in fog can range from 10 to 20 microns; the diameter highly dependent on the location of the fog event on the planet, as mentioned earlier.

Different types of fog exist in nature. Radiation fog forms closer to the ground. This happens when the air closer to the ground surface radiatively cools to a temperature below the dew point. The water droplets present in radiation fog are very small in diameter, and the occurrence of this fog is common when there is a huge difference in day and night temperatures, and there is sufficient water content in the air.

In coastal areas, the wind carries water vapour from the ocean surface to cooler surface areas on the landmass, which instigates the condensation phenomena to result in advection fog. Evaporation fog is formed due to the re-condensing of evaporated water vapour upon mixing with dry and cool air. The National Weather Service website (https://www.weather.gov) provides more details of these and other types of fog, and their role in the forecasting of the weather phenomena (Fig. 2.1).

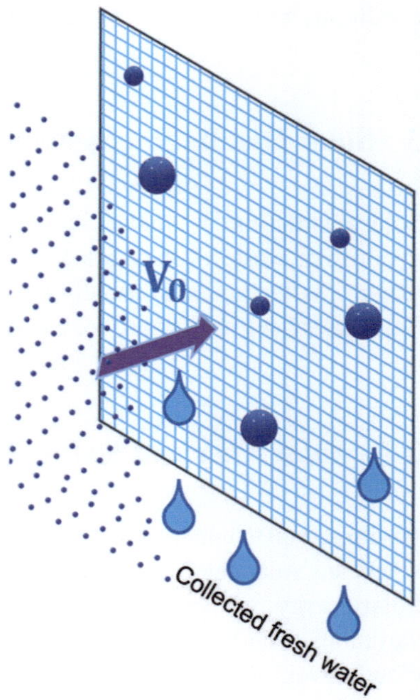

Fig. 2.1 Schematic of a mesh-based fog harvester. Droplets of fog impacting on the mesh grow in size due to droplet coalescence, and slide down the mesh wires due to gravity enabling collection of water from the bottom of the mesh

2.2 Factors Affecting Atmospheric Water Capture

Atmospheric water generators use the principle of capturing fog onto some suitable surfaces that enable the deposition of fog droplets and promote accumulation of droplets of water, which is then collected, after drainage along the collection surfaces. Passive water generators do not force the fog onto the deposition surfaces, and hence, there is no compulsory withdrawal of moisture from the atmosphere. In that sense, this methodology does not conflict the natural balance of the environment while capturing fog. Passive collection of fog involves the deposition of fog (by wind) on surfaces such as wire meshes, or thin cylindrical wires arranged in some form. The yield of water generated, therefore depends primarily upon the geographical location of the water capture device and the quality of surfaces (wettability, in particular) used for fog capture. Additional factors that affect the yield of water collected include the relative humidity, temperature, wind speed of the locality.

Nature has its own way of collecting fog from the atmosphere and use it for the survival of its flora and fauna in many arid and semi-arid regions of the world. The Namib desert beetle (Spirig et al. 2021) is a classic example often used by many researchers to highlight the importance of surface topology on fog harvesting. Certain plants such as the Geophytes (Vogel and Muller-Doblies 2011), use their leaves for extracting the fog to make water for its survival. Plants closer to the ground surface utilise dew formation on leaves that occurs in regions for environmentally favourable conditions unlike fog, wherein droplet movement is the distinguishing characteristic.

3 Solar Distillation Systems

Solar distillation is a means by which solar energy is utilised to evaporate a fixed quantity of wastewater or brackish water, with the objective of generating potable water. The technique is usually envisaged in a portable device using solar energy. Solar stills have been in use for a very long time since the time of Aristotle (Fig. 2.2).

3.1 Solar Flux and Spatio-Temporal Distribution

It is important to understand the availability of solar radiation at the location of interest where any solar device is installed. This requirement applies for solar distillation also. Solar energy received at the surface of the earth is subject to spatial and temporal variations. The spatial variability depends on weather-related phenomena such as cloud-cover, aerosols, and water vapour content—more so, with respect to cloud cover. It also depends upon the limits of integration (hourly, daily, monthly, yearly) that relate to the overall availability. Major countries of the world have networks of measuring instruments and satellites (polar or geo-stationary) that help to provide

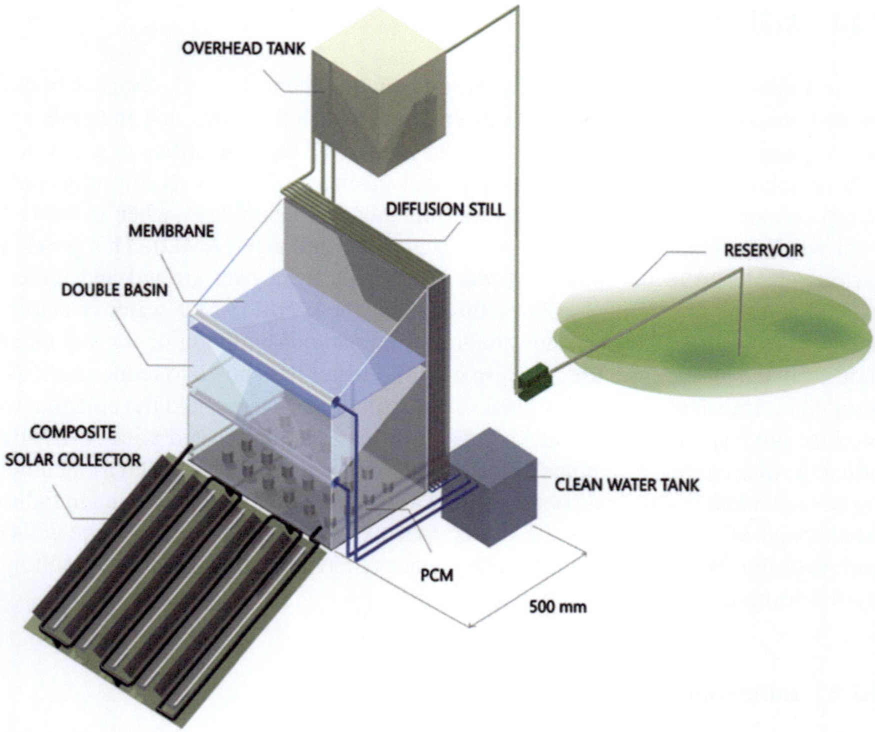

Fig. 2.2 Schematic of a solar thermal desalination method using solar still. Augmentation devices such as thermosyphon and heat pipes can also be used

solar irradiation data on a suitable time-scale basis, using a combination of field measurements and satellite imagery. This data is often available from meteorological departments. In predicting the variability, it is often prudent to take re-course to historical data over several years, and in this regard, it is possible that AI-based predictions could realistically provide an estimate of accurate variability of solar irradiance, within reasonable predictable variations. The average yearly global solar irradiance over the Indian sub-continent (Tyaji 2009) is estimated to be about 7000 MJ/m^2.

3.2 *Potential Industrial Applications of Solar Distillation*

We believe the following two major industrial applications of solar distillation are important, if actively pursued, it will have the potential for enabling significant generation of freshwater from wastewater sources. These include: (i) acid-mine drainage and (ii) industrial wastewater recovery.

3.2.1 Acid-Mine Drainage

Critical minerals are required for the ever-expanding technological advances of our modern society. This means countries endowed with metal ores and minerals are actively engaged in the mining of ores and minerals. Huge quantities of excavated leftover material in mine tailings on mine sites (abandoned as well as currently operating) contain trace quantities of several minerals and sulphides—when in contact with water generates what is known as acid-mine drainage (AMD). This poses a serious challenge to the quality of water available from over-ground and underground sources in mining localities throughout the world. AMD water is acidic, and sulphur-rich: this water is unsuitable for human consumption or for any other industrial and agricultural use. The use of AMD water has shown to reduce agricultural yields drastically. Well-designed solar distillation systems hold the potential to produce potable water from brackish water or water contaminated with AMD, by utilising solar energy. Examples of water contaminated with AMD such as metallic copper is evident from (Chakravarthy 2013), for a mine discharge happening in India. Abandoned Mine Drainage in the United States (Abandoned Mine Drainage 2024) and Australia (Harries 1997) provides evidence of environmental pollution resulting from acid-mine drainage.

3.2.2 Industrial Wastewater Recovery

It is estimated (UN World Water Report 2017) that the combined agriculture, municipal, and industrial wastewater released yearly to the environment is about 56%. Out of this, about 80% of wastewater is released to the environment without any treatment. As discussed in Chap. 1, this untreated wastewater has the potential to contaminate overground and underground supplies of freshwater. It is therefore important to treat industrial wastewater and explore the potential of freshwater recovery. Considering that all wastewater is made up of 99% water, scientific advances for recovering freshwater from wastewater is becoming important in recent times, as evidenced from the number of scientific articles (Nadel-Bach et al. 2021) published in this area. This article also provides a review of active and passive methods employed in solar stills to make the process cost-effective, and efficient.

Industrial, agricultural, and municipal wastewater consist of several chemical ingredients, and not just sodium chloride. Research into water quality of condensate water obtained using solar stills from the use of agricultural and industrial wastewater requires further exploration. It has become evident that not many research papers seem to focus on this area and is something worth pursuing in the future. No environmental issues appear to have been reported against the use of solar stills for water purification, and hence, this is a completely environmentally friendly choice for converting brackish wastewater into potable water, especially for captive sources in remote and regional communities.

4 Low-Temperature Solar Still and Productivity Indices

The productivity index of solar stills is a function of environmental, design, and operational factors. The environmental parameters include location-specific factors such as the availability of solar radiation, and local weather conditions. Design parameters have a very strong effect on the yield of freshwater produced by the device. Research literature provides information on both active and passive means of enhancement of solar still performance. Excellent reviews on this topic are available (Shadi et al. 2016; Elsheikh et al. 2023; Rufuss et al. 2016).

Among the environmental parameters, an enhancement in potable water yield from solar stills has been observed for an increase in solar radiation flux, ambient temperature, relative humidity, and wind velocity, while a decrease in performance is noted during the presence of cloud cover in the locality, as well as dust on the transparent cover of the solar still. The design parameters or the engineering design of the still itself can have a significant effect on the freshwater yield. These factors take into consideration the evaporation area available for water, the depth of vapour space within the solar still, the angle of the condensing surface, surface features of the condensing surface, thermal storage materials, the use of solar tracking for the still, as well as the insulation of the various components of the still. The use of solar stills in thermal desalination and the enhancement of solar still performance are covered later under a separate chapter. Operational factors mainly relate to the maintaining the cleanliness of the transparent covers, wastewater containment systems and piping, removal of the deposited salts, as well as the removal of clogged filters and membranes.

In spite of several years of advancements in the design of solar stills, very few of them have gone onto a prototype or commercial—captive type of water generation system. Most of these ones are indeed confined to custom designs.

5 Closure

The chapters so far covered have highlighted the need for the development of sustainable water harvesting technologies that will serve to augment existing water generation methods. The scientific principles that underpin these technologies require consideration at this stage, and hence, these principles will now constitute the remainder of the chapters, that will enable the reader to be familiar with most concepts of thermodynamics, thermal desalination, atmospheric water generation, and so on.

References

Abandoned Mine Drainage (2024) https://www.epa.gov/nps/abandoned-mine-drainage. Accessed on 22 Apr 2025

Chakravarthy A (2013) Curse of copper. Down to Earth. www.downtoearth.org.in. Accessed on 22 Apr 2025
Elsheikh A, Hammoodi KA, Ibrahim AMM, Mourad AHI, Fujii M, Abd-Elaziem W (2023) Augmentation and evaluation of solar still performance: a comprehensive review. Desalination 574. https://doi.org/10.1016/j.desal.2023.117239
Harries J (1997) Acid mine drainage in Australia: its extent and potential future liability, Department of Climate Change, Energy, the Environment and Water-Australian Government. https://www.weather.gov/ Accessed on 22 Apr 2025
Hutchinson J (1919) The rain tree of Hierro, Canary Islands. (Oreodaphne foetens). Bull Miscell Inform (Royal Botanic Gardens, Kew) 1919(3):153–164 (12 pages). https://doi.org/10.2307/4107704
Jacobson MZ (2005) Fundamentals of atmospheric modelling 2/e. Cambridge University Press, Cambridge, UK
Mohammed Shadi S, Abujazar S, Fatihah AR, Rakmi MZS (2016) The effects of design parameters on productivity performance of a solar still for seawater desalination: a review. Desalination 385:178–193
Nadal-Bach J, Bruno JC, Farnos J, Rovira M (2021) Solar stills and evaporators for the treatment of agro-industrial liquid wastes: a review. Renew Sustain Energy Rev 142:110825. https://doi.org/10.1016/j.rser.2021.110825
Nelson RA (2003) Air Wells, fog fences and dew ponds. http://www.rexresearch.com/airwells/airwells.html
Rufuss D, Iniyan S, Davies P, Dsilva Winfred Rufuss D, Suganthi L (2016) Solar stills: a comprehensive review of designs, performance and material advances. Renew Sustain Energy Rev 63:464–496. https://doi.org/10.1016/j.rser.2016.05.068
Spirig R, Vogt R, Feigenwinter C (2021) Droplet size distribution, liquid water content and water input of the seasonally variable, nocturnal fog in the Central Namib Desert 0169–8095
The United Nations World Water Development Report (2017) Wastewater–The Untapped Resource, UN World Water Assessment Programme, Italy
Tyaji AP (ed) (2009) Solar radiant energy over India, India meteorological department, Ministry of Earth Sciences, Government of India
Vogel S, Muller-Doblies U (2011) Desert geophytes under dew and fog: the 'curly-whirlies' of Namaqualand (South Africa), Flora–morphology. Distribut Funct Ecol Plants 206(1):3–31. https://doi.org/10.1016/j.flora.2010.01.006
Wipperman A (2020) How ancient water conservation methods are reviving in India. Global Centre on Adaptation

Thermodynamics of Atmospheric Water Harvesting

Malay K. Das

1 Introduction

Atmospheric water harvesting (AWH) refers to the collection of processes where liquid water is collected from ambient air (Tu and Hwang 2020). Figure 1 shows a schematic representation of typical AWH technologies. There are generally two different pathways: (a) active or passive cooling and (b) sorption followed by cooling (Nikkhah et al. 2023). In active cooling, we use a standard refrigeration cycle to cool the ambient air (Eslami et al. 2018). The condensed water, generated via cooling, is then collected. Since the refrigeration system adds to the cost as well as the carbon footprint of the process, the goal here is to devise efficient refrigeration technologies, often using renewable energy sources (Tashtoush and Alshoubaki 2023). The refrigeration-based systems are usually unsuitable for arid regions due to the high cost of refrigeration. Conversely, in passive cooling, the main goal is to collect naturally condensed water, such as fog and dew. While natural cooling processes make such technologies cost-effective, relying on natural processes often leads to uncertainties in output. Finally, sorption-based processes use desiccant-type materials to trap water vapor from the atmosphere, producing locally high-humidity regions (Li et al. 2022). The vapor is then cooled using active or passive cooling techniques to produce liquid water. While sorption-based techniques are particularly suitable for arid regions due to their lower cooling cost, higher material costs for such systems may preclude scaling up of technology (Bilal et al. 2022). Irrespective of the technologies used, condensation of water vapor and consequent collection of liquid water constitute the key processes behind AWH (Xiang et al. 2024). Thermodynamics play a major role in understanding the phase change behavior of water and thus designing the AWH systems (Kocher and Menon 2023).

M. K. Das (✉)
IIT Kanpur, Kanpur, India
e-mail: mkdas@iitk.ac.in

R. Narayanaswamy et al. (eds.), *Sustainable Water Harvesting*, Springer Tracts in Mechanical Engineering, https://doi.org/10.1007/978-981-95-3373-2_3

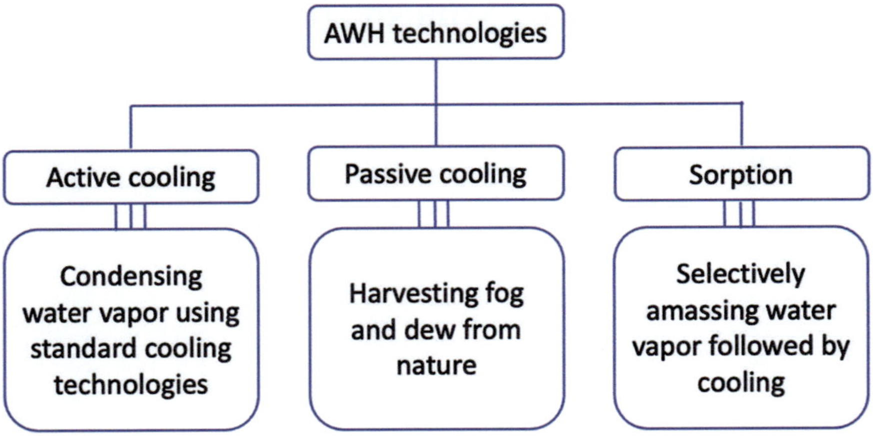

Fig. 1 Typical pathways for atmosphere water harvesting. Each includes condensation of water vapor present in the ambient air

2 Simple Compressible System, Equilibrium, Pure Substance, and Phases

A thermodynamic *system* is a region in space or a fixed amount of matter that we wish to investigate. *Surroundings* consist of everything outside the system. A real or imaginary *boundary* separates the system and the surroundings. The system contains a *working fluid* that is not subjected to gravitational, electrical, magnetic, or surface tension effects. The working fluid is also assumed to be not in motion. A thermodynamic system containing such an idealized working fluid is known as a *simple compressible system.* In thermodynamics, we study a system in *equilibrium.* In the context of thermodynamics, equilibrium implies a state of stability when no heat mass or momentum transfer happens within the system. Thermodynamic calculations rely on the measurement of thermodynamic *properties* such as density and enthalpy entropy. Particularly important are *intensive properties* that are independent of the mass of the working fluid contained in the system. Some of the important intensive properties include specific volume, specific enthalpy, specific entropy, etc. In equilibrium, the state of a simple compressible thermodynamic system is uniquely specified by two independent intensive properties.

Further in thermodynamics, *pure substance* refers to substances with a fixed chemical composition throughout the thermodynamic system. Examples include water, oxygen, hydrogen, and various other elements and compounds. A mixture of multiple pure substances may also be treated as a pure substance if the mixture exhibits a homogeneous chemical composition throughout the thermodynamic system. For instance, although air is a mixture of different gases, in many thermodynamic applications, air is considered to be a pure substance. Such an idealization of air is possible since, in such applications, air maintains a homogeneous chemical composition throughout.

If two or more pure substances separate themselves, forming an interface, each substance is referred to as a *phase* (Zemansky and Dittman 1997). A thermodynamic system that contains working fluid with different phases is called a *multiphase* system. Since most oils do not mix homogeneously with water, the oil–water system is a typical multiphase thermodynamic system. As the oil–water system exhibits different chemical compositions within each phase, such a mixture is not a pure substance. However, a multiphase system containing ice, liquid water and steam may still be considered a pure substance since each phase has the same chemical composition.

3 Phase Equilibrium

Thermodynamics is primarily governed by two laws: the first law and the second law. A detailed discussion of these laws can be found elsewhere (Cengel and Boles 2019). In a simple compressible thermodynamic system of fixed mass, the application of the first and the second laws of thermodynamics provides:

$$\delta Q - PdV = dU \tag{1}$$

$$dS \geq \frac{\delta Q}{T} \tag{2}$$

where Q is the heat transfer to the system with internal energy U, pressure P, volume V, temperature T, and entropy S. Combining Eqs. 1 and 2:

$$dS \geq \frac{PdV + dU}{T} \Rightarrow dU + PdV - TdS \leq 0 \tag{3}$$

Since *Giggs free energy* is defined as $G = H - TS$, where H is the enthalpy of the system, we may write:

$$(dG)_{P,T} = dH - TdS = d(U + PV) - TdS = dU + pdV - TdS \tag{4}$$

Thus, combining Eqs. 3 and 4, we have:

$$(dG)_{P,T} \leq 0 \tag{5}$$

Equation 4 is applicable to all thermodynamic systems undergoing a constant P, constant T process. A constant P, constant T process continues spontaneously as long as $(dG)_{P,T} < 0$ and reaches equilibrium at $(dG)_{P,T} = 0$. $(dG)_{P,T} > 0$ is a thermodynamic impossibility since such a process violates the laws of thermodynamics. Applying Eq. 5 to a phase change system yields the criterion of phase equilibrium, as shown in Fig. 2. A phase change process continues as long as $(dG)_{P,T} < 0$ reaching the thermodynamic equilibrium at $(dG)_{P,T} = 0$.

Fig. 2 Criterion for thermodynamic equilibrium in a phase change system. The idea is applicable to all thermodynamic systems

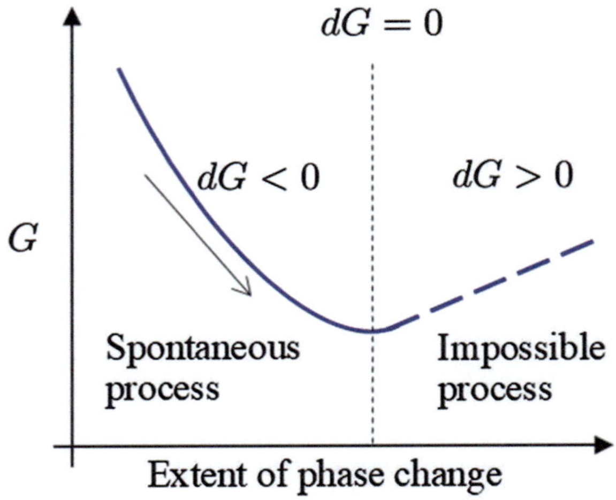

Let us consider a fixed-mass system containing a mixture of liquid water and water vapor. Assume masses of liquid and vapor are m_f and m_v, respectively. If the specific Gibbs energy of liquid and vapor are g_f and g_v, respectively, The Gibbs energy of the liquid–vapor mixture is:

$$G = m_f g_f + m_v g_v \tag{6}$$

Consider a small mass of liquid dm_f is converted to vapor through a constant P, constant T process. The vapor mass increases by dm_v and applying the mass conservation principle $dm_v = -dm_f$. The change in G, in this process, is:

$$(dG)_{P,T} = g_f dm_f + g_v dm_v = \left(g_f - g_v\right) dm_f \tag{7}$$

Thus, the condition of phase equilibrium is:

$$(dG)_{P,T} = 0 \Rightarrow g_f = g_v \tag{8}$$

Therefore, a two-phase mixture remains at equilibrium if the specific Gibbs energies of the two phases match with each other (Sonntag et al. 2002). Thus, a liquid–vapor phase change process proceeds with vapor being converted to liquid ($dm_f > 0$) if $g_f < g_v$. Similarly, a liquid–vapor phase change process proceeds with liquid being converted to vapor ($dm_f < 0$) if $g_f > g_v$.

4 Liquid–Vapor Phase Change in Water

Consider a thermodynamic system of a certain mass of liquid water at 25 °C in a piston-cylinder system as shown in Fig. 3a. The weight of the piston dictates the water pressure. For simplicity, neglecting the piston weight, we assume that the water remains at 1 atm pressure (101 kPa). If we now slowly supply heat to the cylinder, the water temperature will rise up to 100 °C (Fig. 3b). During this heating process, there will be a negligibly small increase in the specific volume of water. As the liquid water attains 100 °C temperature, further addition of heat does not immediately increase the water temperature, rather, liquid–vapor phase change begins at 100 °C temperature. Thus, the temperature remains constant at 100 °C until the entire liquid water is converted to water vapor. The temperature 100 °C is called the *saturation temperature* (T_{sat}) of water at 1 atm pressure. At 1 atm pressure, liquid water at 100 °C is called *saturated liquid* and water vapor at 100 °C is called *saturated vapor*. Thus, at 1 atm and 100 °C, the thermodynamic state of water is a mixture of saturated liquid and saturated vapor. The mass fraction of saturated vapor in this mixture is known as the dryness fraction or *quality*. Again, at 1 atm pressure liquid water below 100 °C temperature is known as subcooled or *compressed liquid*, while vapor above 100 °C is called *superheated vapor*. To move from saturated to superheated state, at 1 atm pressure, further addition of heat is required, as shown in Fig. 3c.

The constant-pressure (1 atm, 101 kPa) liquid–vapor phase change of water, described above, is illustrated by the *T*-*v* diagram in Fig. 4. Here, the heating begins at *A* where the water remains as a compressed liquid. During the process *AB*, the water temperature increases due to *sensible heat* addition. The phase change happens during the process *BC*, where the temperature remains constant at the saturation temperature 100 °C despite heat addition. While point *B* indicates saturated liquid, water becomes saturated vapor at point *C*. The added heat in process *BC* is referred to as the latent heat. As the process continues beyond point *C*, superheated steam is generated via further addition of sensible heat.

Repeating the constant-pressure phase-change experiments described above, we generate a family of curves shown in Fig. 5. Scrutinizing Fig. 5, we find that the saturation temperature increases with pressure, as shown in Fig. 6. Furthermore, Fig. 5 shows that the specific volume of saturated liquid (v_f) and the specific volume of

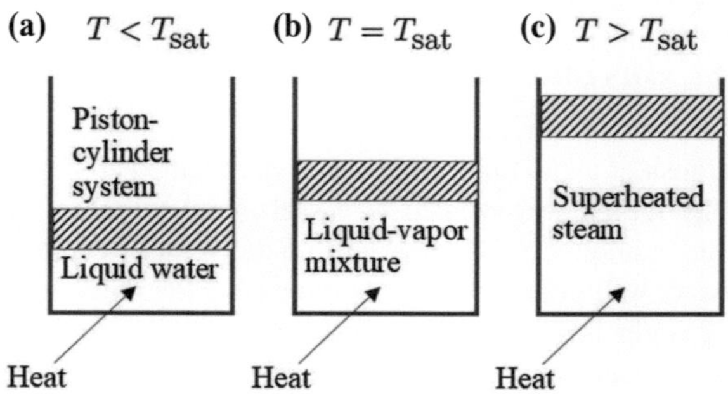

Fig. 3 Constant pressure liquid–vapor phase change of water: **a** compressed liquid, **b** a mixture of saturated liquid and vapor, and **c** superheated steam

Fig. 4 T-v diagram of constant pressure (101 kPa) liquid–vapor phase change of water

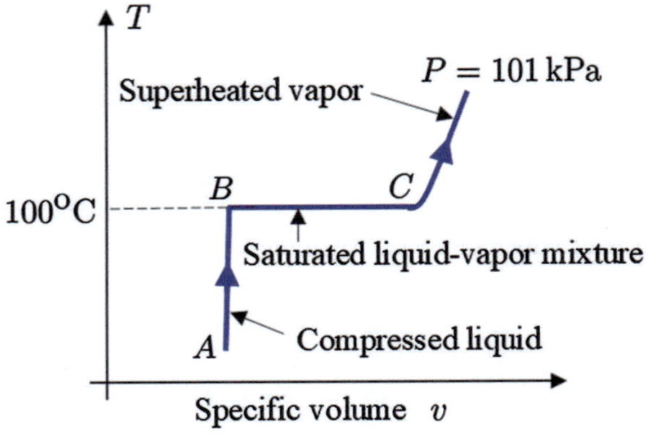

Fig. 5 Liquid–vapor phase diagram of water

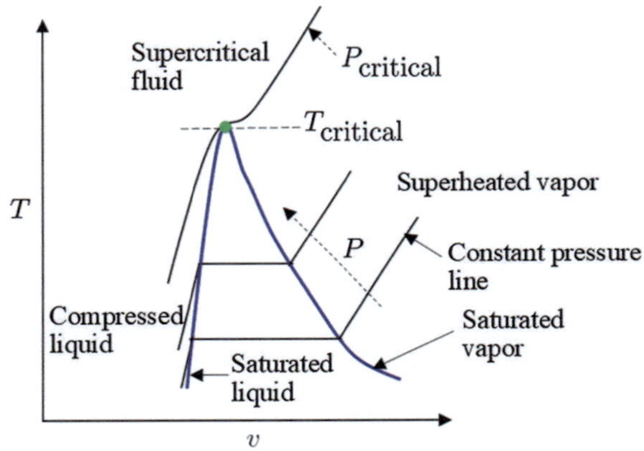

saturated vapor (v_g) come close as the pressure increases. Finally, at critical pressure $P_{critical} = 22.0$ MPa the saturation temperature reaches the critical temperature $T_{critical} = 323$ °C where v_f and v_g match exactly. The constant pressure line at $P_{critical}$ shows an inflection point. Beyond the critical pressure, the distinction between liquid and vapor ceases to exist, and the fluid is marked as supercritical.

5 Gas and Vapor Mixture

Consider a gas-filled rigid vessel maintained at a constant temperature, as shown in Fig. 7. First, we evacuate the vessel completely (Fig. 7a), and then we introduce a very small amount of liquid water into the vessel, keeping the temperature constant. The water will evaporate slightly increasing the pressure in the vessel. Since the amount of water is very small, it is likely that the entire water will turn into superheated vapor, and the pressure in the vessel will be less than the saturation pressure of

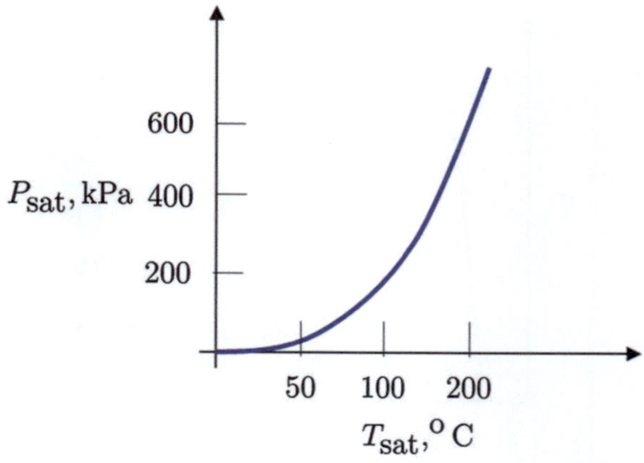

Fig. 6 Liquid–vapor phase diagram of water

water at the specified temperature of the vessel (Fig. 7b). If we keep on injecting more liquid water into the vessel, the water will continue evaporating, increasing the vessel pressure. The evaporation process will continue until the vessel pressure equates to the saturation pressure of water at the vessel temperature (Fig. 7c). Further addition of liquid water will not lead to evaporation. Rather, some water vapor may now condense. The vessel pressure will remain constant at the saturation pressure and water inside the vessel will remain as a mixture of saturated liquid and saturated vapor. If we keep on injecting more liquid water into the vessel, the vapor will continue to condense keeping vessel pressure constant. The process will continue until the entire water in the vessel condenses. Once the vessel is filled with liquid water, only a small amount of liquid water can be injected further since liquid water is virtually incompressible. During this phase of liquid water injection, the vessel pressure will increase considerably, and the state of water in the vessel becomes compressed.

During the above process, the vessel temperature was kept constant. Similar outcomes emerge if we repeat the experiments at various temperatures, as shown in Fig. 8. Figure 8 indicates three isothermal processes *AB*, *CD*, and *EF*. Each process starts from the same pressure ($P_A = P_C = P_E$). Further, each process starts from superheated state and terminates at the saturated vapor state. Additionally, in each

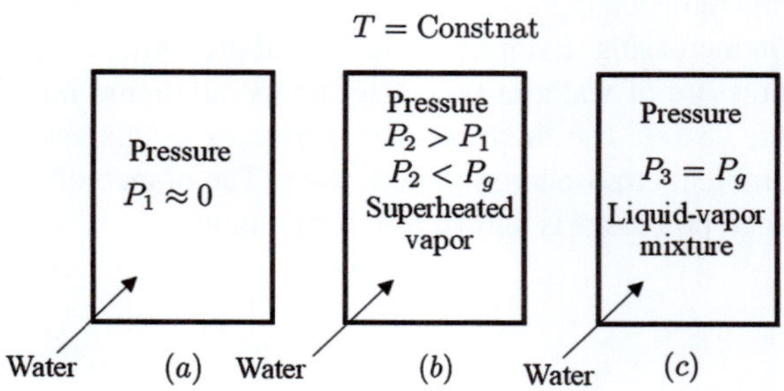

Fig. 7 Isothermal phase change of water

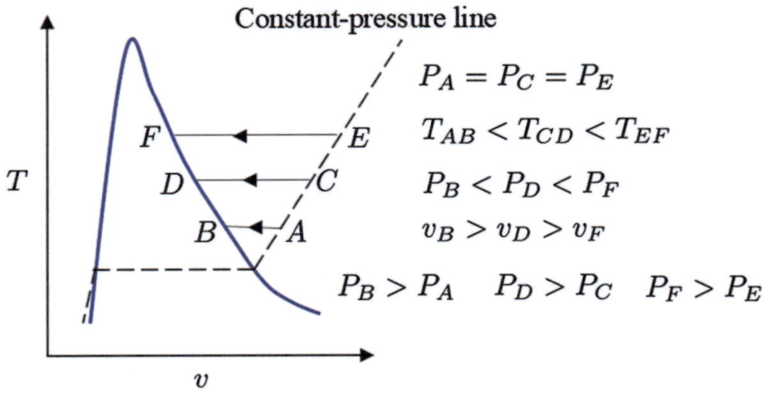

Fig. 8 Isothermal phase change of water at various temperatures

process, the final pressure is greater than the initial pressure and the final specific volume is less than the initial specific volume. Clearly, each process indicates water vapor accumulation in the system due to water injection from the surroundings. Finally, we observe that the specific volume at the end of the process decreases with the increase in temperature. In other words, a higher-temperature environment can contain more water vapor.

While all the above experiments start from an evacuated vessel, similar outcomes may also be achieved if the vessel contains some air initially. If we introduce a small amount of liquid water in an isothermal vessel containing dry air, the water soon evaporates. If we keep the vessel temperature constant and introduce more liquid water into the vessel, water will continue evaporating, increasing the vessel pressure. Eventually, we reach a state where evaporation stops, and further injection of water keeps water in the liquid state. In other words, the results from these ai-filled vessel experiments are in consonance with the experimental results of the evacuated vessel.

In a vessel containing a mixture of dry air and water vapor, each constituent, according to *Gibbs-Dalton law*, exerts pressure as if it occupies the entire vessel. The pressure of each constituent, while occupying the entire vessel, is called the *partial pressure* of the constituent in the mixture. Initially when the vessel contains only dry air, the partial pressure of water in the vessel is zero. As we keep on injecting more and more water into the vessel, the partial pressure of water vapor, in the vessel keeps on increasing. Eventually, the partial pressure of water vapor reaches the saturation pressure of water at the vessel temperature and liquid–vapor mixture is formed in the vessel. The thermodynamic process in this vessel mimics the processes in the ambient atmosphere in many ways. The branch of thermodynamics that deals with such processes is known as Psychrometry.

6 Psychrometry

Psychrometry refers to the branch of thermodynamics that study the processes involving atmospheric air where atmospheric air is treated as a mixture of dry air and water vapor. While at first glance, such an assumption about the nature of air looks like an oversimplification, in many applications, such as air-conditioning or atmospheric water harvesting, experiments justify the above assumption. Furthermore, while dry air is a mixture of several gases, we usually treat dry air as an ideal gas in psychrometric studies and ignore the contributions of individual constituent gases. Additionally, in psychrometric processes, the variations of temperature and pressure are such that the composition and thermophysical properties of dry air may be assumed constant. The amount of water vapor, however, varies in atmospheric air due to the phase change of water. For instance, condensation of water vapor forming dew, fog, and rain, evaporation of water from water bodies as well as transpiration of water vapor from plants change the concentration of water vapor in atmospheric air.

In ambient conditions, water usually exists as superheated vapor. For instance, at an ambient temperature of 298 K saturation pressure of water is 3.17 kPa. Thus, at 298 K ambient air, water vapor pressure may reach a maximum of 3.17 kPa. Maximum water vapor pressure is reached only when water vapor is in thermodynamic equilibrium with liquid water. Such a thermodynamic equilibrium is achieved only when water vapor is in contact with liquid water. Therefore, in general, the water vapor pressure in ambient air is much lower than the saturation pressure of water at ambient temperature. Under such low-pressure conditions, superheated or even saturated vapor can be reasonably approximated as an ideal gas (Moran et al. 2015). Treating the water vapor as an ideal gas is a fundamental assumption in psychrometry. Once both dry air and water vapor are considered as ideal gases, we can apply Dalton's law of partial pressure as follows:

$$P = P_a + P_s \tag{9}$$

where P is the atmospheric pressure, P_a and P_s are the partial pressures of dry air and superheated vapor, respectively.

6.1 *Psychrometric Quantities*

Consider $P_s = 1.7$ kPa at an ambient temperature of 40 °C. The saturation pressure of water at 40 °C is 7.4 kPa. Thus, the water vapor at $P_s = 1.7$ kPa is superheated, shown as point *A* in Fig. 9. If we now put a glass of water in the ambient air and slowly cool the glass by placing ice cubes in the water, eventually, we will observe condensation of atmospheric water vapor on the outer side of the glass. The process is called dew formation and the temperature at which the dew formation begins is

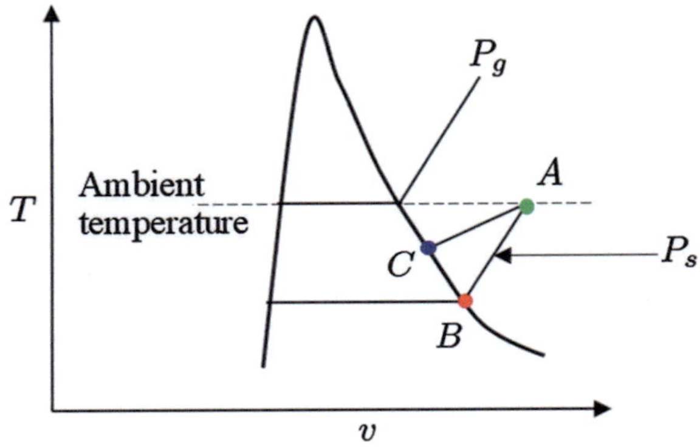

Fig. 9 Water vapor at ambient condition (**a**), at the dew point (**b**), and at adiabatically saturated state (**c**) in a *T*-*v* diagram

known as the dew point (T_{dp}), shown as point *B* in Fig. 9. The cooling of air due to the presence of the cold water-filled glass is a constant-pressure process, shown as the process *AB* in Fig. 9. Therefore, the dew point T_{dp} is the saturation temperature of water at Ps = 1.7 kPa, which is 15 °C.

While T_{dp} is an important quantity in psychrometric processes, several other psychrometric quantities are also of practical importance in atmospheric water harvesting. The first one is the specific (or absolute) humidity, defined as:

$$\omega = \frac{m_{\text{s}}}{m_{\text{a}}} \tag{10}$$

where m_{s} and m_{a} are the masses of water vapor and dry air, respectively. As per our previous argument of treating both dry air and water vapor as ideal gases, we may write:

$$\omega = \frac{P_s V}{R_s T}\frac{R_a T}{P_a V} = \frac{P_s}{P_a}\frac{R_a}{R_s} \tag{11}$$

Here, *V* and *T* are the volume and temperature of the moist air, respectively. Also, R_a and R_s are the gas constants of air and water, respectively. Further, we know that:

$$\frac{R_a}{R_s} = \frac{R_u/M_a}{R_u/M_s} = \frac{M_s}{M_a} = \frac{18}{28.96} = 0.622 \tag{12}$$

While R_{u} is the universal gas constant, M_{a} and M_{s} are the air and water molar masses, respectively. Thus, combining Eqs. 2–4, we derive:

$$\omega = 0.622\frac{P_s}{P_a} \tag{13}$$

Apart from the dew point and specific humidity, relative humidity is an important parameter in designing the atmospheric water harvesting processes. As mentioned before, water vapor in the atmosphere is generally superheated. However, atmospheric water harvesting becomes easier if the water vapor is saturated. Relative humidity provides an idea of how much water vapor would be necessary to saturate the vapor at the atmospheric temperature. Precisely, relative humidity is defined as the ratio of the actual mass of water vapor in a given volume of moist air to the mass of water vapor to saturate at the ambient temperature. For some moist air of volume V and temperature T, the relative humidity (ϕ) is defined as:

$$\phi = \frac{m_s}{m_{s,\,sat}} \tag{14}$$

while m_s is the mass of water vapor in the moist air of volume V and temperature T, $m_{s,\,sat}$ is the mass of water vapor in volume V and temperature T if the vapor is saturated. If P_g is the saturation pressure of water at temperature T, as shown in Fig. 9, we may write:

$$\phi = \frac{P_s V}{R_s T} \frac{R_s T}{P_g V} = \frac{P_s}{P_g} \tag{15}$$

While dew point, specific humidity, and relative humidity are powerful tools for AWH, directly measuring these quantities could be expensive and time-consuming. Thus, in practice, we rely on temperature measurements to indirectly determine the values of humidities. One way to estimate humidities is to conduct an adiabatic saturation experiment. While in Fig. 9, the process AC depicts the adiabatic saturation process in a T-v diagram, Fig. 10 shows the schematic of the adiabatic saturation experiments. The experiment is conducted in a long, thermally insulated setup where water vapor in ambient air saturates via liquid water supplied from outside, as shown in Fig. 10. In general, ambient air enters the system with relative humidity $\phi_1 < 1$. As the liquid water enters the system and is exposed to the flowing air, the liquid water evaporates. During evaporation, water absorbs latent heat from the surroundings. Thus, the water temperature drops, and provided the setup length is long enough, the relative humidity at the exit would be $\phi_2 = 1$. The key feature of this experiment is that the temperatures at the water inlet and the air exit are the same. The experiment starts by supplying water at a temperature T_1. As the exit temperature of the air drops below T_1, we adjust the water inlet temperature to the air exit temperature. As the process continues, the air exit temperature and, consequently, the water inlet temperature stabilizes to T_2, the adiabatic saturation temperature. As the air exit temperature stabilizes, given sufficient length of the setup, the water vapor reaches the saturated state ($\phi_2 = 1$) at the exit.

Thermodynamic analysis of the above process invokes mass conservation as follows:

$$\dot{m}_a = \dot{m}_{a1} = \dot{m}_{a2} \tag{16}$$

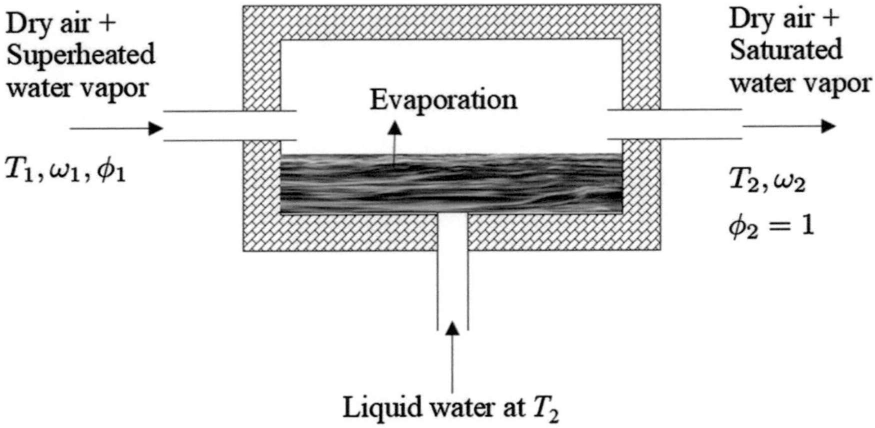

Fig. 10 Superheated water vapor, present in ambient air, is undergoing adiabatic saturation process

$$\dot{m}_{w1} + \dot{m}_f = \dot{m}_{w2} \tag{17}$$

where $\dot{m}_a$ is the mass flow rate of dry air, which remains the same at the inlet ($\dot{m}_{a1}$) and at the outlet ($\dot{m}_{a2}$). Further, $\dot{m}_{w1}, \dot{m}_{w2}$ are water mass flow rates at the air inlet and exit, respectively. Assuming ω_1, ω_2 be the specific humidities at the air inlet and exit, respectively, Eq. 17 is written as:

$$\begin{aligned}\dot{m}_{w1} + \dot{m}_f &= \dot{m}_{w2} \Rightarrow \dot{m}_a\omega_1 + \dot{m}_f \\ &= \dot{m}_a\omega_2 \Rightarrow \dot{m}_f = \dot{m}_a(\omega_2 - \omega_1)\end{aligned} \tag{18}$$

Setting h_1, h_2 and h_{f2} to be the specific enthalpies at the air inlet, air exit, and water inlet, respectively, the energy conservation equation is written as follows:

$$\begin{aligned}\dot{m}_a h_1 + \dot{m}_f h_{f2} &= \dot{m}_a h_2 \Rightarrow \dot{m}_a h_1 + \dot{m}_a(\omega_2 - \omega_1)h_{f2} \\ &= \dot{m}_a h_2 \Rightarrow h_1 + (\omega_2 - \omega_1)h_{f2} = h_2\end{aligned} \tag{19}$$

If c_p is the specific heat of dry air and h_{s1} is the specific enthalpy of water vapor at the air inlet, while h_{g2} is the specific enthalpy of saturated water vapor at the air outlet, Eq. 19 can be written as:

$$\begin{aligned}c_p T_1 + \omega_1 h_{s1} + (\omega_2 - \omega_1)h_{f2} &= c_p T_2 + \omega_2 h_{g2} \Rightarrow \omega_1 \\ &= \frac{c_p(T_2 - T_1) + \omega_2 h_{g2}}{h_{s1} - h_{f2}}\end{aligned} \tag{20}$$

In the adiabatic saturation experiment, we measure T_1 and T_2 to calculate ω_1 using Eq. 20. Since ω_1 is unknown, h_{s1} is also unknown. However, we assume $h_{s1} \approx h_{g1}$, where h_{g1} is the enthalpy of saturated vapor at a saturation temperature of T_1. Now, assuming negligible pressure drop of air as it flows through the system, the moist air pressure at the outlet is the ambient pressure P. Using Eq. 5, we can find the value of ω_2, and thus, Eq. 20 can be written as:

$$\omega_1 = \frac{c_p(T_2 - T_1) + 0.622\frac{P_{g2}h_{g2}}{P - P_{g2}}}{h_{g1} - h_{f2}} \tag{21}$$

Here, P_{g2} is the saturation pressure of water at the temperature T_2. Once we measure T_1 and T_2, we can find c_p, P_{g2}, h_{f1}, and h_{g2} using standard thermodynamic Tables. Further, knowing the ambient pressure P, we can evaluate the specific humidity (ω_1) of air. Knowing the temperature T, pressure P, and specific humidity ω_1, any thermodynamic property of moist air can be evaluated using standard thermodynamic Tables. Thus, at a given ambient pressure, any property of moist air is a function of ambient temperature and adiabatic saturation temperature.

While the above discussion provides a relatively easier framework for finding the properties of moist air from temperature measurements only, finding the adiabatic saturation temperature requires conducting a sophisticated experiment, as discussed above. The experiment can be simplified by using a wet-bulb thermometer, as shown in Fig. 11. In the experimental arrangement, shown in Fig. 11, the thermometer bulb is in contact with the wet wick that is kept wet by dipping in the liquid water. As the air flows over the damp wick, due to evaporation, the air temperature drops below the ambient temperature. The temperature thus measured is called wet-bulb temperature (T_{wb}). Experiments suggest that the T_{wb} can well approximate adiabatic saturation temperature.

Figure 12 shows the simple handheld device called a sling psychrometer that simultaneously measures the ambient and wet-bulb temperatures. In this apparatus, an ordinary thermometer and a wet-bulb thermometer are placed over a single frame. The frame is connected to a handle via a hinge and can be rotated manually to establish airflow over the moist wick. While the ordinary thermometer measures the

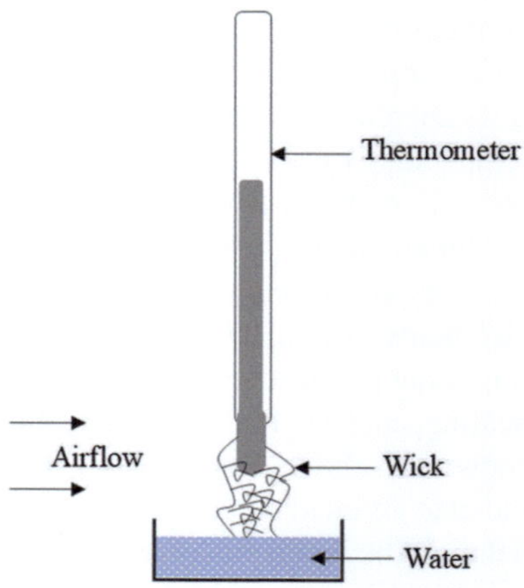

Fig. 11 Apparatus for measuring wet-bulb temperature. Airflow over the wick brings the local air temperature close to the adiabatic saturation temperature

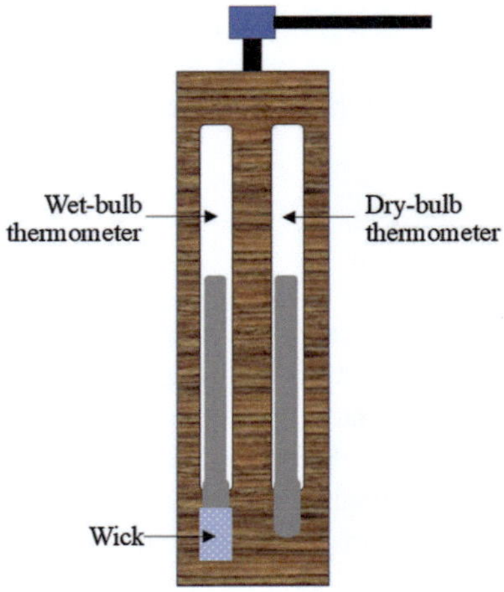

Fig. 12 Sling psychrometer simultaneously measures dry-bulb (ambient) and wet-bulb (adiabatic saturation) temperatures

ambient temperature, which is called the dry-bulb temperature (T_{db}), the wet-bulb thermometer measures the T_{wb}. Once the T_{db} and T_{wb} are measured, specific humidity can be calculated using Eq. 21. Knowing the ambient pressure, temperature, and specific humidity necessary thermodynamic properties for psychrometric application can be easily calculated.

6.2 *Psychrometric Chart*

At a given ambient pressure, the thermodynamic state of atmospheric air can be determined from two independent properties. Thus, in practice, we use a two-dimensional chart containing the necessary properties of atmospheric air. Schematically shown in Fig. 13, such a chart is called a psychrometric chart. While dry-bulb temperature forms the x-axis of the chart, specific enthalpy remains in the y-axis. The chart contains the lines of constant relative humidity, specific volume, enthalpy, and wet bulb temperature. While few sample lines are shown in Fig. 13, families of such lines are plotted in an actual chart. In practice, after measuring the dry- and wet-bulb temperatures using a sling psychrometer, we can locate the corresponding atmospheric condition as a point on the psychrometric chart leading to the evaluations of the remaining properties.

Apart from finding various thermodynamic properties, the psychrometric chart also helps to visualize important psychrometric processes. In the context of AWH, important psychrometric processes are combinations of four fundamental processes: heating, cooling, humidification, and dehumidification (Tian et al. 2023). Figure 14 shows the above four processes. While heating and cooling (processes AB, AC) indicate an increase or decrease of T_{db}, keeping the ω constant, humidification and dehumidification (processes AD, AE) increase or decrease the ω, keeping the T_{db} constant (Eastop and McConkey 2002).

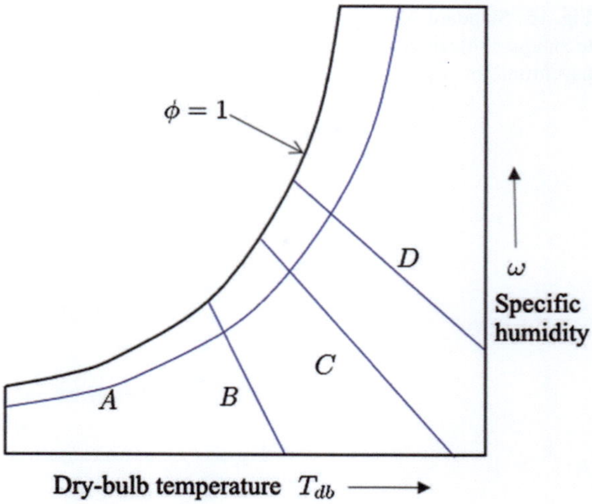

Fig. 13 Schematic of a typical psychrometric chart. **a** Line of constant relative humidity, **b** Line of constant specific volume, **c** Line of constant specific enthalpy, and **d** Line of constant wet-bulb temperature

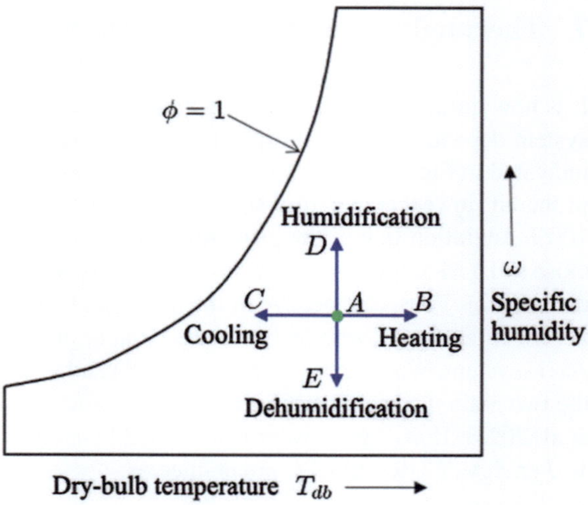

Fig. 14 Four fundamental psychrometric processes shown in a psychrometric chart

For harvesting water from the atmosphere, we must bring the water vapor to 100% saturation and collect the condensate. Figure 15 summarizes the techniques typically used in AWH. Setting point *A* to be the ambient condition, the process *AB* indicates the cooling of water vapor to 100% saturation without adding any water, a phenomenon observed in dew formation. Similarly, the process *AC* follows a constant T_{wb} line indicating the adiabatic saturation process. Such a process becomes useful in water harvesting from cooling towers. Finally, sorption-based processes (*AD*) first increase the ω using appropriate materials and structures, followed by cooling water vapor to 100% saturation (*DE*) (Hanikel et al. 2020).

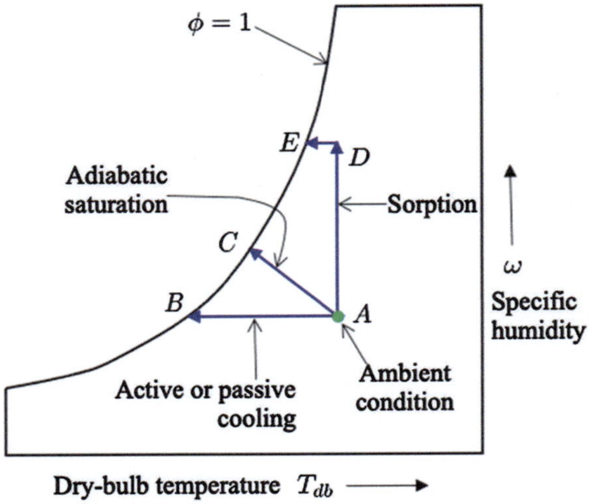

Fig. 15 Standard AWH techniques shown in a psychrometric chart

7 Thermodynamic Analyses of Standard AWH Techniques

It is now quite clear that at a given ambient pressure, the performance of an AWH system depends strongly on specific humidity and temperature (Bagheri 2018). As indicated in Fig. 13, while higher specific humidity implies higher moisture content in the air, lower temperature signals the closeness of atmospheric condition to the 100% saturation line in the psychrometric chart. Having an atmospheric condition close to 100% saturation means less energy expenditure for water removal (Kandeal et al. 2022). Thus, irrespective of the technology used, the thermodynamic concepts described above provide us few basic tools to judge the overall performance of an AWH system. Water harvesting rate (WHR) and unit power consumption (UPC) are the two main thermodynamic performance indicators for an AWH system (Nikkhah et al. 2023). If m_w kg of water is produced by a AWH system in Δt sec consuming W J energy, WHR and UPC are defined as:

$$\text{WHR} = \frac{m_w}{\Delta t} \ \text{UPC} = \frac{W}{m_w} \tag{22}$$

Clearly, high WHR and low UPC are desirable for an AWH system.

7.1 *AWH via Active Cooling*

Harvesting atmospheric water using active cooling usually requires the incorporation of a standard refrigeration cycle to cool the atmospheric air such that the vapor reaches the 100% saturation line (Ahrestani et al. 2023). Typical refrigeration systems

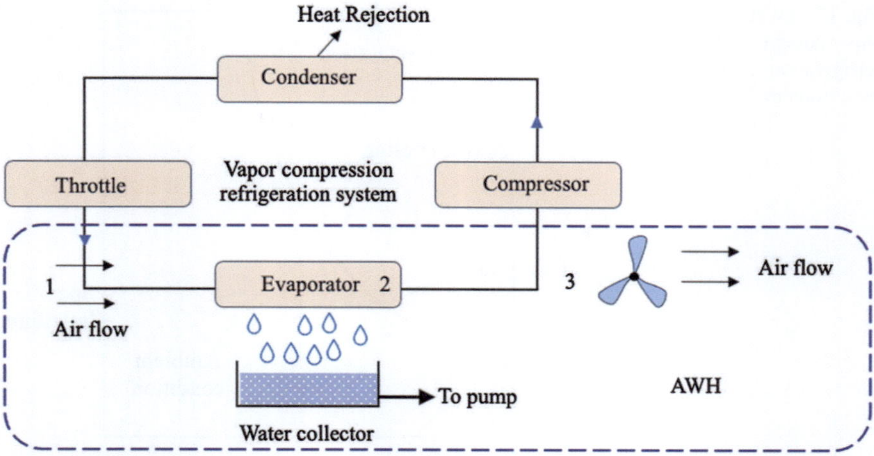

Fig. 16 AWH system using vapor compression refrigeration

include vapor compression refrigeration, vapor absorption refrigeration, thermoelectric cooling, etc. Figure 16 shows the schematic of an AWH system that uses vapor compression refrigeration for cooling of atmospheric air. The refrigeration system includes four major components: compressor, condenser, throttle, and evaporator. A fan establishes atmospheric airflow over the evaporator bringing the atmospheric water vapor to 100% saturation. Air enters the system at 1, reaches 100% saturation at 2, and leaves the system at 3. The condensed water is then collected and pumped to storage facilities (Gido et al. 2016).

Figure 17 shows the above process on a psychrometric chart. Ambient air enters at 1 and undergoes cooling at the evaporator. Condensation starts when the air reaches 100% saturation (point 2). Since the condensate is removed from the air, the specific humidity of the air is dropped while the air remains at 100% saturation (process 2–3). Finally, air leaves the evaporator at 3. Air, leaving at 3, can still be used for other purposes, such as air-conditioning.

Once we locate points 1, 2, 3 on the psychrometric chart, if the mass flow rate of moist air ($\dot{m}_a$) entering the evaporator is measured as a function of time (t), thermodynamic quantities, such as WHR and UPC, can be easily calculated. The heat transfer rate at the evaporator is:

$$\dot{Q} = \dot{W}.\mathrm{COP} = \dot{m}_a(h_1 - h_2) \tag{23}$$

Here, $\dot{W}$ is the electrical power consumption rate of the vapor compression system and COP is the coefficient of performance of the refrigeration system. Thus, from Eq. 23, we find:

$$\dot{W} = \frac{\dot{m}_a(h_1 - h_2)}{\mathrm{COP}} \tag{24}$$

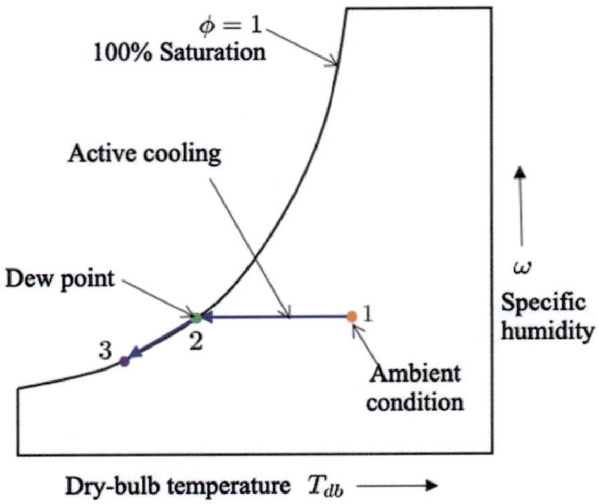

Fig. 17 AWH system using vapor compression refrigeration shown on a psychrometric chart

Now, the total energy consumption between the time interval t_1 to t_2 is obtained by integrating Eq. 24.

$$W = \int_{t_1}^{t_2} \dot{W}\, dt = \frac{(h_1 - h_2)}{\text{COP}} \int_{t_1}^{t_2} \dot{m}_a dt = \frac{m_a(h_1 - h_2)}{\text{COP}} \tag{25}$$

where m_a is the total mass flow of ambient air between the time interval t_1 to t_2.

Again, knowing the specific humidities at 2 and 3, from the psychrometric chart, the mass flow rate of condensate is:

$$\dot{m}_w = \dot{m}_a(\omega_2 - \omega_3) \tag{26}$$

Thus, the total mass of liquid water collection in the time interval t_1 to t_2 is:

$$m_w = \int_{t_1}^{t_2} \dot{m}_w dt = \int_{t_1}^{t_2} \dot{m}_a(\omega_2 - \omega_3) dt = m_a(\omega_2 - \omega_3) \tag{27}$$

Thus, the thermodynamic performance parameters of the AWH system are:

$$\begin{aligned} \text{WHR} &= \frac{m_w}{\Delta t} = \frac{m_a(\omega_2 - \omega_3)}{t_2 - t_1} \\ \text{UPC} &= \frac{W}{m_w} = \frac{\frac{m_a(h_1 - h_2)}{\text{COP}}}{m_a(\omega_2 - \omega_3)} = \frac{h_1 - h_2}{\text{COP}(\omega_2 - \omega_3)} \end{aligned} \tag{28}$$

Thus, high value of $\omega_2 - \omega_3$ and low value of $h_1 - h_2$ would yield superior thermodynamic performance. While the value of $\omega_2 - \omega_3$ primarily depends on the design of the refrigeration system, $h_1 - h_2$ is largely dictated by the ambient conditions.

From a thermodynamic viewpoint, we may assign $\omega_2 - \omega_3 \approx \omega_2$ and thus, the limiting performance of the AWH can be stated as:

$$\mathrm{WHR}_{\max} = \frac{m_a \omega_2}{t_2 - t_1} \; \mathrm{UPC}_{\min} = \frac{h_1 - h_2}{\mathrm{COP}\omega_2} \tag{29}$$

Clearly, the $\mathrm{WHR}_{\max}$ can be increased, and $\mathrm{UPC}_{\min}$ can be decreased by increasing ω_2 and reducing h_1–h_2. As the psychrometric chart suggests, this can be achieved by raising the ω_1. Thus, we may use adiabatic saturation or sorption-based techniques to improve the system performance, as shown in Fig. 15 (Feng et al. 2022).

7.2 AWH via Adiabatic Saturation

Adiabatic saturation, as discussed before, saturates the water vapor by adding liquid water to the system. At first glance, such a technique does not seem to be quite efficient since the final goal of AWH is to produce liquid water. However, the adiabatic saturation process is routinely used in cooling towers where evaporative cooling brings down the cooling water temperature to T_{wb}. During this process, cooling towers produce exhaust fog that can be harvested using standard fog harvesting techniques. Another idea is to supply saline water to adiabatically saturate the water vapor present in the ambient air. The liquid water generated this way may not be readily potable and may need desalination. Figure 18 shows the process on a psychrometric chart. Here, process 1–2 shows adiabatic saturation that may be occurring in a cooling tower that is an auxiliary to another engineering system. The thermodynamic analyses of such processes can be accomplished using the formulations shown above.

7.3 AWH via Sorption-Based Techniques

A closer look at the psychrometric chart reveals that high-temperature, low-humidity condition increases $h_1 - h_2$. Such a situation, often encountered in arid regions, is thus not suitable for AWH vis active cooling (Aleem et al. 2023). In such cases, we must rely on passive cooling, such as fog or dew harvesting. It is also possible to use sorption-based AWH techniques in arid regions (Shukla et al. 2024). Sorption-based techniques, typically used in arid regions, use appropriate materials that increase the specific humidity and consequently reduce the dew point (Gentile et al. 2022). Materials such as activated carbon, silica gen, zeolites, and metal-oxide frameworks are hydrophilic materials that efficiently absorb moisture from the ambient atmosphere

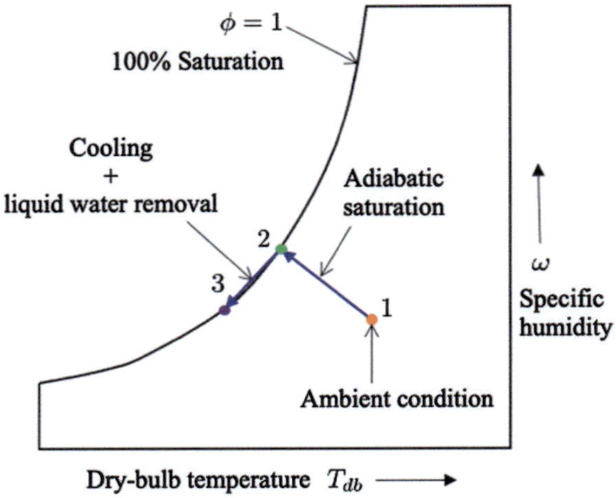

Fig. 18 AWH, via adiabatic saturation, on a psychrometric chart. Water vapor is saturated between 1 to 2. Liquid water is collected between 2 to 3

(Kim et al. 2020). At an elevated temperature, such materials release water vapor that is condensed to produce liquid water (Pan et al. 2020). The heating is usually accomplished via solar radiation since most arid regions have a sufficient supply of solar energy (Zhang et al. 2023). Figure 19 shows the process on the psychrometric chart. Here, ambient air humidity is raised by an adsorbent in the process 0–1. Once the air reaches 1, we may use an active cooling as discussed before. Thus, air enters the evaporator at 1. The process then follows the same pathway as discussed in the previous section. The thermodynamic parameters, such as the WHR and UPC, can also be calculated using the formulation shown in the previous section (Kalmutzki et al. 2018).

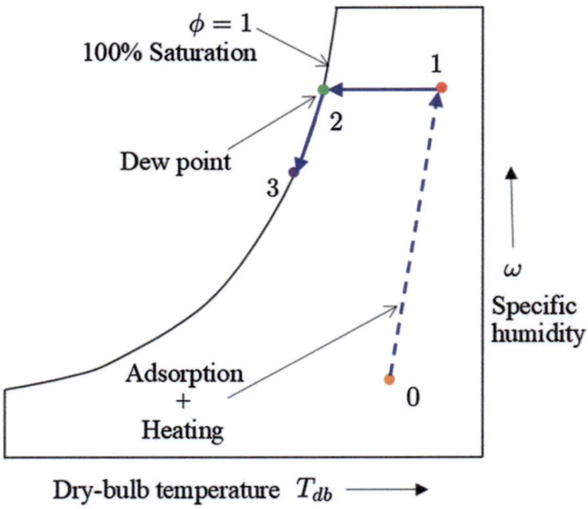

Fig. 19 Sorption-based AWH system on a psychrometric chart. Water vapor is adsorbed between 0 to 1. Water vapor is then cooled using a standard vapor compression or similar refrigeration cycle

If the system undergoes active cooling without adsorption, air will enter the evaporator at 0. As Fig. 19 suggests, in the case of active cooling, the dew point would have been very low, leading to a large enthalpy drop in the evaporator. Large enthalpy drop increases the UPC of the AWH system. Thus, active cooling is not the best method for high temperature, low humidity ambient (Salek et al. 2018). It is generally agreed that active cooling may prove inefficient for ambient conditions with $\phi < 0.2$ (Rao et al. 2022). In such cases, adsorption followed by active cooling is a better idea. The goal of adsorption, in any case, is to increase the dew point temperature, leading to a decrease in the enthalpy drop across the evaporator and thus improving the thermodynamic performance (Logan et al. 2020).

Using the procedures discussed above, any AWH system can be thermodynamically analyzed. Thermodynamic analysis provides an inexpensive way to find the limiting performance of the system (Qureshi and Zubair 2003). Details design of the AWH system, of course, requires coupling of transport phenomena with the thermodynamics.

8 Closure

In this chapter, we have discussed the thermodynamics of atmospheric water harvesting (AWH). Since AWH requires condensing superheated water vapor to liquid water, phase change thermodynamics is central to the thermodynamic analysis of AWH. Apart from the phase equilibrium and phase change thermodynamics, the chapter also discusses the psychrometric principles in detail. Finally, examples show the efficacy of thermodynamic analysis in quantifying the limiting performance of AWH systems.

A review of the present state of AWH research indicates two main research directions, namely material development, and system optimization. In both cases, thermodynamics analysis could prove to be an efficient and inexpensive tool to quantify the limiting performances of novel materials and innovative AWH system designs. We will now move on to look at the principles of evaporation in the next chapter.

Nomenclature

COP	Coefficient of performance
c_p	Specific heat at constant pressure (J kg^{-1} K^{-1})
dm	Change in mass (kg)
g	Specific Gibbs energy (J kg^{-1})
G	Gibbs energy (J)
h	Specific enthalpy (J kg^{-1})
H	Enthalpy (J)
m	Mass (kg)

$\dot{m}$	Mass flow rate (kg s^{-1})
M	Molar mass (kg $kmol^{-1}$)
P	Pressure (Pa)
Q	Heat transfer (J)
$\dot{Q}$	Heat transfer rate (W)
R	Gas constant (J kg^{-1} K^{-1})
R_u	Universal gas constant (J K^{-1} mol^{-1})
S	Entropy (J K^{-1})
t	Time (s)
T	Temperature (K)
UPC	Unit power consumption (J kg^{-1})
V	Volume (m^3)
v	Specific volume (m^3 kg^{-1})
W	Energy (J)
$\dot{W}$	Power (W)
WHR	Water harvesting rate (kg s^{-1})

Greek Symbols

ϕ	Relative humidity
ω	Specific humidity (kg kg^{-1})

Subscripts

A	Dry air
Critical	Critical point of water
Db	Dry bulb
Dp	Dew point
F	Liquid
Max	Maximum
Min	Minimum
S	Superheated
Sat	Saturation
V	Vapor
W	Liquid water
Wb	Wet bulb

Abbreviations

AWH Atmospheric water harvesting

References

Ahrestani Z, Sadeghzadeh S, Motejadded Emrooz HB (2023) An overview of atmospheric water harvesting methods, the inevitable path of the future in water supply. RSC Adv 13:10273–10307

Aleem M, Sultan M, Farooq M et al (2023) Evaluating the emerging adsorbents for water production potential and thermodynamic limits of adsorption-based atmospheric water harvesting systems. Int Commun Heat Mass Transf 145. https://doi.org/10.1016/j.icheatmasstransfer.2023.106863

Bagheri F (2018) Performance investigation of atmospheric water harvesting systems. Water Resour Ind 20:23–28. https://doi.org/10.1016/j.wri.2018.08.001

Bilal M, Sultan M, Morosuk T et al (2022) Adsorption-based atmospheric water harvesting: a review of adsorbents and systems. Int Commun Heat Mass Transf 133. https://doi.org/10.1016/j.icheatmasstransfer.2022.105961

Cengel YA, Boles MA (2019) Thermodynamics: an engineering approach, 9th edn. McGraw Hill Education

Eastop TD, McConkey A (2002) Applied thermodynamics for engineering technologists

Eslami M, Tajeddini F, Etaati N (2018) Thermal analysis and optimization of a system for water harvesting from humid air using thermoelectric coolers. Energy Convers Manag 174:417–429. https://doi.org/10.1016/j.enconman.2018.08.045

Feng A, Akther N, Duan X et al (2022) Recent development of atmospheric water harvesting materials: a review. ACS Materials Au 2:576–595

Gentile V, Bozlar M, Meggers F, Simonetti M (2022) Liter-scale atmospheric water harvesting for dry climates driven by low temperature solar heat. Energy 254. https://doi.org/10.1016/j.energy.2022.124295

Gido B, Friedler E, Broday DM (2016) Assessment of atmospheric moisture harvesting by direct cooling. Atmos Res 182:156–162. https://doi.org/10.1016/j.atmosres.2016.07.029

Hanikel N, Prévot MS, Yaghi OM (2020) MOF water harvesters. Nat Nanotechnol 15:348–355

Kalmutzki MJ, Diercks CS, Yaghi OM (2018) Metal–organic frameworks for water harvesting from air. Adv Mater 30

Kandeal AW, Joseph A, Elsharkawy M et al (2022) Research progress on recent technologies of water harvesting from atmospheric air: a detailed review. Sustain Energy Technol Assess 52. https://doi.org/10.1016/j.seta.2022.102000

Kim H, Rao SR, LaPotin A et al (2020) Thermodynamic analysis and optimization of adsorption-based atmospheric water harvesting. Int J Heat Mass Transf 161. https://doi.org/10.1016/j.ijheatmasstransfer.2020.120253

Kocher JD, Menon AK (2023) Addressing global water stress using desalination and atmospheric water harvesting: a thermodynamic and technoeconomic perspective. Energy Environ Sci 16:4983–4993

Li AC, Zhang L, Zhong Y et al (2022) Thermodynamic limits of atmospheric water harvesting with temperature-dependent adsorption. Appl Phys Lett 121. https://doi.org/10.1063/5.0118094

Logan MW, Langevin S, Xia Z (2020) Reversible atmospheric water harvesting using metal-organic frameworks. Sci Rep 10. https://doi.org/10.1038/s41598-020-58405-9

Moran MJ, Shapiro HN, Boettner DD, Bailey MB (2015) Principles of engineering thermodynamics. Wiley

Nikkhah H, Azmi WMBW, Nikkhah A et al (2023) A comprehensive review on atmospheric water harvesting technologies: from thermodynamic concepts to mechanism and process development. J Water Process Eng 53

Pan T, Yang K, Han Y (2020) Recent progress of atmospheric water harvesting using metal-organic frameworks. Chem Res Chin Univ 36:33–40

Qureshi BA, Zubair SM (2003) Application of exergy analysis to various psychrometric processes. Int J Energy Res 27:1079–1094. https://doi.org/10.1002/er.933

Rao AK, Fix AJ, Yang YC, Warsinger DM (2022) Thermodynamic limits of atmospheric water harvesting. Energy Environ Sci 15:4025–4037

Salek F, Moghaddam AN, Naserian MM (2018) Thermodynamic analysis and improvement of a novel solar driven atmospheric water generator. Energy Convers Manag 161:104–111. https://doi.org/10.1016/j.enconman.2018.01.066

Shukla A, Sunil, Thakur AD, Raj R (2024) Experiment and modelling of an improvised atmospheric water harvester for arid and semi-arid conditions. Appl Therm Eng 242. https://doi.org/10.1016/j.applthermaleng.2024.122486

Sonntag RE, Borgnakke C, Van Wylen GJ (2002) Fundamentals of thermodynamics. John Wiley & Sons

Tashtoush B, Alshoubaki A (2023) Atmospheric water harvesting: a review of techniques, performance, renewable energy solutions, and feasibility. Energy 280

Tian S, Geng Y, Huang L et al (2023) Fundamental and perspectives of thermo-responsive materials for dehumidification and water harvesting. J Clean Prod 412

Tu R, Hwang Y (2020) Reviews of atmospheric water harvesting technologies. Energy 201

Xiang T, Xie S, Chen G et al (2024) Recent advances in atmospheric water harvesting technology and its development. Mater Horiz

Zemansky MW, Dittman RN (1997) Heat and thermodynamics, 7th edn. McGraw-Hill Higher Education

Zhang Q, Tu R, Liu M (2023) Performance analyses and optimization studies of desiccant wheel assisted atmospheric water harvesting system under global ambient conditions. Energy 283. https://doi.org/10.1016/j.energy.2023.128477

Fundamentals of Evaporation

Manish Bhendura and K. Muralidhar

1 Introduction

Evaporation is a liquid-to-vapor phase transition process that proceeds with the absorption of latent heat of vaporization from the liquid phase. It is one of the essential steps in temperature-driven water desalination systems, wherein brackish water is heated and evaporated, and the generated water vapor is transported to a relatively colder surface for condensation as the distillate (Tiwari and Sahota 2017). The condensate purity is sufficiently high and serves the purpose as potable water. In the context of atmosphere water harvesting (AWH) technologies, water evaporates from natural water reservoirs, such as rivers and ponds, partially or fully saturating the atmospheric air with vapor (Tu and Hwang 2020). Moisture thus available may be suitably condensed to become a source of clean water. Figure 1 classifies AWH techniques based on water–vapor saturation in the ambient air as per sorption–desorption cycle, dropwise condensation, and fog collection (Lu et al. 2022). In other words, evaporation is the first and the rate-limiting step in thermally driven water distillation and AWH systems.

The vaporization process is a multi-faceted phenomenon that generates pure vapor, induces a cooling effect in the liquid body, and causes interface movement owing to a variety of factors including loss of liquid mass. The gaseous phase above the water body is set into buoyancy-driven motion. Simultaneous, the liquid medium is also set in motion from factors such as thermo-capillary flow, buoyancy-driven flow, interfacial deformation, and motion of the three-phase contact line at the bounding walls. The coupled nature of these phenomena complicates the analysis of evaporation, especially when the molecular interaction among the three phases is considered.

M. Bhendura
IIT, Roorkee, India

K. Muralidhar (✉)
IIT Kanpur, Kanpur, India
e-mail: kmurli@iitk.ac.in

R. Narayanaswamy et al. (eds.), *Sustainable Water Harvesting*, Springer Tracts in Mechanical Engineering, https://doi.org/10.1007/978-981-95-3373-2_4

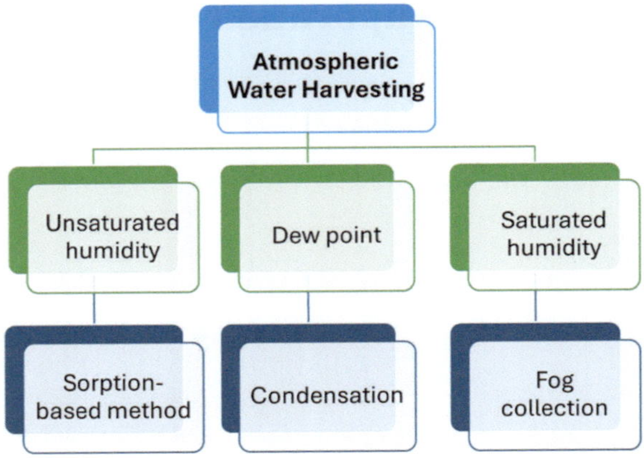

Fig. 1 A classification chart of atmospheric water collection systems based on humidity. In the sorption-based approach, a desiccant absorbs water vapor from air and is subsequently heated to release water vapor and condense it as water. Alternatively, a surface relatively colder than the ambient nucleates droplets that cyclically grow and depart, producing a steady stream of water. Meanwhile, fog collectors capture suspended water droplets in flowing air on mesh screens, not involving any phase change or energy input

Therefore, a comprehensive analysis jointly with experimental validation becomes essential to develop understanding of evaporation rate and its dependence on the process variables.

This chapter presents a discussion of the evaporation process in the context of three specific applications: basin-type solar distillation, membrane distillation, and droplet evaporation. The working medium is water or an aqueous solution unless otherwise stated; temperatures are below the boiling point and phase-change by boiling is not considered. Section 2 describes three distinct approaches to determining the evaporative mass flux of a warm water pool at a given temperature surrounded by ambient air. The implementation of these models leads to numerical analysis of an air–water system along with experimental validation as described in Sect. 3. Evaporation of a pool of water is extended to membrane-controlled evaporation, wherein a combination of Knudsen diffusion, ordinary diffusion, and Poiseuille flow defines vapor transport through the membrane pores (Sect. 4). An extreme limiting geometry of a pool of water with a deformed interface is a droplet whose evaporation is described in Sect. 5. Mechanisms of interest here are evaporation-induced internal liquid flow owing to surface tension gradients and contact line motion. The possibility of an analytical solution for diffusion-limited droplet evaporation is also discussed. The chapter concludes with a discussion on thin-film evaporation and the available modeling approaches.

2 Evaporation Models: Air–Water System

In evaporation of a water body, the air–water interface generates a vapor mass flux transported into air and induces heat transport in both air and water. The vapor flux depends upon thermodynamic state variables—temperature and partial pressure of vapor in air, flow velocities in air and water, and thermophysical properties, mainly the mass diffusion coefficient of vapor in air. Mathematically, three approaches have been proposed in the literature to determine the evaporation rate: ab initio, kinetic theory of gases, and empirical correlations. These are described below in the following sub-sections.

2.1 Ab Initio Model Including Stefan Flow

From first principles, the evaporation rate of a body of water is the summation of diffusive and convective vapor fluxes (Misyura 2018; Vlasov 2020). Following these studies, the expression for the total evaporative vapor flux is obtained as (Fig. 2):

$$\dot{m}_{\text{evap}} = \dot{m}_{w,\text{diff}} + \dot{m}_{w,\text{conv}} \tag{1}$$

The diffusive flux is evaluated using Fick's law of mass diffusion at the air–water interface and is expressed in the following form (Bird et al. 2006):

$$\dot{m}_{w,\text{diff}} = -\rho_a D_{a-w} \nabla Y_w \tag{2}$$

Here, ρ_a is the density of air in kg/m^3, D_{a-w} is the binary mass diffusion coefficient of moisture in air in units of m^2/s, and Y_w is the mass fraction of water vapor in air.

The convective vapor flux is caused by bulk gaseous flow, equivalent to the diffusive flux of air, and is referred as Stefan flow. Figure 2 illustrates induced Stefan

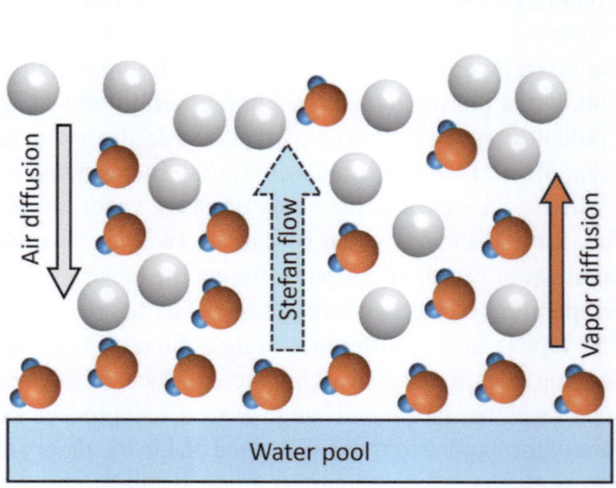

Fig. 2 Physical interpretation of Stefan flow induced by vapor diffusion into air; H_2O molecules in the vapor phase are schematically shown in red and blue color while air molecules are in white

flow caused by the diffusion of air and water–vapor. As water evaporates and vapor diffuses into air, an equal and opposite air flux is transported toward the interface. Since the interface is impermeable to air, Stefan flow is induced near the interface. This convective flow is expressed as:

$$\dot{m}_{w,\text{conv}} = \rho v_s Y_w \tag{3}$$

Here, ρ is the density of the air-vapor mixture in units of kg/m^3 and v_s is the Stefan velocity in m/s. Since the Stefan flux is equivalent to the diffusive flux of air, equating these fluxes determines the Stefan velocity. Substituting the Stefan velocity into Eqs. (3) and (1) leads to the augmented evaporative mass flux at the interface, expressed as:

$$\dot{m}_{\text{evap}} = \frac{1}{1 - Y_{w,\text{sat}}}(-\rho_a D_{a-w} \nabla Y_w) \tag{4}$$

The mass fraction of vapor in vapor-saturated air at the interface is denoted as $Y_{w,\text{sat}}$.

2.2 Kinetic Theory Approach

The ab initio approach assumes the air–water interface thickness to be zero and temperature variation is continuous between the two bulk phases, namely, liquid and vapor. However, a temperature jump is observed in highly resolved experiments across the liquid–vapor interface, which continuum models fail to account for. The non-equilibrium model addresses this temperature discontinuity, wherein two microlayers are assumed between the bulk phases, namely interface transition and Knudsen layer, as depicted in Fig. 3 (Aursand and Ytrehus 2019). The transition layer close to the liquid phase is of the order of a few molecular diameters and exhibits a rapid change in density and temperature. This layer is followed by the Knudsen layer of few molecular mean free path lengths and extends to within the bulk vapor phase. It exists in a non-equilibrium state and involves low temperature gradient across its thickness. The kinetic theory of gases models the microscopic variation and temperature discontinuity in these layers.

In the classical kinetic theory of gases, vapor molecules emerging from the interface have velocities given by the Maxwellian distribution (Van Carey 2020). The statistics of this distribution including the mean value and variance depend upon the temperature of the two bulk phases. The temperature discontinuity between liquid and vapor causes a difference in mass fluxes related to the distribution of molecules moving toward and away from the interface. In this respect, the net mass flux determines whether the phase change process will proceed as evaporation or condensation. One of the expressions derived based on kinetic theory to estimate the interfacial mass flux is given by Schrage-Mills (Aursand and Ytrehus 2019) and is written as follows:

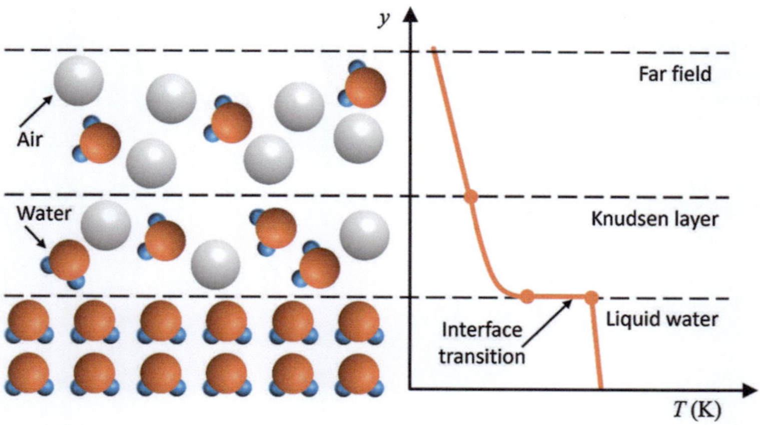

Fig. 3 Schematic of two microlayers formed at the air–water interface, namely interface transition and Knudsen layer and variation of temperature across the interface for an evaporating body of water (graph on the right is not drawn to scale)

$$\dot{m}_{\text{evap}} = \frac{2\alpha_e}{2-\alpha_e}\left[\frac{p_{w,\text{sat}}}{\sqrt{2\pi R_w T_w}} - \frac{p_\infty}{\sqrt{2\pi R_w T_\infty}}\right] \tag{5}$$

where, α_e is the accommodation coefficient (dimensionless), $p_{w,\text{sat}}$ is saturation pressure (N/m^2) at liquid temperature T_w (K), R_w is the gas constant of water (J/kg-K), p_∞ is vapor pressure (N/m^2), and T_∞ is the vapor temperature (K).

The accommodation coefficient is the fraction of molecules that changes its initial liquid or vapor phase out of the total number of molecules striking or coming out of the liquid–vapor interface (Persad and Ward 2016). Its value varies from 0 to 1; however, its exact value for polar liquids such as water and methanol is a subject of debate with 0.01 being preferred.

2.3 *Empirical Correlations*

The ab initio and kinetic theory approaches are computationally expensive, particularly for larger systems such as water pools, because of a prerequisite in the form of determination of vapor concentration in the bulk gaseous phase. In such cases, empirical correlations are proposed for specific applications involving evaporation, wherein the evaporation rate is a function of partial vapor pressure difference between the water surface and ambient air. A generic expression for evaporation rate can be written as (Sartori 2000):

$$\dot{m}_{\text{evap}} = (a + bV)\left(p_{w,\text{sat}} - p_v\right)^n \tag{6}$$

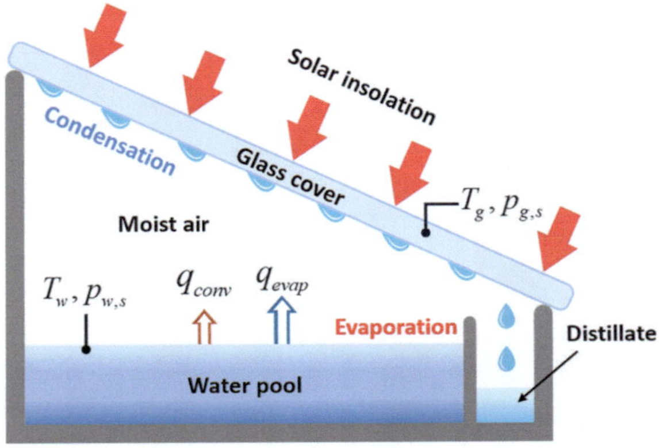

Fig. 4 Pictorial depiction of a basin-type passive solar desalination unit, highlighting pool evaporation of water and condensation at glass cover; respective temperatures and saturation pressures are indicated

Here, a, b, and n are empirical constants; V is the ambient velocity in m/s; $p_{w,\text{sat}}$ and p_v are saturated vapor pressure at the water surface and partial vapor pressure in air, respectively. The vapor pressures are in units of N/m^2.

A solar desalination system also incorporates a water pool inside a thermally insulated basin covered by a relatively colder glass cover at the top (Fig. 4). Water in the apparatus is heated using solar insolation, and the vapor generated is collected beneath the glass cover, where it condenses and is collected as potable water. For these systems, evaporation is modeled in terms of evaporative heat flux and defined in the following manner (Tiwari and Sahota 2017):

$$q_{\text{evap}} = h_{\text{evap}}\left(T_w - T_g\right) \tag{7}$$

In this expression, the evaporative heat transfer coefficient is symbolized as h_{evap} and has units of W/m^2 K. Symbols T_w and T_g are the temperatures of water surface and glass cover in units of K, respectively.

In evaporation of a water pool, air velocity acquired by buoyant convection impacts the evaporation rate as it dictates vapor transport from the water surface to the glass cover. An expression relating the convective and evaporative coefficients inside a solar basin is derived as follows (Tiwari and Sahota 2017):

$$h_{\text{evap}} = 16.273 \times 10^{-3}(h_{\text{conv}})\left(\frac{p_{w,\text{sat}} - p_{g,\text{sat}}}{T_w - T_g}\right) \tag{8}$$

Here, h_{conv} is convective heat transfer coefficient arising from air velocity in units of W/m^2 K.

For air enclosed inside the basin, simultaneous heat and vapor transport leads to buoyant convection. The applicable heat transfer coefficient of convective heat transport is determined using the following empirical correlation proposed by Dunkle in the context of a solar distillation system (Tiwari and Sahota 2017):

$$h_{\text{conv}} = 0.884 \times \left[\left(T_w - T_g\right)\right.$$

$$+\frac{\left(p_{w,\text{sat}}-p_{g,\text{sat}}\right)\left(T_w+273.15\right)}{268900-p_{w,\text{sat}}}\Bigg]^{1/3} \tag{9}$$

In Eqs. (8–9), $p_{w,\text{sat}}$ and $p_{g,\text{sat}}$ represent the saturation pressure of vapor in N/m^2 at the water surface and glass cover temperature (in Kelvin units) T_w and T_g, respectively.

The simple and direct nature of Dunkle's model (Eqs. 8–9) has made it popular and is extensively utilized in the design and analysis of solar distillation systems on various scales. However, the empirical correlation (Eq. 8) is obtained for free convection between two parallel plates without considering active evaporation and variation in length scales such as wall inclination. Kumar and Tiwari (1996) addressed these shortcomings and adopted the Nusselt number approach, wherein the constants are not predetermined but estimated first with linear regression analysis using experimental data. The Nusselt number dependence on the buoyancy parameter, namely Grashof number is expressed as:

$$\text{Nu}=\frac{h_{\text{conv}}L_c}{k_a}=C(\text{Gr Pr})^n \tag{10}$$

Here, L_c is characteristic length in m, k_a is thermal conductivity of moist air in W/m K, Gr is Grashof number, Pr is Prandtl number, while C and n are empirical constants. The dimensionless numbers are defined as follows:

$$\text{Grashof number}:\ \text{Gr}=\frac{\beta g L_c^3\left(T_w-T_g\right)\rho_a^2}{\mu_a^2} \tag{11}$$

$$\text{Prandtl number}:\ \text{Pr}=\frac{\mu_a C_{p,a}}{k_a} \tag{12}$$

In this formulation, the water distillation output in kg is calculated in terms of the evaporative heat flux using the following expression:

$$\begin{aligned} m_{\text{dist}} &= \frac{q_{\text{evap}}}{h_{fg}}\times A_w\times t \\ &= \frac{h_{\text{evap}}\left(T_w-T_g\right)}{h_{fg}}\times A_w\times t \end{aligned} \tag{13}$$

where, t and A_w are time (in seconds) and evaporative surface area (in m^2), respectively. Substituting Eqs. (8) and (10) into Eq. (13) leads to the expression below:

$$\begin{aligned} m_{\text{dist}} = &\frac{0.01623}{h_{fg}}\times\frac{k_a}{L_c}\times C(\text{Gr Pr})^n \\ &\times\left(p_{w,\text{sat}}-p_{g,\text{sat}}\right)\times A_w\times t \end{aligned}$$

$$= R \times C(\text{Gr Pr})^n \tag{14}$$

Here

$$R = \frac{0.01623}{h_{fg}} \times \frac{k_a}{L_c} \times \left(p_{w,\text{sat}} - p_{g,\text{sat}}\right) \times A_w \times t \tag{15}$$

Equation (14) can be rewritten in compact form as:

$$Y = AX + B \tag{16}$$

In this modified expression, $Y = \ln\left(m_{\text{dist}}/R\right), X = \text{Gr}\cdot\text{Pr}, A = \ln(C)$, and $B = n$. The parameters A and B are determined using linear regression analysis, while X and Y are obtained from the experimental values of water distillate and temperatures of water surface and glass cover. The expressions that calculate the parameters are based on the method of least squares and obtained as follows:

$$B = \frac{N(\sum XY) - (\sum X)(\sum Y)}{N(\sum X^2) - (\sum X)^2} \tag{17}$$

$$A = \frac{\sum Y}{N} - B\frac{\sum X}{N} \tag{18}$$

In Eqs. (17–18), N is the number of experimental observations.

This approach resolves the inadequacies of Dunkle's model (Eq. 9) and includes the effect of basin dimensions, operating temperature ranges, and orientation of the glass cover. Kumar and Tiwari (1996) showed a better agreement between the experimental and model values (~20%) for the solar basin having a double-sloping glass cover and three temperature ranges compared to the Dunkle model where the agreement with experiments was ~48%.

3 Water Pool Evaporation: Computational Modeling

Simplified thermal analysis of a basin-type solar desalination unit may be performed using the lumped system approach, wherein characteristic equations can be derived from energy balance of the water body and glass cover (Tiwari and Tiwari 2006; Matrawy et al. 2015). These equations relate time-varying temperatures to the appropriate external and internal heat transfer rates (e.g., radiative, convective, and evaporative) and thermal losses to the ambient. This basic energy-balance model is valuable for preliminary analysis. However, its accuracy is limited since the model does not account for spatial temperature variation, changes in buoyancy-driven flow, and temperature stratifications in both fluids. The shortcomings of lumped system analysis can be addressed using the continuum approach, which includes physical phenomena neglected in lumped analysis. As a result, the detailed modeling approach allows for a deeper understanding of the performance of solar stills though at the expense of computational cost.

3.1 Spatially Distributed Modeling and Governing Equations

As stated previously, the solar still is a thermally insulated basin, partially filled with water and exposed to a relatively colder glass cover at its top. A computational geometry containing two fluid domains of equal heights, i.e., air and water, is utilized as a case study to formulate the mathematical model for simultaneous heat and moisture transport in an enclosed air–water system. The geometry is axisymmetric as it mimics a circular basin of the solar desalination unit and is shown in Fig. 5. Water is assumed to be preheated to relax the complexities of radiative heat transfer, while the top boundary is considered to be flat and isothermal instead of tilted transparent glass. Buoyancy-driven flow patterns in air and water are modeled using the differential equations of mass conservation, momentum, and thermal energy, as stated below (Akkus et al. 2020; Bhendura et al. 2023):

$$\nabla \cdot \vec{v}_m = 0 \tag{19}$$

$$\rho_m \frac{D\vec{v}_m}{Dt} = -\nabla p_m + \mu_m \nabla^2 \vec{v}_m + \rho_m \vec{g} \tag{20}$$

$$\frac{DT_m}{Dt} = \alpha_m \nabla^2 T_m \tag{21}$$

Here, the index m denotes two phases: $m = 1$ for air, while $m = 2$ for water; $\vec{v}$ is the velocity vector (m/s), T is the temperature (K), p is the pressure (N/m^2), $\vec{g}$ is the

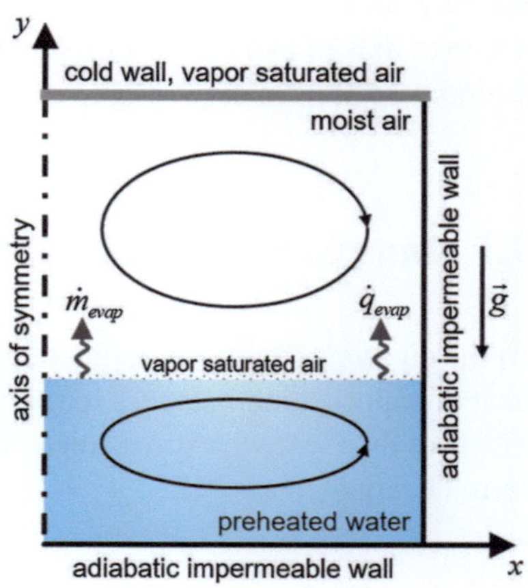

Fig. 5 Computational domain for air–water system in an axisymmetric framework, mimicking convection-driven evaporation in a basin-type solar distillation system

gravitational acceleration (m/s^2), ρ is the density (kg/m^3), μ is the absolute viscosity (Pa s), and α is the thermal diffusivity (m^2/s).

In addition, the water–vapor transport in air is modeled using the conservation law of species transport, expressed as follows (Bird et al. 2006; Bhendura et al. 2023):

$$\frac{D\omega}{Dt} = D_{a-w}(T, p)\nabla^2\omega \tag{22}$$

Here, ω represents absolute humidity in units of kg of water per kg of dry air (kg/kg) and D_{a-w} is the binary mass diffusion coefficient of water vapor in air (m^2/s). This coefficient varies with the local temperature and pressure as per the following relationship (Bird et al. 2006):

$$D_{a-w}(T, p) = 1.87 \times 10^{-10}\frac{T^{2.072}}{p} \tag{23}$$

In this equation, temperature and pressure are in units of Kelvin and atm, respectively.

3.2 *Initial and Boundary Conditions*

As the water is preheated, the initial water temperature is T_h. Air is assumed to be at a relatively lower temperature T_c, with ϕ, the relative humidity. The top boundary is maintained at the cold temperature T_c, while the right-side and lower boundaries are thermally insulated. As the vapor produced due to evaporation condenses over the top boundary, the gas phase at the top boundary is assumed to be vapor-saturated at temperature T_c. The condensate is taken to be appropriately removed but is not explicitly represented in the governing equations. The right and lower boundaries are assumed to be impermeable to vapor transport. In addition, no-slip conditions of velocity are enforced at the top, right, and bottom boundaries and the left boundary is one of axis of symmetry. Pressure may be stipulated as zero, a datum value, at the midpoint of the cavity.

3.3 *Interfacial Conditions*

At the air–water interface, temperature is assumed to be continuous, while the latent heat of vaporization balances the net heat flux crossing the interface from each side. The heat flux balance expression at the interface is now represented in the following form (Janajreh et al. 2017):

$$\sum_m [-\hat{n} \cdot (k_m \nabla \cdot T_m)] = \dot{m}_{\text{evap}} h_{fg} \tag{24}$$

In this expression, the evaporation rate $\left(\dot{m}_{\text{evap}}\right)$ is evaluated using one of the approaches discussed in Sect. 2.

Moreover, evaporation ensures the interface to be saturated with water vapor at the interfacial temperature due to molecular diffusion. The absolute humidity of saturated air is evaluated using the following definition applicable for ideal gases (Demirel 2007).

$$\omega_{\text{in,sat}} = \frac{M_w}{M_a} \times \frac{p_{\text{in,sat}}}{p - p_{\text{in,sat}}} \tag{25}$$

Here, M_w and M_a are the molecular weights of water and air in units of kg/kmol, respectively, while $p_{\text{in,sat}}$ is the saturation pressure of water at the interfacial temperature. An empirical correlation from steam tables relating saturation pressure and temperature is expressed as (Tiwari and Sahota 2017):

$$p_{\text{in,sat}} = \exp\left(25.317 - \frac{5144}{T_{\text{in}}}\right) \tag{26}$$

The interfacial temperature (T_{in}) in Kelvin results in a saturation pressure in units of N/m^2.

3.4 Validation Against Interferometry Experiments

Bhendura et al. (2023) validated the mathematical model formulated in Sect. 3 against the interferometric experiments. The experimental apparatus contains two identical circular cavities (diameter 120 mm and height 60 mm) (Fig. 6a and b). One of them, the test cavity is filled with preheated water to a depth of 30 mm while the reference cavity is entirely filled with warm water. These cavities are thermally insulated and placed in the path of light beams. Initially, the top of both cavities is covered with insulated plates, and to begin the experiments, the test cavity cover is replaced by a cooler constant temperature surface. The constant temperature is maintained using a water bath that circulates water through a cuboidal copper box in a serpentine path created by internal fins.

As the constant temperature surface is placed over the test cavity, the air–water system cools, and the water vapor condenses beneath the cold surface. It leads to simultaneous heat and moisture transport in air and evaporative cooling of water. The cooling down of the water body is visualized using a Mach–Zehnder interferometer (MZI). The interferometer operates in the wedge fringe mode, producing equally spaced and vertically erect fringes in water. After placing the cold surface over the test cavity, these fringes bend gradually near the air–water interface, representing changes

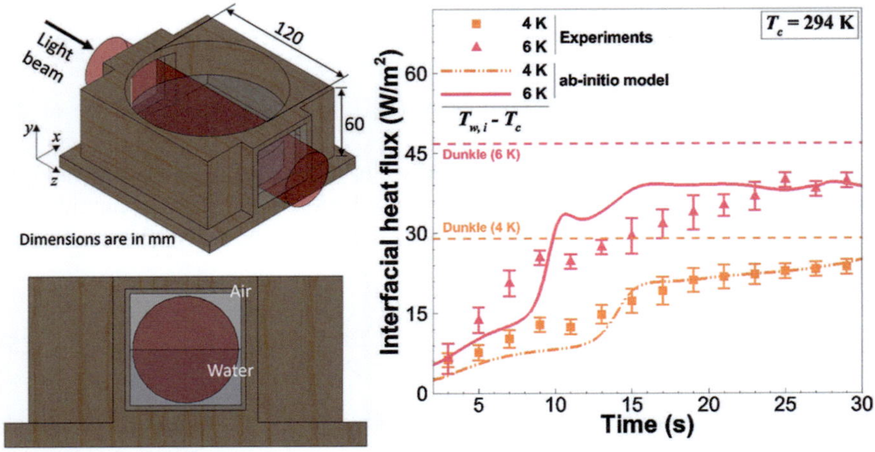

Fig. 6 **a** Isometric drawing and **b** front view of the test cavity partly-filled with water with a coherent light beam propagating through the cavity; **c** time-varying heat fluxes measured using laser interferometry and compared with estimated values using ab-initio and Dunkle models. (Reproduced with permission from Bhendura et al. 2023. © 2023 Elsevier B.V.)

in the interfacial temperature gradient. Naylor and Duarte (1999) have derived the following expression to relate temperature gradient with the fringe bending angle:

$$\left.\frac{\mathrm{d}T}{\mathrm{d}y}\right|_{\mathrm{in}} = \frac{\lambda / L}{\mathrm{d}n / \mathrm{d}T}\left(\frac{-1}{d \times \tan\theta}\right) \tag{27}$$

Here, λ is the wavelength of the light beam in m, L is the length of beam propagation through the test cavity in m, n is refractive index, $\mathrm{d}n/\mathrm{d}T$ is a thermo-optic constant in units of K^{-1}, d is the initial gap between two parallel fringes in units of m, and θ is the fringe bending angle at the interface in units of radians.

Temperature gradient in water (Eq. 27) is further utilized to evaluate the interfacial heat flux in water using Fourier's law of heat conduction $\left(q_{w,\mathrm{in}} = -k_w \cdot (\mathrm{d}T/\mathrm{d}y)_{\mathrm{in}}\right)$. This heat flux is the sum of evaporative heat flux and heat flux in air and is also equal to the interfacial heat flux in water. In numerical simulation of the air–water system enclosed in a circular cavity, the interfacial heat flux in water is calculated after numerically solving the governing equations (Eqs. 19–22) with appropriate initial, boundary, and interfacial conditions (Eqs. 24–25). Figure 6c illustrates the time-varying interfacial heat flux for two initial temperature differences, i.e., 4 and 6 K, where the cold temperature boundary at the top of the test cavity is maintained at 294 K while the initial water temperatures are 298 and 300 K, respectively. The ab initio model (continuous lines) shows a good match with experimental values (symbols). The model further captures the rise in interfacial heat flux at a later time, being attributed to the delayed onset of natural convection in air.

For the given initial temperature difference, the Dunkle model overpredicts the interfacial heat fluxes, shown as dotted lines in Fig. 6c. The predictions of the Dunkle model can be improved after re-calibrating the coefficients of Eq. (10), using the methodology discussed in Sect. 2.3. Improvements in empirical modeling can also be obtained my machine learning techniques (Bhendura et al. 2022).

4 Evaporation Through a Hydrophobic Membrane

Membrane distillation is a thermally driven system containing a semi-permeable hydrophobic membrane that only allows water vapor molecules to pass through but restricts the movement of liquid water molecules (Khayet 2011). The membrane also separates the feed and permeate sides of a channel of the process reactor. Heated water flows at the feed side and evaporates through the membrane pores; the vapor thus produced is transported to the permeate side through the membrane, where it condenses, and pure water is collected. This transmembrane evaporation is modeled with the presumption that evaporative mass flux is proportional to the difference in vapor saturation pressures on the two faces of the membrane and is formulated as follows (Janajreh et al. 2017):

$$\dot{m}_{\text{evap}} = C_m\left(p_{w,\text{sat}} - p_{a,\text{sat}}\right) \tag{28}$$

Here, C_m is the membrane mass transfer coefficient (also called, permeability) expressed in units of m^{-1} s while $p_{w,\text{sat}}$ and $p_{a,\text{sat}}$ are vapor saturation pressures on water and air sides of the membrane at the respective temperatures.

Membrane permeability is directly related to its characteristics, specifically, porosity, tortuosity, pore size, and thickness and operating conditions in terms of temperature and pressure, thus determining the mode of vapor flow through the membrane pores. Three modes of vapor transport have been identified in the literature: Knudsen diffusion, Ordinary diffusion, and Poiseuille flow (Kadi et al. 2020). As the geometric parameters of the membrane may have a non-uniform distribution, the resulting flow might be simultaneously governed by more than one mechanism. The modes of transport are then suitably combined to predict the membrane coefficient in such a way that there is overall match with experiments. The coefficients applicable in Eq. (28) for each of the modes are calculated as follows (Kadi et al. 2020):

$$\text{Knudsen diffusion}:\ C_{m,k} = 1.064 \times \frac{\varepsilon r_p}{\tau \delta_m}\sqrt{\frac{M_w}{RT_m}} \tag{29}$$

$$\text{Ordinary diffusion}:\ C_{m,o} = \frac{\varepsilon p D_{a-w}}{p_a \tau \delta_m} \times \frac{M_w}{RT_m} \tag{30}$$

$$\text{Poiseuille flow}: \; C_{m,p} = 0.125 \times \frac{\varepsilon r_p^2}{\tau \delta_m} \times \frac{M_w p_{m,\mathrm{sat}}}{R T_m \mu_w} \tag{31}$$

In Eqs. (29–31), the membrane properties: porosity, tortuosity, pore size, and thickness are denoted using ε, τ, r_p, and δ_m, respectively; the operating conditions: average temperature across the membrane, corresponding saturation pressure, partial air pressure, and total pressure are represented using $T_m, p_{m,\mathrm{sat}}, p_a$, and p, respectively; the thermophysical properties of water vapor: M_w is molecular weight, μ_w is absolute viscosity, D_{a-w} is mass diffusivity of vapor in dry air, and R is the universal gas constant. These coefficients are judiciously combined as discussed in the following section.

4.1 Validation Against Interferometry Experiments

The interferometry experiments discussed in Sect. 3 have been extended to study evaporation through a hydrophobic membrane (Bhendura et al. 2023). The membrane utilized in these experiments is made of polytetrafluoroethylene (PTFE) with a thickness of 160 μm, pore size of 0.4 μm, porosity of 80%, and an equilibrium contact angle of $115^\circ \pm 5^\circ$. This membrane is placed over water, which is preheated and filled inside the test cavity to a depth of 30 mm while filling the reference cavity entirely with water at the same temperature as the test cavity. The cold temperature surface is placed over the test cavity to create an overall temperature difference. The induced buoyant flow in water is recorded in the form of interferograms in the wedge fringe setting. Subsequently, these interferograms are analyzed to measure the interfacial heat flux (Eq. 27). Fringe analysis described in the previous section is employed for the present study as well.

Numerical modeling of an enclosed water-membrane-water system incorporates governing equations (Eqs. 19–22) from Sect. 3.1 with an additional conservation equation of thermal energy for the membrane (solid). Bulk vapor flow through the membrane pores is neglected and the diffusion equation of thermal energy is expressed as:

$$\frac{\partial T_s}{\partial t} = \alpha_{e,s} \nabla^2 T_s \tag{32}$$

Here, $\alpha_{e,s}$ is the effective thermal diffusivity of the membrane in units of m^2/s.

The initial and boundary conditions for the present simulation are similar to those discussed in Sect. 3.2, but the interfacial conditions are altered in the presence of a membrane. At the membrane-air interface, temperature and heat flux continuity are imposed along with a vapor mass flux (Eq. 28), while at the membrane-water interface, temperature itself is continuous. The heat flux imbalance related to the evaporative mass flus is given as (Janajreh et al. 2017):

$$\left(\sum_{i=w,s} (-\hat{n} \cdot (k_i \nabla \cdot T_i)) \right) = \dot{m}_{\text{evap}} h_{fg} \tag{33}$$

Effective membrane permeability in the vapor mass flux expression (Eq. 28) is evaluated using the following combination of mass transfer coefficients (Kadi et al. 2020):

$$C_m = \left(\frac{1}{C_{m,k}} + \frac{1}{C_{m,o}} \right)^{-1} + C_{m,p} \tag{34}$$

A comparative plot of interfacial heat flux measured using the interferometry experiments and the numerical values for two initial temperature differences (4 and 6 K) between water and the top cover (maintained at 294 K) is shown in Fig. 7a. The experimental values are shown with related uncertainty estimated using the repeatability approach. Similar to open surface evaporation, discussed in Sect. 3.4, the surge in interfacial heat flux is caused by the transition from diffusive transport in air to buoyancy-driven convective transport.

Additionally, the numerical study also characterizes individual vapor transport mechanisms through the membrane. Figure 7b demonstrates that among the three transport processes represented by Eqs. (29–31), Knudsen diffusion has the highest vapor flux, whereas Poiseuille flow has the lowest. However, as the harmonic sum of Knudsen and ordinary diffusion is employed to determine the overall mass flux (Eq. 34), ordinary diffusion-driven by pressure difference across the pores proves to be the dominant mode of vapor transport through the membrane.

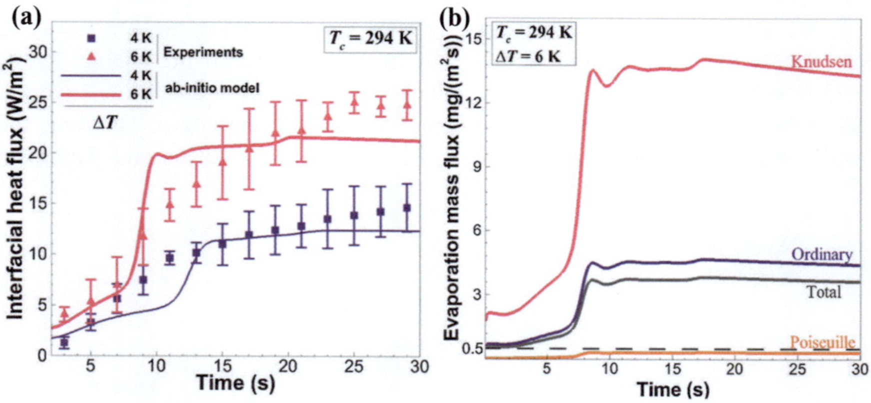

Fig. 7 **a** Variation in the instantaneous surface-averaged interfacial heat flux with time in membrane-assisted evaporation of water for two initial temperature differences of 4 and 6 K with symbols representing experimental values while continuous lines are from numerical simulations and **b** contribution of the three vapor transport mechanisms through a hydrophobic membrane to the overall evaporative mass flux. (Reproduced with permission from Bhendura et al. 2023. © 2023 Elsevier B.V.)

5 Droplet Evaporation and Substrate Cooling

In contrast to the flat air–water interface in pool evaporation, the curved interface of a sessile droplet causes additional phenomena that influence evaporation rates. These include evolution of the droplet shape as the liquid mass shrinks due to evaporation, stick–slip motion of a three-phase contact line caused by pinning, surface-tension-driven flow such as Marangoni and capillary flow, and changes in vapor saturation pressure owing to large curvature. These effects become increasingly dominant as the droplet size reduces and further influence the droplet evaporation rate. Figure 8 schematically represents the induced internal flow within the droplet, the changes in the evaporation rate across the interface, and the movement of the three-phase contact line.

5.1 *Modes of Droplet Evaporation*

The shape of a droplet placed on a substrate depends upon its wettability with respect to the substrate and the combined forces of surface tension and gravity. The shape, in turn, determines the internal pressure of the droplet. Wettability is prescribed in terms of the apparent contact angle; for example, a contact angle <90° is considered hydrophilic, 90°–120° is hydrophobic, and 120°–180° defines a superhydrophobic surface (Zang et al. 2019). The initial shape of a sessile droplet will evolve in time as the droplet evaporates. Two independent modes of evaporation are known to determine the resulting change in shape; these are referred as the constant contact radius (CCR) and constant contact angle (CCA) modes (Birdi and Vu 1993).

In the CCR mode, the wetted area remains constant while the contact angle decreases during evaporation. In the CCA mode, in contrast, the contact radius recedes, but the contact angle remains unchanged. Figure 9a and b illustrate the change in the interface profiles, the contact radius, and the contact angle as the droplet evaporates in CCR and CCA modes, respectively (Wilson and Ambrosio

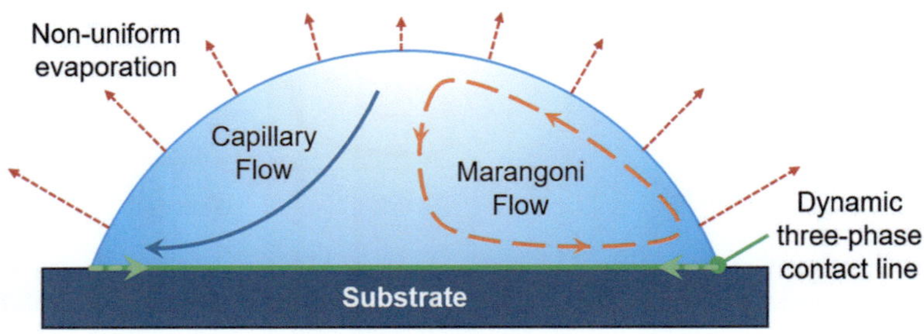

Fig. 8 Schematic representation of sessile droplet evaporation from a heated substrate into a cooler environment with induced internal liquid flow, non-uniform evaporation rate over the interface, and three-phase contact line motion

2023). However, in many instances, experiments reveal that the droplet evaporates in a mixed mode, being a combination of the two extreme modes. In a commonly observed mixed mode, the contact motion displays stick–slip (SS) movement (Dash and Garimella 2013), wherein the CCR mode is followed by the CCA mode, after attaining a critical contact angle, as shown in Fig. 9c. Stick–jump (SJ) mode is another commonly observed mixed mode of droplet evaporation depicted in Fig. 9d, which follows a series of CCR steps separated by rapid jump phases (Debuisson et al. 2016). This mode consists of several critical contact angles at which jumps takes place.

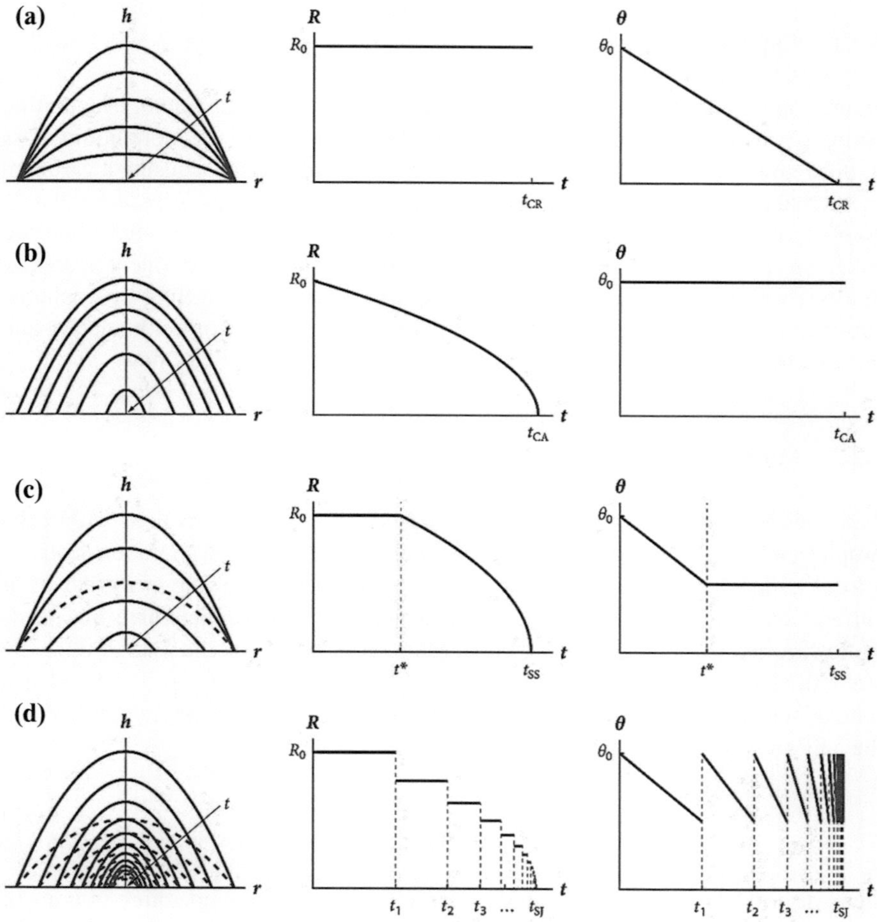

Fig. 9 Changes in interface shape, contact radius (R), and instantaneous contact angle (θ) for individual evaporation modes of a sessile droplet placed on a hydrophobic surface; **a** CCR mode, **b** CCA mode, **c** SS mode, and **d** SJ mode. The dashed lines in SS and SJ modes denote the critical contact angles at which CCR-CCA transition takes place. Abbreviations are explained in the text. (Figure reproduced from Wilson and Ambrosio 2023. Licensed under CC BY 4.0)

5.2 Evaporation-Induced Flow in Droplet

Evaporation of a sessile droplet induces internal liquid motion attributed to complex phenomena such as non-uniform evaporation rate, temperature and concentration distribution within the droplet and along the interface, and surface-tension-driven minimization of interfacial area. The induced flow is categorized into four types based on the prevalence of the stated phenomena: capillary flow, Marangoni flow, buoyancy-driven flow, and flow coupled with a three-phase contact line.

5.2.1 Capillary Flow

Here, evaporative mass flux varies along the interface with maximum evaporation taking place at the three-phase contact line, thus losing relatively more liquid mass in this region to vapor. This lost mass distorts the droplet shape, causing a variation in internal pressure (also called, Laplace pressure). Thus, to attain pressure equality, compensatory flow from the top of the droplet originates toward the contact line and is known as capillary flow (Gelderblom et al. 2022). This flow is prevalent in configurations with a pinned contact line (CCR mode) on partially wetting hydrophobic substrates. Thus, the local changes in evaporation rate at the drop-air boundary and the area minimization characteristic of the interface cause capillary flow.

5.2.2 Marangoni Flow

The cooling effect of evaporation re-distributes the temperature field within the droplet and along the interface. The non-uniform nature of evaporation leads to a local temperature differential over the curved interface, resulting in a gradient of surface tension. These interfacial gradients generate a tangential force that leads to flow vortices in the liquid, known as Marangoni flow (Xu and Luo 2007). The strength of this flow field is characterized by Marangoni number. It is defined as a ratio of surface tension induced by a temperature gradient along the free surface and the adhesive force and is formulated as:

$$\mathrm{Ma} = -\frac{\mathrm{d}\sigma}{\mathrm{d}T}\frac{L_c \Delta T}{\mu \alpha} \tag{35}$$

Here, $\mathrm{d}\sigma \big/ \mathrm{d}T$ denotes surface tension coefficient relative to temperature in units of N/m K, L_c is the characteristic length scale in m, μ is dynamic viscosity in Pa s, α is thermal diffusivity in m^2/s, and ΔT is temperature difference across the liquid layer in K.

Apart from temperature, surface tension is further influenced by the concentration of dissolved substances such as dissolved salts, electrolytes, colloids, and surfactants

(Nguyen and Stebe 2002). Thus, flow induced by the surface tension gradient originating from local concentration differences caused by evaporation is characterized by the following expression:

$$\mathrm{Ma}_C = -\frac{\mathrm{d}\sigma}{\mathrm{d}C}\frac{L_c \Delta C}{\mu \alpha} \tag{36}$$

where, $\mathrm{d}\sigma / \mathrm{d}C$ is the surface tension gradient with respect to solute concentration in N/kg m^2, and ΔC is the solutal concentration difference in kg/m^3. Symbol Ma_C is referred as the solutal Marangoni number.

5.2.3 Buoyancy-Driven Flow

In addition to temperature gradient over the interface, evaporative cooling further sets up temperature gradients parallel and perpendicular to gravity within the body of the droplet. These gradients result in density differentials that lead to internal buoyancy-driven convection. Invariably, convective motion takes place vertically upward along the axis of the drop, while descending along the periphery toward the three-phase contact line. Since buoyancy force scales with the cube of the length scale parallel to the gravity vector, the associated strength of the convective field is strong in drops with large heights, such as a droplet in the Cassie-Baxter state. However, for droplets with capillary length scales, surface tension forces predominate the gravitational; yet one may expect buoyancy to be significant, at least in the early stages of evaporation and counteract Marangoni flow. In a droplet arising from a binary mixture of liquid phases, density difference as well as differences in vapor pressure will be responsible for buoyant as well as Marangoni convection. The strength of gravity-induced convection is measured in terms of the Rayleigh number, which is the ratio of convective to diffusive transport rates and expressed in the following form:

$$\mathrm{Ra} = \frac{g\beta \Delta T L_c^3}{\nu \alpha} \tag{37}$$

In this formula, g is gravitational acceleration (m/s^2), β is the thermal expansion coefficient (K^{-1}), ΔT is the overall temperature difference, L_c is the relevant length scale (m), ν is the kinematic viscosity of liquid (m^2/s), and α is the thermal diffusivity of liquid (m^2/s).

A theoretical study on the strength of evaporation-induced flow in a millimeter-sized water droplet suggests that Marangoni flow of thermal origin would dominate capillary flow, except on a partially wetting substrate, specifically near the contact line (Hu and Larson 2005). Regardless of the dominance (based on Marangoni number), these two flows assist each other. However, experiments show that the accidental presence of a surfactant monolayer at the interface counteracts thermal Marangoni flow and weakens thermal advection (Cammenga et al. 1984). A similar analysis based on the ratio of Marangoni and Rayleigh numbers for water indicates that surface-tension-driven flow dominates the buoyancy-driven flow at room conditions. However, in contaminated water, the surfactants negatively affect the surface tension gradient, and natural convection takes precedence despite the large ratio of Marangoni to Rayleigh numbers (Diddens et al. 2021).

5.3 *Effect of Curvature on Saturation Pressure: Kelvin's Effect*

In a flat interface, the imbalance of cohesive and adhesive forces causes surface tension; however, on a curved interface, the net resulting force is upscaled. This is because of the pressure difference across the interface, called the Laplace pressure, that varies with the local interfacial radius of curvature. The expression relating Laplace pressure to the interface radius is derived as follows (Van Carey 2020):

$$\Delta P_L = \sigma\left(\frac{1}{R_1} + \frac{1}{R_2}\right) \tag{38}$$

where, σ denotes surface tension (N/m), while R_1 and R_2 are the three-dimensional principal radii of curvature in units of (m). Laplace pressure further affects the saturation pressure. The expression for the change in saturation pressure due to the curvature in a droplet, referred as Kelvin's effect, is expressed as (Semenov et al. 2012):

$$\ln\left(\frac{p_{\text{sat}}}{p_{\text{sat},0}}\right) = \frac{2\sigma V_m}{rRT} \tag{39}$$

The saturation pressure of liquid in N/m^2 at flat and curved interfaces are symbolized as $p_{\text{sat},0}$ and p_{sat}, respectively, V_m is the molar volume (m^3/mol), and r is the droplet radius (m).

Equation (39) shows that the deviation in saturation pressure is inversely related to droplet radius. A sample calculation at 25 °C shows that a water droplet with a diameter of 0.1 μm experiences a 1% change in saturation pressure. Thus, Kelvin's effect can be neglected for droplet sizes larger than 0.1 μm.

5.4 *Effects of Substrate Properties*

The molecular interaction between a liquid drop and substrate is determined by the substrate characteristics, such as roughness, surface energy, and thermal conductivity. These interactions affect the wettability of the liquid relative to the substrate, pinning of the three-phase contact line, modes of evaporation, and evaporation-induced internal flow. In addition, inhomogeneity in surface composition or surface roughness leads to contact angle hysteresis, wherein the advancing contact angle differs from the receding contact angle. This hysteresis further influences the transition of the mode of evaporation (CCR to CCA), thus, the stick–slip motion of the contact line and, eventually, the patterns of solute deposits during evaporation. Moreover, droplet evaporation absorbs the latent heat from its surroundings, which lowers the substrate temperature, in-turn, reducing the evaporation rate itself (Sobac and Brutin 2012). Hence, in experiments, the substrate temperature is required to be maintained constant by external heating, which promotes Marangoni flow, affects the stick–slip behavior of the contact line and finally, the solute deposition patterns (Zang et al. 2019).

5.5 *Modeling Droplet Evaporation*

In droplet evaporation, estimating the evaporation rate is quite a challenge because the evaporation process is coupled with vapor transport rates into the surrounding air, internal liquid flow, evaporative cooling, three-phase contact line motion, and conjugate heat transport through the substrate. These complexities led to the formulation of simplified models of droplet evaporation, while expressions for the evaporation mass flux are determined by solving these models either by analytical or numerical tools. In these models, two basic assumptions are made for the droplet shape, applicable for smaller values of Bond number $\left(\mathrm{Bo} = \rho g R^2 / \sigma\right)$ and Capillary number $\left(\mathrm{Ca} = \mu u / \sigma\right)$. These assumptions ensure that surface tension dominates the gravity and viscous forces, respectively. Thus, the sessile droplet is modeled as a hemispherical cap during its placement over both non-wetting and partially-wetting substrates. The following sections describe a few of the widely accepted evaporation models derived for sessile droplets.

5.5.1 Diffusion-Limited Evaporation Model

Evaporation of a sessile droplet in ambient conditions is a slow process. Experiments show that for partially wetting (hydrophilic) substrates, the CCR mode of evaporation is more prolonged than the CCA mode. Hence, the contact line is presumed to be pinned for modeling slow evaporation, while the evaporation rate is considered quasi-steady. As the vapor transport in air above the droplet is assumed to be diffusive, the Laplace equation for vapor concentration $\left(\nabla^2 c = 0\right)$ is solved to derive an expression of a time-dependent evaporation rate. The expression for the evaporation rate as a function of radial coordinate is expressed as (Popov 2005):

$$J(\alpha) = \frac{c_{\mathrm{sat}} - c_{\infty}}{R/D}\left(\frac{\sin\theta}{2} + \sqrt{2}(\cos h\alpha + \cos\theta)^{3/2}\right.$$

$$\times \int_0^\infty P_{-1/2+i\tau}(\cos h\alpha)\frac{\tau \cos h\theta\tau \tan h[(\pi-\theta)\tau]}{\cos h\pi\tau}\mathrm{d}\tau \Bigg) \tag{40}$$

where, α is the toroidal coordinate, $P_{-1/2+i\tau}$ is the Legendre functions of the first kind, R is the wetting radius in m, D is vapor diffusivity in m^2/s, c_{sat} is the saturated vapor concentration in kg/m^3, c_∞ is the vapor concentration at far-field in kg/m^3, θ is the contact angle in radians, and τ is the dummy integration variable.

Figure 10 depicts the local evaporative mass flux along the interface for four different contact angles obtained using Eq. (40). In these plots the principal axes and the local evaporation rates are scaled with droplet volume. For contact angle of 10° (Fig. 10a), a theoretical singularity occurs at the contact line, while for 90°, uniform distribution of evaporative mass flux is observed.

In addition, Hu and Larson (2002) solved the Laplace equation of concentration distribution using the finite element method (FEM) and derived a simplified expression for evaporative mass flux as given below:

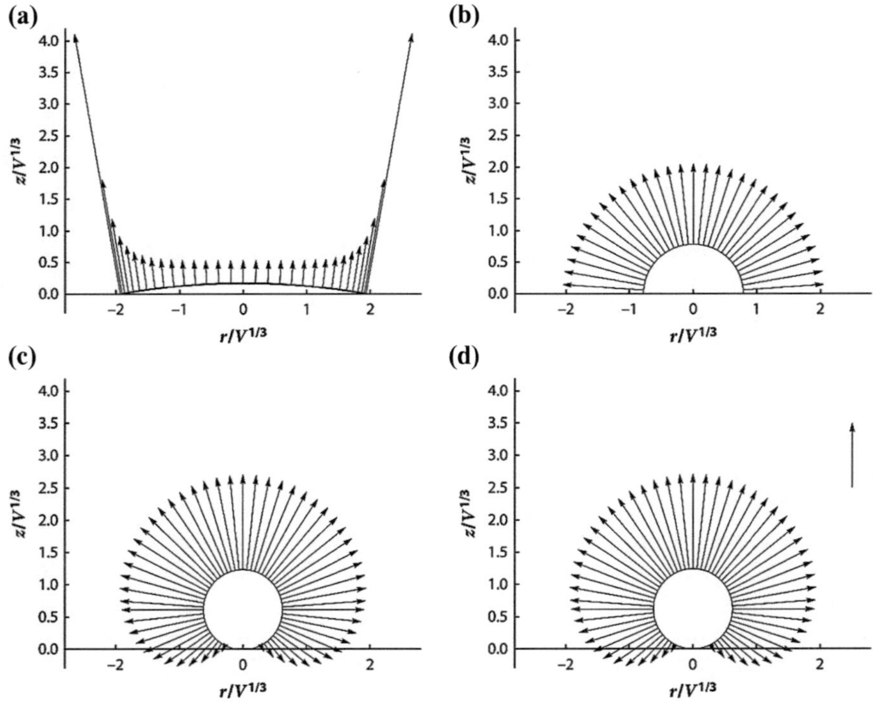

Fig. 10 Scaled evaporation rates over the air–liquid interface for four contact angles; **a** 10°, **b** 90°, **c** 170°, and **d** 180°. The contact radii (abscissa) and heights (ordinate) of the droplets are scaled with the cube root of droplet volume (V), while the local evaporation rates (proportional to the length of the arrows) are scaled as $JV^{1/3}/D(c_{sat}-c_\infty)$. (Figure reproduced from Stauber et al. 2015.

$$J = \frac{Dc_{\text{sat}}(1-H)}{R}(0.27\theta^2 + 1.30)$$
$$\left(0.6381 - 0.2239 \times (\theta - \pi/4)^2\right)(1 - \hat{r}^2)^{-\lambda(\theta)} \quad (41)$$

Here, H is relative humidity, $\hat{r}$ is the non-dimensional radial distance $(\hat{r} = r/R)$, and λ is the fitting parameter representing non-uniformity of evaporation rate $(\lambda(\theta) = 0.5 - \theta/\pi)$.

Equations (40) and (41) are derived with isothermal conditions, i.e., the droplet temperature is assumed to be uniform, and evaporative cooling, i.e., cooling of the droplet owing to a consumption of latent heat of evaporation is neglected. Thus, the estimated evaporation rates are slightly higher than the experimentally observed evaporation rates. Nguyen et al. (2018) included the effect of interfacial cooling and solved the coupled Laplace equations for temperature $(\nabla^2 T = 0)$ and vapor concentration $(\nabla^2 c = 0)$. The revised expression obtained for evaporation flux is given as:

$$J(\alpha) = \frac{c_e - c_\infty}{R/D}\Big(\sqrt{2}(\cos h\alpha + \cos\theta)^{3/2}$$
$$\times \int_0^\infty E_c P_{-1/2+i\tau}(\cos h\alpha)\left[\frac{\cos h[(\theta-\pi)\tau]\sin\theta}{2(\cos h\alpha + \cos\theta)}\right.$$
$$\left. -\tau \sin h[(\theta - \pi)\tau]\right]d\tau\Big) \quad (42)$$

In this expression, c_e is the saturated vapor concentration at the temperature of the droplet edge, while E_c is defined as follows:

$$E_c = \frac{\cos h\theta\tau}{\cos h\pi\tau \cos h[(\theta-\pi)\tau]} \times \frac{\tau F \cot h\theta\tau - \frac{1}{3}\frac{dF}{d\theta}}{\left(\tau F \cot h\theta\tau - \frac{1}{3}\frac{dF}{d\theta}\right) - E_0\left(\tau F \tan h\theta\tau - \frac{1}{3}\frac{dF}{d\theta}\right)} \quad (43)$$

where $F = \sin\theta\tau/(\sin h\pi\tau \sin\theta)$ and $E_0 = (dc/dT)h_{fg}D/k_l$. E_0 is the evaporative cooling number, while k_l is the thermal conductivity of the liquid phase in W/m K.

The analytical solutions predict the variation in evaporation rate over the curved interface with increasing values from the apex of the droplet to the edge of the contact line. As these solutions are obtained with pure diffusive heat and mass fluxes, their accuracy is limited to slow-evaporating droplets. For fast evaporation, the internal and external convection patterns in air and liquid dominate heat and mass fluxes. Hence, a comprehensive approach needs to be developed which incorporates the effects of evaporative cooling, thermocapillary convection, natural convection, and substrate characteristics, for proper understanding of droplet evaporation.

5.5.2 Numerical Models

The limitations of analytical solutions originate from diffusive modeling of heat and vapor transport in liquid and surrounding gas. These are addressed by including the acceleration and convective transport terms in numerical models. In the continuum framework, these models consist of two sets of conservation equations for mass, momentum, thermal energy, and vapor for the liquid droplet and the gaseous medium above the droplet (Akkus et al. 2020). These two sets of equations are coupled using appropriate interfacial conditions to incorporate evaporative cooling and thermocapillary convection. The governing equations for heat and vapor transport are similar to the equations reported in Sect. 3.1. Likewise, the evaporative mass flux is calculated using the superposition of diffusive and convective vapor transport near the air–water interface, namely, Eq. (4). Lastly, the effect of evaporative cooling on the substrate is modeled using unsteady thermal energy transport in solids (Eq. 32).

In addition to heat and mass movement, loss of liquid mass loss continuously reshapes the droplet. Therefore, the interface must be tracked simultaneously, particularly during large evaporation rates. The following numerical techniques are available for interface tracking: volume of fluids (VOF), level-set, phase-field, and arbitrary Lagrangian–Eulerian (ALE). These methods have proven to be effective in certain cases, though they could suffer from numerical instability, mass imbalance, numerical diffusion, and the need for tuning parameters. Numerical approaches are comprehensive and inclusive but are constrained by the time bounds to achieve convergence and the related extremely high computing cost. These limits are further exacerbated for liquid droplets containing salts or contaminants that generate additional physico-chemical phenomena.

6 Thin-Film Evaporation

On a hydrophilic substrate, a liquid layer will tend to spread out and achieve a contact angle of $<90°$. In the case of low surface tension liquids placed over a high-surface energy substrate, the liquid spreads entirely over the substrate and forms a thin film of sub-micrometer thickness with a contact angle close to zero (Van Carey 2020). These thin films form adjoining the three-phase contact line, for example, an extended liquid meniscus in a wicking structure or a capillary tube or a precursor film of a sessile droplet. The extended meniscus is divided into three regions, namely, intrinsic meniscus or bulk region, evaporating thin film region, and adsorbed or non-evaporating region, as sketched in Fig. 11. Among these three regions, the thin film ensures maximum evaporation rates within the extended meniscus due to low thermal resistance.

As the thickness of the extended meniscus is in the submicron range, molecular interactions among the three phases in the form of weak forces adjust the meniscus profile, suggesting macroscopically a change in the internal pressure. Examples of intermolecular forces are dispersion forces, electrostatic forces, steric forces, and

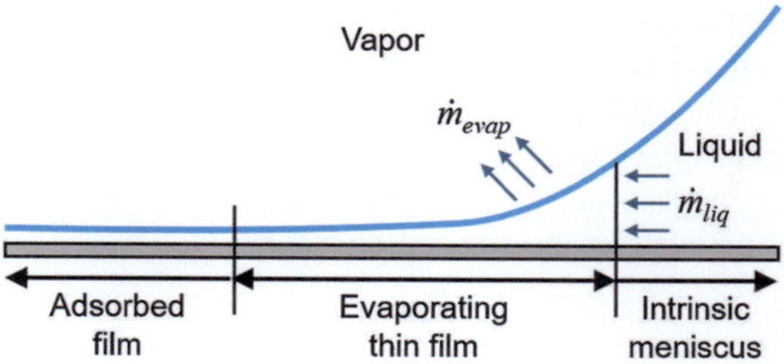

Fig. 11 Sketch of an extended meniscus shown with three regions formed adjacent to the three-phase contact line; these regions are the adsorbed film, evaporating thin film, and intrinsic meniscus

chemical bonding (Wang et al. 2009). The pressure difference between the extended meniscus and the bulk liquid caused by molecular interactions is referred as disjoining pressure. At the macroscopic level, the disjoining pressure is entirely a function of film thickness. It is negligible in the bulk liquid but gradually increases along the meniscus where the film thickness decreases and eventually attains a maximum value in the non-evaporating zone. Further, the disjoining pressure adsorbs the last segment of the meniscus on the substrate, suppressing evaporation despite the lowest thermal resistance. However, negative disjoining pressure in the adsorbed film pulls the liquid from the intrinsic meniscus to the thin film, thus assisting the overall evaporation rate.

The evaporative mass flux of a thin film is predicted using the kinetic theory of evaporation (Sect. 2.2) after including the impact of capillary and disjoining pressures on the saturation vapor pressure at the interfacial temperature. For a flat interface in bulk fluid, the evaporation rate is estimated using Eq. (5), wherein the interfacial vapor pressure is considered to be the vapor saturation pressure. In a thin film, the adjacent vapor pressure is altered in the presence of capillary and disjoining pressures, manifesting as a reduction in vapor saturation pressure. Therefore, the liquid must be slightly superheated beyond saturation for thin-film evaporation. The augmented vapor pressure adjacent to the thin film is derived as follows (Ranjan et al. 2011):

$$\ln\left[\frac{p_{v,\mathrm{in}}(T_{\mathrm{in}})}{p_{\mathrm{sat}}(T_{\mathrm{in}})}\right] = \frac{p_{v,\mathrm{in}}(T_{\mathrm{in}}) - p_{\mathrm{sat}}(T_{\mathrm{in}}) - (p_d + p_c)}{\rho_l T_{\mathrm{in}} R/M} \tag{44}$$

Here, $p_{v,\mathrm{in}}$, p_{sat}, p_d, and p_c are the pressures of vapor saturation at thin-film interface, vapor saturation at thick-film interface, disjoining, and capillary pressures in N/m^2, respectively. The symbols T_{in}, R, M, and ρ_l represent the interfacial temperature (K), the universal gas constant (J/kmol K), the molecular weight (kg/kmol), and the density of liquid (kg/m^3), respectively. The vapor saturation pressure adjacent to the flat interface of a thick film is calculated using the Clausius–Clapeyron equation (Van Carey 2020):

$$p_{\mathrm{sat}}(T_{\mathrm{in}}) = p_{\mathrm{sat,ref}}\left(T_{\mathrm{sat,ref}}\right) \exp\left[\frac{h_{fg} M}{R}\left(\frac{1}{T_{\mathrm{sat,ref}}} - \frac{1}{T_{\mathrm{in}}}\right)\right] \tag{45}$$

The reference temperature is in units of Kelvin and the corresponding saturation pressure in N/m^2, being, respectively, denoted as $T_{\text{sat,ref}}$ and $p_{\text{sat,ref}}$.

The left-hand side of Eq. (44) is approximated using the Taylor series expansion and the interfacial vapor pressure is substituted into Eq. (5) to incorporate the effect of capillary and disjoining pressures on the evaporative mass flux. The resulting expression for the evaporation flux is as follows (Akkuş et al. 2017):

$$\dot{m}_{\text{evap}} = \frac{2\alpha_e}{2-\alpha_e}\sqrt{\frac{M}{2\pi RT_{\text{in}}}}\left[\frac{Mp_v h_{fg}}{RT_v T_{\text{in}}}(T_{\text{in}} - T_v) - \frac{p_v V_l}{RT_{\text{in}}}(p_c + p_d)\right] \tag{46}$$

Here, V_l is the molar volume of liquid, while variables with subscript 'v' are related to the bulk vapor phase.

The capillary and disjoining pressures in Eq. (46) are estimated using the following expressions (Ahmed and Pandey 2019):

$$p_c = \sigma \frac{\mathrm{d}^2\delta/\mathrm{d}x^2}{\left\{1 + (\mathrm{d}\delta/\mathrm{d}x)^2\right\}^{3/2}} \tag{47}$$

$$p_d = \frac{\overline{A}}{\delta^3}\frac{1 + 3b\delta/2\lambda}{(1 + b\delta/2\lambda)^2} \tag{48}$$

In these expressions, σ is surface tension (N/m), δ is thin-film thickness (m), $\overline{A}$ is Hamaker constant (J), b is empirical factor (= 5.32), and λ is characteristic wavelength (= 100 nm).

6.1 *Theoretical Modeling of Thin-Film Shapes*

The thin film region in Fig. 11 is modeled for its heat-transfer characteristics, incorporating the evaporative flux at the interface, compensatory liquid flow into the film, and variable film thickness. Liquid flow within the thin film is assumed to be steady, laminar, and unidirectional, while the interface is considered to be shear-free (Wayner et al. 1976).

In the thin-film region, both capillary and disjoining pressures are present and the pressure balance across the liquid–vapor interface is expressed as follows:

$$p_v = p_l + p_d + p_c \tag{49}$$

Substituting Eqs. (47) and (48) into Eq. (49), and neglecting the gradient in vapor pressure leads to the following third-order differential equation for the film thickness (Ahmed and Pandey 2019):

$$\frac{d^3\delta}{dx^3} = \frac{3\left(\frac{d\delta}{dx}\right)\left(\frac{d^2\delta}{dx^2}\right)^2}{1 + \left(\frac{d\delta}{dx}\right)^2} - \frac{1}{\sigma}\left(\frac{dp_d}{dx} + \frac{dp_l}{dx}\right)\left(1 + \left(\frac{d\delta}{dx}\right)^2\right)^{3/2} \tag{50}$$

Moreover, lubrication theory is utilized to model liquid flow within the thin film. The corresponding gradient of liquid pressure is derived in the following form (Ahmed and Pandey 2019):

$$\frac{dp_l}{dx} = \mu_l\left(\frac{d^2u_l}{dy^2}\right) \tag{51}$$

Since the process is assumed to be steady, the loss of liquid mass due to evaporation is compensated by the liquid mass flux $\left(\dot{m}_l = \int_0^\delta \rho_l u_l dy\right)$. Hence, the gradient of liquid mass flux along the curvature of the thin film is written as:

$$\frac{d\dot{m}_l}{dx} = -\dot{m}_{\text{evap}} \tag{52}$$

In the present formulation, the interfacial temperature (T_{in}) is determined by balancing the evaporative heat flux with the heat conduction across the thin film, leading to the following expression (Ahmed and Pandey 2019):

$$T_{\text{in}} = T_w - \frac{\dot{m}_{\text{evap}} h_{fg} \delta}{k_l} \tag{53}$$

Hence, the film thickness profile, liquid pressure, and evaporative mass flux are determined by solving Eqs. (50–52) with simplified boundary conditions.

The above discussion highlights the importance of governing equations that model thin-film evaporation. The applicable boundary conditions are not explicitly available and one may take recourse to experiments. However, the model is a good starting point for characterizing flow and transport in thin films, though it needs continuous validation against experiments. The formulation given above does not include the adsorbed region. The exclusion of micro-scale surface tension forces further limits its accuracy. These factors need to be addressed in future research.

7 Closure

The present chapter outlines phenomena involved in the evaporation of a liquid pool with and without membrane and their extension to a droplet and an extended meniscus. The discussions further include the mathematical approaches to predict the evaporation rate in specific applications. In water pool evaporation, a comprehensive model for an air–water system with three independent expressions of evaporation rates is validated against interferometric experiments. This combined experimental and numerical study is further extended to determine the evaporative flux through a membrane. Additionally, droplet evaporation with induced fluid motion within the curved interface is discussed. Evaporation rates are derived for the diffusion-limited model of evaporation. Lastly, the concept of disjoining pressure and its impact on the evaporation of thin films is reviewed.

Nomenclature

C_m	Mass transfer coefficient (m^{-1} s)
C_p	Specific heat at constant pressure (J/kg K)
d	Fringe gap (m)
D_{a-w}	Mass diffusion coefficient (m^2/s)
g	Gravitational acceleration (m/s^2)
Gr	Grashof number
h_{conv}	Convective heat transfer coefficient (W/m^2 K)
h_{evap}	Evaporative heat transfer coefficient (W/m^2 K)
h_{fg}	Latent heat of vaporization (J/kg)
J	Evaporation rate (kg/m^2 s)
k	Thermal conductivity (W/m K)
L_c	Characteristic length (m)
Ma	Marangoni number
$\dot{m}_{\text{evap}}$	Evaporation mass flux (kg/m^2 s)
M	Molecular weight of water (kg/kmol)
n	Refractive index
Nu	Nusselt number
p	Pressure (N/m^2)
p_c	Capillary pressure (N/m^2)
p_d	Disjoining pressure (N/m^2)
Pr	Prandtl number
R	Universal gas constant (J/kmol K)
Ra	Rayleigh number
r_p	Pore radius of membrane (m)
T	Temperature (K)
u	x-Velocity component (m/s)
v	y-Velocity component (m/s)
v_s	Stefan velocity (m/s)
Y	Mass fraction

Greek Alphabets

α	Thermal diffusivity (m^2/s)
α_e	Accommodation coefficient
β	Thermal exp. coefficient (1/K)
δ	Thickness of membrane (m)
δ_m	Thin film thickness
ε	Porosity
λ	Wavelength of light beam (m)
μ	Viscosity (Pa s)
ν	Kinematic viscosity (m^2/s)
ρ	Density (kg/m^3)
σ	Surface tension (N/m)
τ	Tortuosity
θ	Contact angle (rad)
ω	Humidity (kg/kg)

Subscript

a	Air
c	Cold
g	Glass cover
h	Hot
in	Interface
sat	Saturation
v	Vapor
w	Water

References

Ahmed S, Pandey M (2019) New insights on modeling of evaporation phenomena in thin films. Phys Fluids 31(9):139. https://doi.org/10.1063/1.5112139

Akkuş Y et al (2017) Two-dimensional computational modeling of thin film evaporation. Int J Therm Sci 121:237–248. https://doi.org/10.1016/j.ijthermalsci.2017.07.013

Akkus Y, Çetin B, Dursunkaya Z (2020) A theoretical framework for comprehensive modeling of steadily fed evaporating droplets and the validity of common assumptions. Int J Therm Sci 158(6):106529. https://doi.org/10.1016/j.ijthermalsci.2020.106529

Aursand E, Ytrehus T (2019) Comparison of kinetic theory evaporation models for liquid thin-films. Int J Multiphase Flow 116:67–79. https://doi.org/10.1016/j.ijmultiphaseflow.2019.04.007

Bhendura M, Muralidhar K, Khandekar S (2022) An improved correlation of evaporation rate of a water pool derived using combined ANN-GA. Comput Therm Sci Int J 15(2):1–19. https://doi.org/10.1615/computthermalscien.2022043453

Bhendura M, Muralidhar K, Khandekar S (2023) Influence of a hydrophobic membrane on evaporation rate of water placed in a top-cooled circular cavity. Desalination 568:37. https://doi.org/10.1016/j.desal.2023.117037

Bird RB, Stewart WE, Lightfoot EN (2006) Transport phenomena, 2nd edn. Wiley, New York

Birdi KS, Vu DT (1993) Wettability and the evaporation rates of fluids from solid surfaces. J Adhesion Sci Technol 7(6):485–493. https://doi.org/10.1163/156856193X00808

Cammenga HK et al (1984) On Marangoni convection during the evaporation of water. J Colloid Interf Sci 98(2):585–586. https://doi.org/10.1016/0021-9797(84)90187-5

Dash S, Garimella SV (2013) Droplet evaporation dynamics on a superhydrophobic surface with negligible hysteresis. Langmuir 29(34):10785–10795. https://doi.org/10.1021/la402784c

Debuisson D et al (2016) Stick–jump (SJ) evaporation of strongly pinned nanoliter volume sessile water droplets on quick drying, micropatterned surfaces. Langmuir 32(11):2679–2686. https://doi.org/10.1021/acs.langmuir.6b00070

Demirel Y (2007) Nonequilibrium thermodynamics : transport and rate processes in physical, chemical and biological systems. Elsevier, Amsterdam

Diddens C, Li Y, Lohse D (2021) Competing Marangoni and Rayleigh convection in evaporating binary droplets. J Fluid Mech 52:914. https://doi.org/10.1017/jfm.2020.734

Gelderblom H, Diddens C, Marin A (2022) Evaporation-driven liquid flow in sessile droplets. Soft Matter Royal Soc Chem 12:8535–8553. https://doi.org/10.1039/d2sm00931e

Hu H, Larson RG (2002) Evaporation of a sessile droplet on a substrate. J Phys Chem B 106(6):1334–1344. https://doi.org/10.1021/jp0118322

Hu H, Larson RG (2005) Analysis of the microfluid flow in an evaporating sessile droplet. Langmuir 21(9):3963–3971. https://doi.org/10.1021/la047528s

Janajreh I et al (2017) Numerical investigation of air gap membrane distillation (AGMD): seeking optimal performance. Desalination 424:122–130. https://doi.org/10.1016/j.desal.2017.10.001

Kadi KE, Moustakas K, Janajreh I (2020) Comparison of different mass transfer models for direct contact membrane distillation flux evaluation. Desal Water Treat 194:379–388. https://doi.org/10.5004/dwt.2020.25776

Khayet M (2011) Membranes and theoretical modeling of membrane distillation: a review. Adv Colloid Interf Sci 64:56–88. https://doi.org/10.1016/j.cis.2010.09.005

Kumar S, Tiwari GN (1996) Estimation of convective mass transfer in solar distillation systems. Sol Energy 57(6):459–464. https://doi.org/10.1016/S0038-092X(96)00122-3

Lu H et al (2022) Materials engineering for atmospheric water harvesting: progress and perspectives. Adv Mater 34(12):79. https://doi.org/10.1002/adma.202110079

Matrawy KK, Alosaimy AS, Mahrous AF (2015) Modeling and experimental study of a corrugated wick type solar still: comparative study with a simple basin type. Energy Conver Manag 105:1261–1268. https://doi.org/10.1016/j.enconman.2015.09.006

Misyura SY (2018) Non-isothermal evaporation in a sessile droplet of water-salt solution. Int J Therm Sci 124:76–84. https://doi.org/10.1016/j.ijthermalsci.2017.10.003

Naylor D, Duarte N (1999) Direct temperature gradient measurement using interferometry. Exper Heat Transf 12(4):279–294. https://doi.org/10.1080/089161599269609

Nguyen VX, Stebe KJ (2002) Patterning of small particles by a surfactant-enhanced Marangoni-Bénard instability. Phys Rev Lett 88(16):164501. https://doi.org/10.1103/PhysRevLett.88.164501

Nguyen TAH, Biggs SR, Nguyen AV (2018) Analytical model for diffusive evaporation of sessile droplets coupled with interfacial cooling effect. Langmuir 34(23):6955–6962. https://doi.org/10.1021/acs.langmuir.7b03862

Persad AH, Ward CA (2016) Expressions for the evaporation and condensation coefficients in the Hertz-Knudsen relation. Chem Rev 116(14):7727–7767. https://doi.org/10.1021/acs.chemrev.5b00511

Popov YO (2005) Evaporative deposition patterns: spatial dimensions of the deposit. Phys Rev E 71(3):036313. https://doi.org/10.1103/PhysRevE.71.036313

Ranjan R, Murthy JY, Garimella SV (2011) A microscale model for thin-film evaporation in capillary wick structures. Int J Heat Mass Transf 54(1–3):169–179. https://doi.org/10.1016/j.ijheatmasstransfer.2010.09.037
Sartori E (2000) A critical review on equations employed for the calculation of the evaporation rate from free water surfaces. Sol Energy 68(1):77–89. https://doi.org/10.1016/S0038-092X(99)00054-7
Semenov S et al (2012) Computer simulations of evaporation of pinned sessile droplets: influence of kinetic effects. Langmuir 28(43):15203–15211. https://doi.org/10.1021/la303916u
Sobac B, Brutin D (2012) Thermal effects of the substrate on water droplet evaporation. Phys Rev E Stat Nonlinear Soft Matter Phys 86(2):1–10. https://doi.org/10.1103/PhysRevE.86.021602
Stauber JM et al (2015) Evaporation of droplets on strongly hydrophobic substrates. Langmuir 31(12):3653–3660. https://doi.org/10.1021/acs.langmuir.5b00286
Tiwari G, Sahota L (2017) Advanced solar-distillation systems. Springer, Singapore
Tiwari A, Tiwari GN (2006) Effect of water depths on heat and mass transfer in a passive solar still: in summer climatic condition. Desalination 195(1–3):78–94. https://doi.org/10.1016/j.desal.2005.11.014
Tu R, Hwang Y (2020) Reviews of atmospheric water harvesting technologies. Energy 201:117630. https://doi.org/10.1016/j.energy.2020.117630
Van Carey P (2020) Liquid-vapor phase-change phenomena, 3rd edn. CRC Press, New York
Vlasov VA (2020) Diffusion-kinetic model of liquid evaporation from a Stefan tube: a solution to the Stefan diffusion problem. Int J Heat Mass Transf 163:120379. https://doi.org/10.1016/j.ijheatmasstransfer.2020.120379
Wang H, Pan Z, Chen Z (2009) Thin-liquid-film evaporation at contact line. Front Energy Power Eng China 3(2):141–151. https://doi.org/10.1007/s11708-009-0020-2
Wayner PC, Kao YK, LaCroix LV (1976) The interline heat-transfer coefficient of an evaporating wetting film. Int J Heat Mass Transf 19(5):487–492. https://doi.org/10.1016/0017-9310(76)90161-7
Wilson SK, Ambrosio HMD (2023) Evaporation of sessile droplets. Annu Rev Fluid Mech 55:2022. https://doi.org/10.1146/annurev-fluid-031822
Xu X, Luo J (2007) Marangoni flow in an evaporating water droplet. Appl Phys Lett 91(12):402. https://doi.org/10.1063/1.2789402
Zang D et al (2019) Evaporation of a droplet: from physics to applications. Phys Rep 804:1–56. https://doi.org/10.1016/j.physrep.2019.01.008

Dew and Fog Harvesting

Ramesh Narayanaswamy

1 Background

Atmospheric water harvesting (AWH) makes use of water vapour present in the atmosphere at a geographical location using dew condensation or fog capture, by employing appropriate devices utilising the scientific principles of thermodynamics, fluid mechanics, heat and mass transfer, material science, and aerodynamics. Progressing from ancient to most modern water capture technologies, these systems are perfect exemplars of significant advances in engineering design, materials, surface engineering, manufacturing, and the use of solar energy. There are several advantages for these systems in terms of cost-effectiveness and points of view from sustainability practices, when used as captive units for supplementing freshwater use for domestic or agricultural use.

Water quality of fog water is a topic of interest for chemistry and hydrology specialists, and hence, those aspects are not considered in this monograph. It must be noted that AWH has its own limitations due to environmental factors affecting either dew formation or fog availability, and hence, its suitability for generating water must require careful consideration of the environmental conditions and year-round weather of the locality.

Coastal areas with nearness to major waterbodies such as oceans will have the benefit of fully functioning AWH water capture systems working on an entirely passive mode without the use of any additional energy requirement. However, interior arid zones will benefit from having additional energy supplied to the devices through renewable means (solar energy, for example)—hence, functioning as an overall 'green' system in terms of sustainable operations. In what follows, we describe general aspects of fog and dew collectors, including sorption-based systems, to provide an overview of AWH methodologies in use for freshwater production. Recent

R. Narayanaswamy (✉)
Curtin University, Perth, WA, Australia
e-mail: r.narayanaswamy@curtin.edu.au

R. Narayanaswamy et al. (eds.), *Sustainable Water Harvesting*, Springer Tracts in Mechanical Engineering, https://doi.org/10.1007/978-981-95-3373-2_5

reviews (Raveesh et al. 2021; Wang et al. 2025) on atmospheric water harvesting provide more information on AWH generation and technologies.

2 Fog and Dew Collection on Surfaces

As indicated earlier, atmospheric water harvesting relies on water vapour in the atmosphere to generate freshwater. These ‘air-to-water’ machines are of different types, depending on how water vapour capture and its processing is adopted in the device. Dew collectors are different from fog collectors; dew forms on cold substrates due to heterogeneous condensation, whereas fog is an agglomeration of condensed water vapour due to homogeneous condensation in the air. Dew accumulates on a surface if the temperature of the surface is below the dew point temperature at that geographical location.

Techniques for lowering the surface temperature of the substrate (condensing surface) can be achieved through active or passive means; although passive methods are always traditionally employed, where no amount of cooling happens due to energy consumption. The yield of water from natural dew condensation (Bayens 2018) has an upper thermodynamic limit of about 0.7 to 0.8 L/m^2/night on a clear, cloudless sky with adequate cooling of the condensing surface, through radiation exchange between the condensing surface and the clear sky, occurring in the spectral wavelength window of 8–13 microns.

Advances in surface engineering have led to new design of surfaces for better dew capture, and novel materials for substrates offering as candidates for passive dew condensation. Water yields from dew capture are limited due to a gamut of environmental factors controlling the condensation rate, such as the availability of clear sky, adequate temperature potential, relative humidity, and wind conditions. Despite shortfalls, dew condensation is found to be a sustainable means for converting atmospheric water vapour into freshwater in small amounts. Traditional use of dew harvesting has been for agricultural applications where dew is used in a sense for the ‘self-watering’ of plants with the help of adequate scaffold constructs using hydrophilic materials barricaded around shrubs and plants. Modern sorption-based dew capture systems can generate potable-quality water for domestic consumption.

3 Dew Condensation Using Radiative Cooling

Dew collectors based on condensation of atmospheric water vapour using radiative cooling and novel surfaces require mention here. These devices have been in existence for a very long time, and produce water by condensing water vapour in the air near the condensing surface. The air just above the condensing surface must become supersaturated with water vapour, to allow for heterogeneous condensation on the condensing surface. A mild wind velocity is helpful to backfill water vapour from surroundings into the air layer adjacent to the active part of the condensing surface.

Thermodynamic considerations provide with ideal conditions that are required for water production: one of the main components that allows for condensation is the condensing surface. A perfectly black body condensing surface does not ensure all-time yields, especially when relative humidity and ambient temperatures interfere with the functioning of the radiative heat transport from the condensing surface to the sky, by not ensuring the required drop in the surface temperature of the condensing surface during the night to satisfy dew point considerations. This has led to the evolution of selective surfaces as possible candidates for condensing surfaces that allow only long-wavelength radiation exchange with the sky, leading to modest increase in water yield, provided other favourable environmental conditions exist during the operation of such dew condensers.

Selective coatings are designed and applied to optimise two quantities of interest in heat exchange, to and from the condensing surface: solar reflection during the day to limit the sunlight absorption (in the short wavelength) and maximising heat emission (long wave-length) during the night. Plenty of research articles on the type of coatings applicable for foils and surfaces suitable for dew harvesting is available.

An important example of such a foil is the one developed by Organisation Pour l'Utilisation de la Rose, the International Organisation for Dew Utilisation (OPUR). The polyethylene film that acts as the condensing surface used in the dew collection device developed by OPUR, is embedded with TiO_2 and $BaSO_4$ microspheres, yielding up to 40 g of freshwater generated per square metre per hour, under conditions of clear sky at night and relative humidity greater than 60%. However, it must be borne in mind that dew-based water vapour capturing machines have no freshwater yield when relative humidity falls below about 50%. Excellent articles on dew condensation with radiative cooling (Haechler et al. 2021; Zhu et al. 2023; Liu et al. 2020) demonstrate the potential of this technique.

Zhao et al. (2019) provide an overview of multi-layered surfaces used in dew condensers. The enhancement of freshwater yield has led to the integration of multiple mechanisms such as thermosyphon effect, radiative cooling, and cellulose-water interactions in combination with selective surfaces to function as one unit for water capture, as described in Zhu et al. (2023) to enable all-weather harvesting of dew.

4 Metal Organic Foam (MOF)-Based Water Vapour Capture

A strategy to overcome the limitation of simple dew collection machines is to make use of adsorption-based systems, for example, the use of suitable metal organic foams. However, one has to take recourse to the use of additional energy for de-sorption and condensation to happen in such cases.

MOF-based devices incorporate an adsorption-based component as the primary unit. The unit works by capturing atmospheric water vapour on a suitable porous metal–organic foam-based framework and utilise solar energy for its re-conversion into liquid water. Kim et al. (2017) have developed a device that uses a suitable combination of metal and organic porous metal–organic framework consisting of a microcrystalline powder form of MOF801, $Zr_6O_4(OH)_4$ (fumarate)$_6$. Adsorbed water vapour is then released (through the desorption process) with the help of solar energy obtained from photo-voltaic cells installed on the device. Freshwater yields of up to 2.8 L of water/kg MOF/day at 20% relative humidity and with no concentrating sources for solar radiation has been reported using this device. In order to maximise freshwater yield, RH-specific MOFs can be designed and fabricated for use in a particular locality, based on historical RH profiles. Use of MOF-based water collection has also been reported by Miller et al. (2017) and Fathieh et al. (2018) among others.

5 Fog Water Harvesting

Fog harvesting is different from dew harvesting in that, fog harvesters capture fog onto their surfaces and yield water due to agglomerated deposits of water droplets, falling off due to an imbalance in the forces affecting the growth versus residence time of the captured droplets of water on a suitable mesh-type structure.

Fog droplets have diameters ranging from 1 to 40 microns, and fog fall velocities on the order of 1 to 5 cm/s. These low velocities enable fog to be transported horizontally by low-speed winds. Engineered fog collectors kept orthogonal to the fog trajectory will enable fog droplets to strike the collector surface, which accumulates over time to generate freshwater. Water collection depends on the geographical location of the place (fog availability, fog event), effectiveness of the fog collector, and the environmental conditions. In highly fog-prone areas, it has been reported (Schemenauer and Cereceda 1994) that fog collectors were able to deliver 11,000 L per day of freshwater in the arid coastal desert of North Chile.

It appears that fog catchers operating at various locations in the world use mesh of the Raschel type, or cylindrical wire mesh arranged in a square, or diagonal formation, though no particular reason can be found in the literature regarding the preference of a mesh at a particular geographical location. Fog catcher installations fall into two major categories, namely, Standard Fog Collector (SFC) and Large Fog Collector (LFC). The meshes are assembled on to square frames 1 to 1.5 m square cross-section, and erected vertically so that the fog-capture surface of the mesh is orthogonal to the direction of fog movement. The frames are supported by scaffolds to prevent uprooting of the structure due to excessive wind velocities. These meshes on frames are laid out in a terrain for several kilometres where fog events occur. Water falling off from the wire meshes is suitable diverted to water collection tanks located beneath the mesh in the fog capture device.

Once installed, occasional cleaning of the meshes is required as these meshes are exposed to the environment, in order to remove wind-borne dust and debris. Major contributors of the operational costs include mesh maintenance, mesh replacement, and pumping equipment from collection cisterns for municipal or irrigational use. The cost of fog water produced has been found to be of the order of US $1 per m^3, and is found to be sustainable in terms of the long-run expected of these devices. The feasibility and sustainability of fog harvesting are provided in detail Batisha (2015) including economic calculations.

5.1 Fog Harvesting Case Studies: Chile

Chile is a country located along the south-western border of South America, serving as a western seaboard for the Pacific Ocean. It has a huge variation in the climate system varying from a desert-type climate in the north to tropical and temperate in the middle, and semi-arid in the south. Back in the earlier years, water shortage issues for some parts of the land led to the early development of large-scale development of fog harvesting devices in this geographical region. One of the driest places on earth, the Atacama Desert is located in Chile; however, marine advection fog that pervades the land provides moisture for certain types of plants to grow in the soil.

Quite naturally, since 1987 El Tofo locality has been the site of developmental testing of large fog collectors to take advantage of exceptional fog events. Fog water production using fog meshes has been in operation since March 1992, delivering about 3 L per square metre per day, and totalling about 11,000 L per day from all of the 25 large scale units installed in the location. There are, of course, zero water production days, when adverse climate (for fog) exists. Water quality from the fog harvesters at this location has been proven to satisfy water quality standards established for Chile as well as the requirements of World Health Organisation on water quality.

5.2 Fog Collection Case Studies: Peru

Like Chile, Peru is also located in South America, on the Pacific coast. The sandy desert coastline of Cerro Orara in Peru provides frequent fog events for effective generation of freshwater from fog meshes installed in the locality. Published literature (Peru's Fog Catchers Net Water Supplies 2012) provides evidence of 1700 square metres of mesh installed near Lima providing fog generated freshwater for a school and a recreational park facility. Peru is sandwiched between the Pacific Ocean and Andes mountains, and has very low precipitation of about less than 4 cm per year, but having a very high relative humidity of about 98%. Fog nets have been reported to provide 'crystal-clear' water in this locality.

5.3 Fog Collection Case Studies: Morocco

Morocco lies in the north-west corner of Africa, and close to Sahara Desert. Several fog nets have been installed, to an extent of about 1686 square metres of net, being one among the largest fog harvesters in the world, along the foothills of Anti-Atlas Mountains in Africa. On an average, about 22 L of freshwater are obtained per square metre annually. The potable quality water has been supplied to nearby villages for consumption.

Similar success stories of freshwater obtained from fog meshes have been reported from other parts of the world such as Bolivia, Tanzania, Yemen, and Nepal among others. From the variety of published literature that we surveyed, it appears that several entrepreneurs and Non-Governmental Organisations (NGOs) have in the past, and present shown interest in setting-up fog catcher nets in some or all of these areas. These agencies have also succeeded in realising some quantity of freshwater to be collected out of the fog. What is clearly lacking is a fundamental, scientific understanding on the design of fog meshes, the phenomena of fog adhesion onto the meshes, coalescence of droplets, drainage, as well as methods to maximise the yield, based on sound engineering principles. These aspects of scientific knowledge will be covered in later chapters in this monograph.

6 Impact Assessment of Water Harvesting Projects Using Fog

As long as the technology used for enabling fog to impact on the mesh surfaces does not forcefully draw fog onto the meshes, there is no likelihood of violation of nature's way of working. Does nature intend to balance out fog movements? This is something that we don't know as yet—in having any impact on the environment. Perhaps, this is a topic that is worth monitoring from the point of view of environmental engineering and climate change effects.

References

Batisha AF (2015) Feasibility and sustainability of fog harvesting. Sustain Water Qual Ecol 6:1–10

Bayens D (2018) Dew water. River Publishers, Denmark

Fathieh F, Kalmutzki MJ, Kapustin EA, Waller PJ, Yang J, Yaghi OM, Fathieh F et al (2018) Practical water production from desert air. Sci Adv 4(8):eaat3198. https://www.opur.cloud. Accessed on 22 Apr 2025

Haechler I, Park H, Schnoering G, Gulich T, Rohner M, Tripathy A, Milionis A, Schutzius TM, Poulikakos D, Haechler D et al (2021) Exploiting radiative cooling for uninterrupted 24-hour water harvesting from the atmosphere. Sci Adv 7(23):eabf3978

Kim H, Yang S, Rao SR, Narayanan S, Kapustin EA, Furukawa H, Umans AS, Yaghi OM, Wang EN, Kim T et al (2017) Water harvesting from air with metal-organic frameworks powered by natural sunlight. Science 356(28):430–434

Liu C, Fan J, Bao H (2020) Hydrophilic radiative cooler for direct water condensation in humid weather. Sol Energy Mater Sol Cells 216:110700

Miller JL (2017) Metal–organic framework extracts water from thin air. Phys Today 70(6):16. https://doi.org/10.1063/PT.3.3579

Peru's fog catchers net water supplies (2012) The Guardian. https://www.theguardian.com/global-development/2012/sep/19/peru-fog-catchers-water-supplies

Raveesh G, Goyal R, Tyagi SK (2021) Advances in atmospheric water generation technologies. Energy Conserv Manag 239:114226

Schemenauer RS, Cereceda P (1994) History of fog meshes, fog collection's role in water planning for developing countries. Nat Resour Forum 18(2):91–100

Wang J, Yang Z, Li Z, Fu H, Chen J (2025) Comprehensive review on atmospheric water harvesting technologies. J Water Process Eng 69:106836

Zhu W, Zhang Y, Zhang C, Shan X, Rao AK, Pitts SL, Woodbury TJ, Masnyk TS, Derome D, Warsinger DM, Ruan X, Mauer LJ, Carmeliet J, Li T (2023) Radiative cooling sorbent towards all weather ambient water harvesting. Commun Eng. https://doi.org/10.1038/s44172-023-00082-3

Zhao B, Hu M, Ao X, Chen N, Pei G (2019) Radiative cooling: a review of fundamentals, materials, applications, and prospects. Appl Energy 236:489–513

Filmwise and Dropwise Condensation

Basant Singh Sikarwar and K. Muralidhar

1 Introduction

In the condensation process, vapor turns into the liquid phase during the cooling process (Incropera et al. 2007; Carey 2018). This phase change occurs when the vapor molecules lose enough thermal energy to overcome their kinetic energy, allowing them to form intermolecular bonds and condense into a liquid (Sheng et al. 2016). The phase change process is seen in pure vapor and mixtures of vapor with non-condensable gases. Condensation may occur in the volume of the body of vapor; condensation on subcooled solid surfaces is interesting to the present discussion. The significance of vapor condensation on cooled solid surfaces is seen in several engineering applications, including the following (El Fil et al. 2020):

- *Heat Exchangers*: In heat exchangers, such as those used in HVAC systems, condensation plays a crucial role in efficiently transferring heat from one medium to another. For instance, in a condenser, refrigerant vapor is cooled and condensed over tubes carrying the coolant, thus releasing latent heat.
- *Power Generation*: In steam power plants, steam condenses back into water in the condenser after passing through the turbine. This process helps create a higher pressure difference, thereby enhancing turbine efficiency and allowing for the recovery of water for reuse.
- *Distillation*: Condensation is a key step in distillation processes, where the components of a mixture are selectively separated based on their boiling points.

B. S. Sikarwar (✉)
Amity University, Noida, India
e-mail: bssikarwar@amity.edu

K. Muralidhar
IIT Kanpur, Kanpur, India

R. Narayanaswamy et al. (eds.), *Sustainable Water Harvesting*, Springer Tracts in Mechanical Engineering, https://doi.org/10.1007/978-981-95-3373-2_6

- *Cooling Systems*: In cooling systems, such as those found in refrigerators and air conditioners, condensation helps remove latent heat from the refrigerant, transferring it to the surroundings.
- *Chemical Processes*: In various chemical engineering processes, condensation on selected surfaces is employed to control reaction conditions, recover solvents, and purify products.
- *Meteorology and Climate Control*: Homogeneous condensation plays a crucial role in cloud formation and precipitation, whose understanding is essential for accurate weather forecasting.
- *Water generation from the atmosphere*: Condensation is a crucial step in producing water from the humid atmosphere, offering an innovative solution to address water scarcity.

Understanding and managing condensation is essential for optimizing efficiency, improving performance, and ensuring the longevity of equipment in engineering applications. Condensation of water vapor can be distinguished as follows: (a) Filmwise condensation and (b) dropwise condensation (Khandekar and Muralidhar 2020). Figure 1(a) shows a pictorial view of dropwise condensation compared to filmwise condensation, along with the orders of magnitude of the condensation coefficients. Filmwise condensation is a mode of condensation in which a continuous film of liquid forms and spreads on a cold surface. This mode of condensation is generally less efficient in terms of heat transfer compared to dropwise condensation, Fig. 1(b). However, dropwise condensation is a mode of condensation in which vapor condenses into small, discrete droplets on a surface, rather than forming a continuous film. This type of condensation is highly efficient for heat transfer because the droplets provide smaller thermal resistance between the vapor and the condensing surface when compared to a liquid film.

1.1 Filmwise Condensation

When vapor encounters a cold, hydrophilic surface that is cooler than the vapor's saturation temperature, it begins to condense into a liquid film that covers the surface. Although less efficient compared to dropwise filmwise condensation, it provides a stable and steady wall heat flux. The process of filmwise condensation is described in Fig. 2. It shows vapor condensation in the form of a film on a wetting vertical surface, as well as heat transfer from the vapor to the cooled surface through the liquid film (Incropera et al. 2007). The behavior and thickness of the condensate film are influenced by gravitational forces and surface tension, which affect the overall heat transfer performance.

Despite the complexities associated with film condensation, Nusselt derived the expression for film thickness and heat transfer coefficient in 1916 by making several assumptions. These include the thermophysical properties of the condensate being constant, the shear stress between the liquid–vapor interface being negligible, and

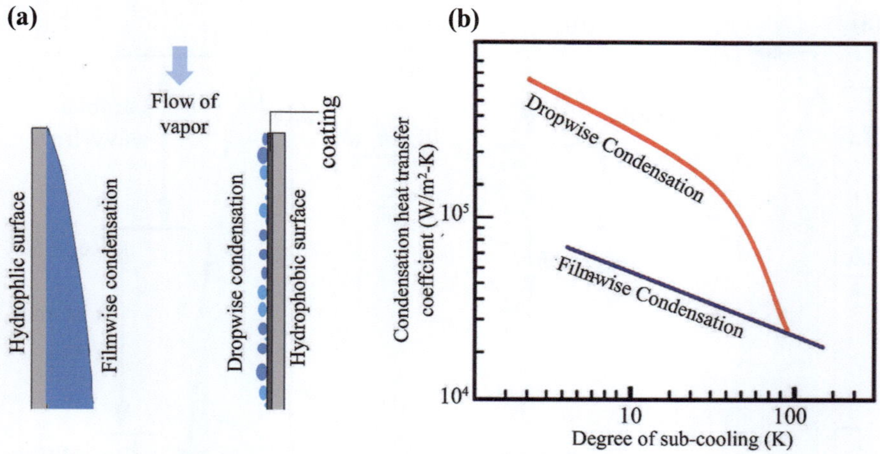

Fig. 1 **(a)** Pictorial view of filmwise and dropwise condensation, respectively, on hydrophilic and hydrophobic surfaces. **(b)** Typical condensation heat transfer coefficient achieved as a function of the degree of subcooling for the dropwise and filmwise modes (Kataria et al. 2024; reproduced with permission from Elsevier)

the uniform temperature of the vapor. Flow within the condensate was taken to be fully developed (Rose 1998, 2011; Beysens 2022). Based on the momentum equation (Eq. 1) and the energy equation (Fig. 2b), condensate film thickness and heat transfer coefficient were derived as in Eqs. 2–6.

$$\frac{\partial^2 u}{\partial y^2} = -\frac{g}{\mu_l}(\rho_l - \rho_v) \tag{1}$$

Integrating Eq. 1 and applying boundary conditions such as $u(0) = 0$ at the wall and $\partial u/\partial y|_{y=\delta} = 0$ at the free surface, the velocity profile within the condensate was obtained as

$$u = \frac{g(\rho_l - \rho_v)}{\mu_l}\left[\frac{y}{\delta} - \frac{1}{2}\left(\frac{y}{\delta}\right)^2\right] \tag{2}$$

Using Eq. 2, the condensate flow rate and hydrodynamic boundary layer thickness can be estimated; details of these calculations are available in the literature (Incropera et al. 2007; Das and Chatterjee 2023). It can be shown that

$$\delta = \left[\frac{4k_l\mu_l(T_{sat} - T_s)x}{g\rho_l(\rho_l - \rho_v)h_{fg}}\right]^{1/4} \tag{3}$$

Rohsenow (2022) recommended using a modified latent heat ($h_{fg}\prime$) in lieu of h_{fg} for improving the sensitivity of Eq. 3:

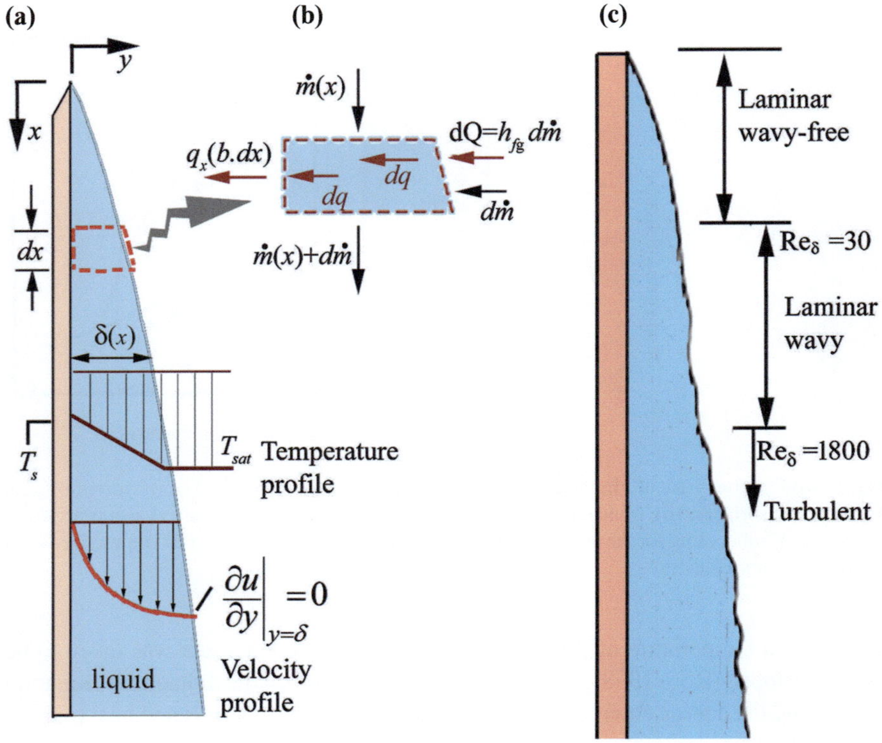

Fig. 2 (**a**) Boundary layer effects related to film condensation on a vertical surface with assumptions associated with Nusselt's analysis, for a vertical plate; (**b**) enlarged view of mass, momentum and energy balance of a small film element for estimating the boundary layer thickness and heat transfer coefficient; and (**c**) flow regimes of filmwise condensation

$$h_{fg}{\prime} = h_{fg} + 0.68C_{pl}(T_{sat} - T_s) \tag{4}$$

The local and average heat transfer coefficients for filmwise condensation can now be written as

$$h = \frac{k_l}{\delta} = \left[\frac{g\rho_l(\rho_l - \rho_v)h_{fg}{\prime}k_l^3}{4\mu_l(T_{sat} - T_s)x}\right]^{1/4} \tag{5}$$

$$\overline{h} = \frac{4}{3}h = 0.943\left[\frac{g\rho_l(\rho_l - \rho_v)h_{fg}{\prime}k_l^3}{4\mu_l(T_{sat} - T_s)L}\right]^{1/4} \tag{6}$$

Equation 5 illustrates that the presence of a liquid film introduces additional thermal resistance, thereby reducing the heat transfer coefficient (Rose 1998; Thome 2015; Das and Chatterjee 2023). In this formulation, the free surface is taken to be flat and surface tension does not play a role. Hence, the behavior and thickness of the condensate film (Eq. 3) are influenced by gravitational forces, significantly impacting overall heat transfer performance. The thermal resistance of the liquid film, which depends on its thickness and thermal conductivity (Eq. 5), is a crucial factor in determining the heat transfer rate. These equations show that thicker films exhibit greater thermal resistance and diminishing heat transfer rates.

Gravity drives the liquid film downward along the surface, and whether this flow is laminar or turbulent is determined by factors such as film thickness, surface texture, and thermophysical properties of the condensate (Kekaula and Chen 2020; Ho and Leong 2021). As the film thickens and flows downward, it may eventually detach from the surface. The flow regime of the condensate film, whether laminar or turbulent, significantly impacts the heat transfer process. In laminar flow, the film moves smoothly over the surface, while in turbulent flow, the film exhibits chaotic and irregular motion, as sketched in Fig. 2c. The transition between these flow regimes is influenced by factors such as surface roughness, condensate properties, and the temperature difference between the vapor and the surface. Understanding these flow regimes is vital for accurately predicting and enhancing heat transfer rates (Incropera et al. 2007; Carey 2018).

One of the primary challenges in filmwise condensation is managing the film thickness to minimize thermal resistance while maintaining a stable condensate layer (Ho and Leong 2021). Excessive film thickness leads to higher thermal resistance and reduced heat transfer rate. Additionally, the downward flow of the condensate film will eventually result in the shedding of the liquid from the surface, which can disrupt the condensation process and further affect heat transfer performance. As most metallic surfaces after manufacture are hydrophilic (Drelich et al. 2011), filmwise condensation is prevalent in many industrial processes, including power generation, refrigeration, and chemical processing (El Fil et al. 2020). Recent research efforts focus on developing surface treatments and coatings to manipulate the film in the form of droplets formed at specific nucleation sites, reduce thermal resistance, and enhance heat transfer rates (Shadmani et al. 2020; Sharma et al. 2020, 2022b; Jain et al. 2023). Additionally, studies are being conducted to better understand the impact of various operating conditions, such as pressure, temperature, and surface orientation on the condensation process (Sikarwar et al. 2012, 2015; Sikarwar 2023). Overall, filmwise condensation remains a critical area of study, with ongoing research aimed at improving control of its thickness and stability to optimize heat transfer performance in industrial systems (Ho and Leong 2021).

1.2 Dropwise Condensation

Dropwise condensation is a process where vapor condenses in the form of discrete droplets on a surface, rather than forming a continuous film of liquid (Vemuri and Kim 2006). Higher heat transfer coefficient occurs in this mode of condensation compared to filmwise condensation, making it of significant interest in various applications, such as in heat exchangers, power generation, and refrigeration systems (Boreyko and Chen 2009; Bonner 2013; Baghel et al. 2019a, b). Dropwise condensation occurs when the surface energy of the condensing surface is low, often achieved by spraying the surface with hydrophobic coatings (Baghel et al. 2020; Bayer 2020). Hydrophobicity facilitates the formation of condensed droplets that grow independently on the surface, rather than spreading into a thin film (Leipertz 2010; Zhang et al. 2022). As these droplets grow, they eventually coalesce and slide-off the surface under the influence of gravity, making room for new droplets to form. The cyclic nature of droplet formation, growth, and departure leads to enhanced heat transfer rate, as thermal resistance associated with the condensate layer is now minimized (Sikarwar 2013; Sikarwar et al. 2013a, b, 2020; Singh et al. 2022) (Fig. 3).

The efficiency of heat transfer in dropwise condensation on hydrophobic surfaces is primarily due to the minimal thermal resistance offered by the discrete droplets. Unlike filmwise condensation, where a continuous liquid film acts as a thermal barrier, heat transfer in dropwise condensation occurs directly through the small droplets, which are in close contact with the condensing surface. Many Researchers specify that dropwise condensation on hydrophobic surfaces can result in heat transfer coefficients that are several times higher than those observed in filmwise condensation. This enhancement is attributed to the continuous renewal of the condensing surface as droplets depart, ensuring that the thermal resistance remains low. Despite its advantages, maintaining dropwise condensation on hydrophobic surfaces over long periods poses significant challenges such as, hydrophobic coatings may degrade due to mechanical wear, high temperatures, or chemical reactions,

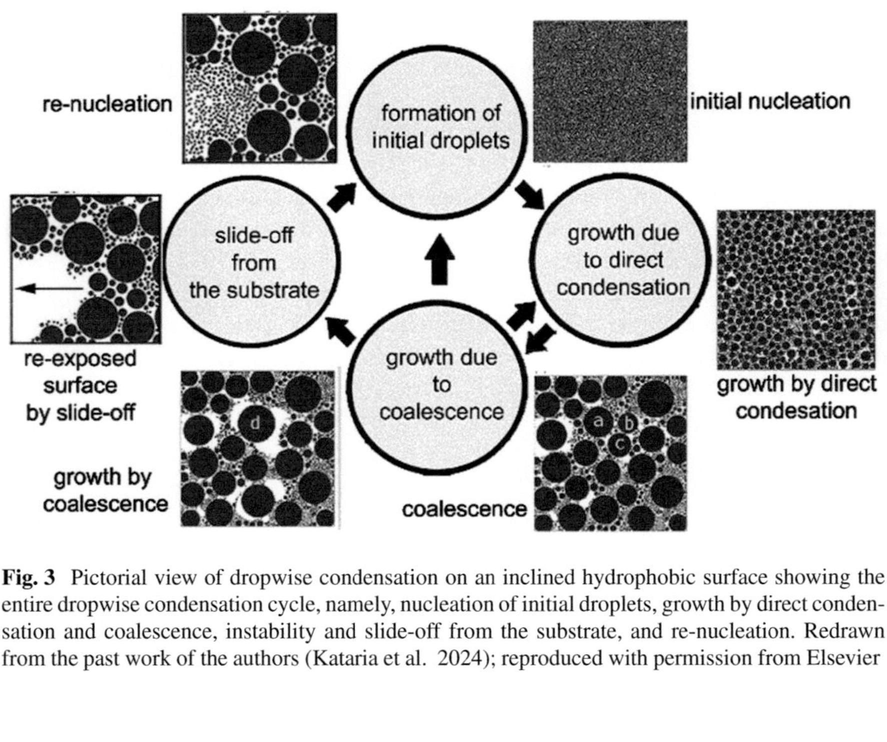

Fig. 3 Pictorial view of dropwise condensation on an inclined hydrophobic surface showing the entire dropwise condensation cycle, namely, nucleation of initial droplets, growth by direct condensation and coalescence, instability and slide-off from the substrate, and re-nucleation. Redrawn from the past work of the authors (Kataria et al. 2024); reproduced with permission from Elsevier

leading to a switch from dropwise to the less efficient filmwise condensation (Kataria et al. 2024).

Several factors influence the efficiency and sustainability of dropwise condensation. Hydrophobic, preferably superhydrophobic, coatings are essential for maintaining dropwise condensation (Wang et al. 2020). These coatings reduce the surface energy, encouraging the formation of droplets. Hierarchical surface roughness (micro- and nanoscale roughness of the surface) plays a crucial role in promoting droplet formation (Miljkovic et al. 2012; Miljkovic and Wang 2013; Nenad et al., 2013a, Nenad et al. 2013b). A hierarchical rough surface with an appropriate hydrophobic coating can further enhance the formation of droplets. In addition, the temperature gradient between the vapor and the condensing surface, as well as bulk pressure, influences the size, frequency, and departure rate of the droplets. The orientation of the surface can affect droplet movement (Leipertz and Fröba 2008). For instance, a vertical surface will promote quicker droplet departure by gravity, a mechanism less effective on horizontal surfaces. The heat transfer coefficient in dropwise condensation is typically much higher than in filmwise condensation because thermal resistance is localized at the droplet footprint over the substrate, where heat flux is localized. The absence of a continuous liquid film on the surface reduces the overall thermal resistance, allowing for higher heat transfer rates.

Several researchers (Vemuri and Kim 2006; Vemuri et al. 2006; Miljkovic and Wang 2013; Jeevahan et al. 2018) have reported heat transfer coefficient in dropwise condensation higher than filmwise by nearly an order of magnitude. Despite its advantages, however, maintaining dropwise condensation over long periods is a challenge (Ahlers et al. 2019; Kataria et al. 2024). The durability of hydrophobic coatings is a significant concern, as these coatings may degrade over time due to factors such as temperature fluctuations, mechanical wear, or chemical reactions (Zheng et al. 2021). Additionally, the cost of applying and maintaining such coatings can be prohibitive. Another limitation is that dropwise condensation is challenging to achieve with certain types of vapors, especially those with a strong affinity for the condensing surface, which can lead to the formation of a continuous film. Recent research (Wang et al. 2022) has focused on developing more durable and efficient hydrophobic coatings, exploring novel materials, and understanding the fundamental physics behind droplet formation and heat transfer mechanisms. The use of nanotechnology has shown promise in enhancing the stability and performance of surfaces for dropwise condensation (Lo et al. 2019). Another emerging area of research is the exploration of hybrid condensation modes, where dropwise and filmwise condensation occur simultaneously, potentially offering a balance between high heat transfer efficiency, stability, and practical realizability (Thomas et al. 2023).

2 Hydrophobic Surfaces

Hydrophobicity is defined by the contact angle made by the surface with water and is typically >90° (de Gennes et al. 2010; Sotoudeh et al. 2023) for a hydrophobic surface, Fig. 4. Hydrophobicity, measured in terms of the contact angle made by the condensate with respect to the condensing surface, is a critical factor that influences droplet formation, growth, and departure (Ma et al. 2022).

On hydrophobic and superhydrophobic surfaces ($\theta > 90°$), water vapor condenses into small, spherical droplets due to the low surface energy. These droplets grow as more vapor condenses, eventually merging with nearby droplets (Chu et al. 2017). Once they reach a critical size, the droplets overcome the adhesive forces between the liquid and the surface, causing them to roll off or slide down the surface under the influence of gravity or other external forces (Sikarwar et al. 2013a, b; Baghel et al. 2019a, b). This self-cleaning mechanism continuously exposes fresh surface areas for new droplets to form, maintaining a small droplet size on average and hence, large heat fluxes (Yilbas et al. 2019).

The hydrophobicity of a surface is influenced by its chemistry, surface roughness and coating of low surface energy materials, as shown in Fig. 5. Surfaces coated with hydrophobic materials like fluoropolymers, silanes, or other low-energy materials are more likely to exhibit dropwise condensation (Zhu et al. 2015). Hierarchical roughness at micro and nano scale enhances the hydrophobicity of a surface by trapping air pockets beneath the droplets while reducing the actual contact area between the liquid and the solid (Tang et al. 2021) (Fig. 6). The drop configuration, known as the Cassie–Baxter state, further promotes dropwise condensation (Singh et al. 2021).

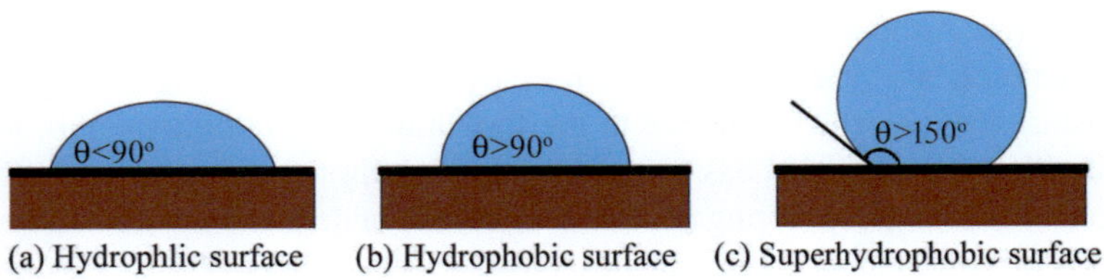

Fig. 4 Schematic representation of hydrophobicity in terms of θ, the contact angle

Fig. 5 Low surface energy material used for enhancing the hydrophobicity of metallic surfaces

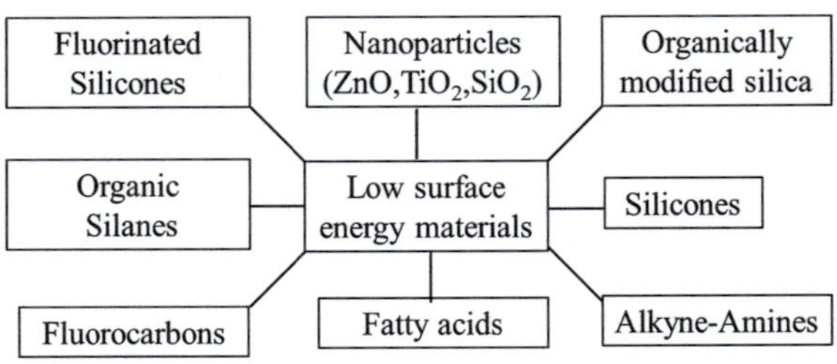

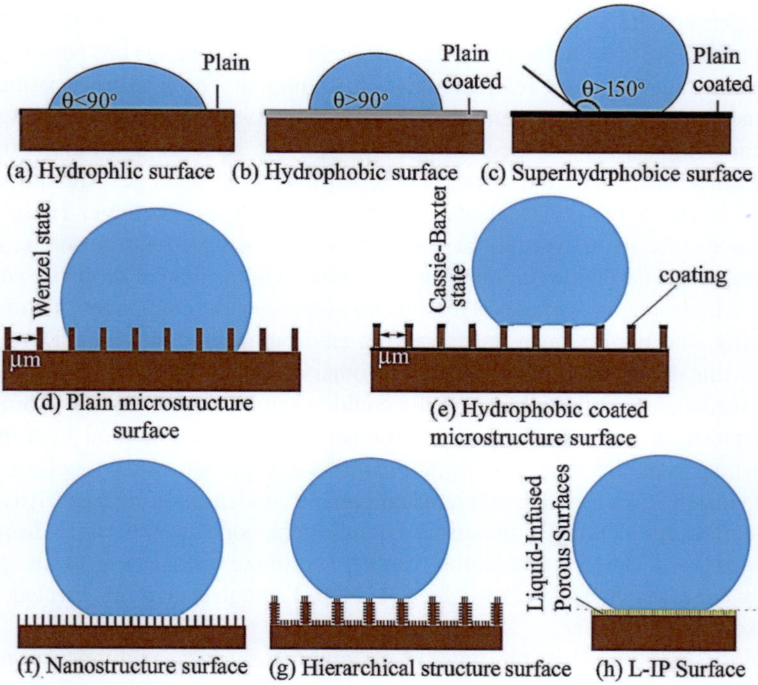

Fig. 6 Droplet state on various chemically and physically textured surfaces. **a** Plain smooth surface having contact angle <90° is known as hydrophilic; **b** low-energy material-coated surface having contact angle $\theta \geq 90°$ is known as hydrophobic; **c** low-energy material-coated surface having contact angle $\theta \geq 150°$ is known as superhydrophobic; **d** plain well-organized microstructured pillars known as a plain structured surface. The droplet sitting on this surface penetrates the inter-pillar gap. The droplet is said to be in the Wenzel state. **e** Microstructured surface coated with a low surface energy material. The water droplet remains in the Cassie state on this surface and does not penetrate the pillar gap. **f** and **g** Nanostructure and hierarchical structured surfaces. These surfaces generally exhibit the Cassie state, and the contact angle is generally ≥120°. Over a nanostructured surface, roughness on the nanoscale is superimposed over micron-order roughness to form a hierarchy of length scales. **h** Liquid-infused porous surfaces (L-IPS) where pores of micron to nanoscale order are created and oil is infused to make them as slippery for droplet movement; from Kataria et al. (2024); reproduced with permission from Elsevier

Droplets need to be efficiently removed from the surface to prevent them from merging into a film, especially on horizontal surfaces where gravity is less helpful. Fabricating surfaces that stay hydrophobic under various conditions requires advanced techniques to balance surface roughness and chemical composition for optimal performance. Despite the challenge to fabricate efficient and durable hydrophobic surfaces for condensation, recent advancements in surface engineering have led to the development of robust hydrophobic surfaces capable of sustaining dropwise condensation. Nanostructured surfaces with precise features have shown promise in enhancing hydrophobicity and maintaining dropwise condensation over longer periods. These surfaces can trap air and reduce the actual contact area between the droplets and the surface, maintaining effectively a low surface energy. Combining hydrophobic and oleophobic (oil-repellent) properties in coatings has been explored to improve durability and resistance to contamination. These hybrid coatings can help maintain dropwise condensation in environments where vapor contains impurities that might otherwise degrade the hydrophobic surface. Innovations in self-healing materials, where the surface can repair itself after damage, offer a potential solution to the durability weakness faced by hydrophobic coatings. These materials can restore their hydrophobic properties after being compromised, ensuring sustained dropwise condensation. Ongoing research in the field of dropwise condensation on hydrophobic surfaces is focused on addressing the challenges of durability, scalability, and performance under varying environmental conditions. Developing surfaces that combine hydrophobicity with other desirable properties, such as corrosion resistance or antimicrobial activity, could expand the applicability of dropwise condensation in more demanding environments. Mimicking natural hydrophobic surfaces, such as the lotus leaf, which exhibits exceptional water repellence, can provide insights into creating efficient and durable condensation surfaces. Advanced computational models that simulate the dynamic behavior of droplets on hydrophobic surfaces under a variety of conditions can help optimize surface designs for specific applications (Jeevahan et al. 2018; Tang et al. 2021; Singh et al. 2021; Fu et al. 2024; Boreyko and Chen 2009).

Table 1 summarizes various types of hydrophobic surfaces with relevant parameters that support drop mobility. Methods and materials reported in the literature to enhance hydrophobicity of a metallic surface for water vapor condensation are summarized in Fig. 7. Additionally, classification of low surface energy materials which are used for modification of surface chemistry for enhancing the hydrophobicity was shown earlier in Fig. 5. While durability of these surfaces is of concern, a combination of physical and texturing techniques that stabilize hydrophobicity of a metallic surface under humid air conditions has been described by these authors.

A combination of physical vapor deposition, chemical etching, electrochemical, and chemical vapor deposition is used to fabricate durable condensing surfaces, proposed in (Baghel et al. 2019a, b; Sharma et al. 2020, 2022a, b; Kataria et al. 2022; Singh et al. 2022; Kataria et al. 2024) shown in Fig. 8.

Typically, condensing surfaces are made of stainless steel, aluminum, and copper. These metallic surfaces have large thermal conductivity but also high surface energy, making them hydrophilic. Special treatment is required to increase their hydrophobicity. Using techniques shown in Fig. 8, hierarchical morphologies on aluminum can be created by altering the roughness in micron-order by plasma etching and coated by a fluoropolymer (Teflon, PTFE) solution followed by silane-based [octa-decyl-tri-chloro-silane (OTS) and hexa-decyl-tri-meth-oxy-silane (HDTMS)] compounds to enhance hydrophobicity. In addition, hydrophobicity can also be enhanced via the incorporation of nanoparticles (e.g., silica, titanium dioxide) into the chosen coatings. Many chemical and mechanical methods alter roughness when coated with a low surface energy material. Before commencing the process of enhancing hydrophobicity of metallic surfaces, these are cleaned and polished by properly selected grade of sandpaper, cleaned in an ultrasonic acetone bath, rinsed with DI water, dipped into

Table 1 Different pattern surfaces with quantifying contact angles and condensation parameters

Surface category	Coating and material	Contact angle (°)	Condensation mode	Droplet dynamics
Superhydrophobic (Singh et al. 2021)	Fluoropolymer	>150	Dropwise	Droplets easily roll off (highly
Hydrophobic rough (Rezaee et al. 2023)	Silica nanoparticles	120–150	Dropwise	Moderate mobility; droplet sliding
Hydrophobic smooth (Parvate et al. 2020)	Teflon	90–120	Dropwise	Low mobility: droplets adhere
Nanostructured (Lee et al. 2020; Tang et al. 2021)	Carbon nanotube (CNT)	150–170	Dropwise	High mobility; minimal droplet retention
Hierarchical (Baba et al. 2021)	Micro/nano combined textures	>160	Dropwise	High droplet mobility
Hydrophilic (Ahmad et al. 2018)	Bare metal surface	<90	Filmwise	Droplets coalesce into a film
Slippery surface (Veronesi et al. 2021)	Infused oil in porous metallic surfaces	>110	Dropwise	High mobility; minimal droplet retention

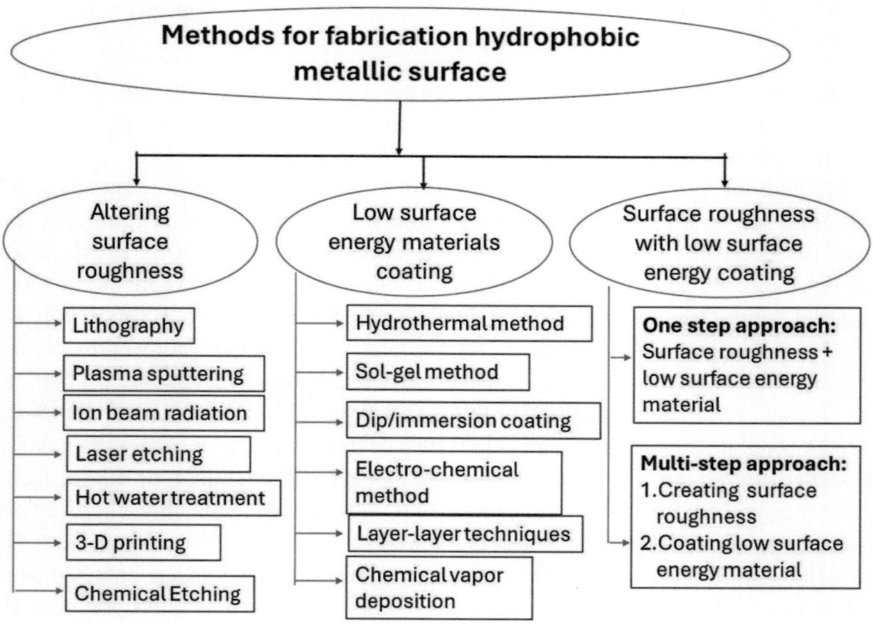

Fig. 7 Techniques for fabricating durable hydrophobic metallic surfaces

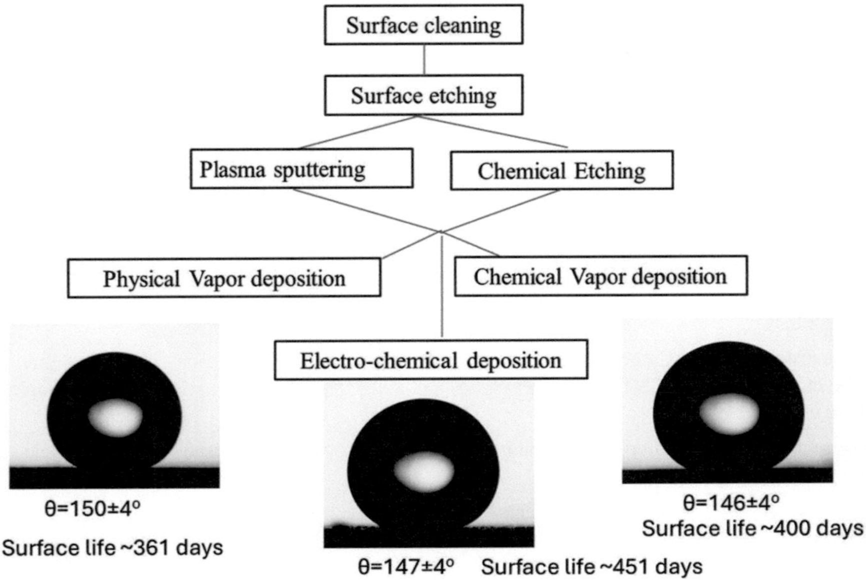

Fig. 8 Hybrid method for the fabrication of a hydrophobic aluminum surface by a combination of two methods—plasma etching and plasma coating (Sharma et al. 2022a; b), chemical etching (Singh et al. 2022), and electrochemical deposition (Baghel et al. 2019a; b). The authors of this chapter also recommend plasma etching along with the chemical vapor deposition process

the etchant for removing oxides, rinsed, and dried with nitrogen. The time period for each process differs according to the composition of chemicals, surface morphology, and grade of metals.

Methods of fabrication can be classified as material subtraction and material addition, 3D printing, vapor deposition, and lithography methods falling in the latter category. In contrast, chemical and electrochemical machining (ECM), electric discharge machining (EDM), and plasma etching are subtractive methods. Most material subtraction methods for fabricating superhydrophobic surfaces involve a multi-step process. The first step is creating surface roughness, followed by chemical or physical modifications to achieve the desired superhydrophobic properties, Fig. 7. However, material addition processes are typically single-step methods where surface roughness is created, and a low surface energy material is simultaneously deposited. In the surface texturing process, mechanical, chemical, or a laser technique is used to create micro- and nanoscale roughness on the metallic surface. This increases surface area and enhances the entrapment of air pockets, that are crucial for achieving hydrophobicity. During the deposition of low surface energy material, a thin, uniform layer of a hydrophobic compound—such as fluoropolymers or silane-based chemicals—is applied to the roughened surface. This coating reduces surface energy, further enhancing water repellence. These refinements are essential for creating a durable, highly hydrophobic surface on metals. Table 2 summarizes

Table 2 Advantages and disadvantages of surface modification used in AWGs for extracting water from moist air

Methods	Advantages	Disadvantages
Layer-by-layer	Fast, low cost, flexible	Less uniform, moderate durability
Chemical vapor deposition	Pure and fine quality coating, controlled coating, complex surfaces	Slow, high cost, high temperature
Physical surface etching	Fast, low cost, excellent mechanical stability	Less control, affects physical and mechanical properties, costly, and not feasible on an industrial scale
Dip-coating	Simple, industrial scale, good durability	Slow, large number of solvents
Chemical surface etching	Simple, one-step method, controlled shape	Slow, severe conditions, small scale, moderate

some of the most widely used methods, along with their pros and cons, for fabricating condensing surfaces.

Both superhydrophobic and superhydrophilic surfaces offer unique advantages for condensation. Fabricating alternative zones of superhydrophobic and superhydrophilic properties will accelerate droplet shedding, as shown in Fig. 9. Over this surface, vapor condenses in the form of droplets in the superhydrophobic zone. In contrast, superhydrophilic surfaces promote thin-film condensation and attract droplets from the superhydrophobic zone by capillary forces. Hence, the superhydrophobic part functions as an efficient vapor condensing zone, and the superhydrophilic portion serves as an effective condensate removal mechanism. Independently, flat surfaces can be replaced with condenser coils, which provide larger surface areas and improved heat exchange efficiency, and the possibility of optimization as in Table 3.

3 Mathematical Modeling of Dropwise Condensation

Figure 3 shows that the dropwise condensation process is a cycle comprising a combination of various subprocesses. These subprocesses—nucleation, growth, coalescence, and departure, involve nanometric sales all the way to the scale of millimeters (Sikarwar et al. 2020). For example, nucleation of a nanometric scale droplet is governed by forces at the atomic cluster level while droplet growth, coalescence, and shedding are governed by continuum scale principles. Several approaches have been used in the literature to simulate dropwise condensation, each with its own strengths and limitations. Empirical approaches rely on experimental data to establish

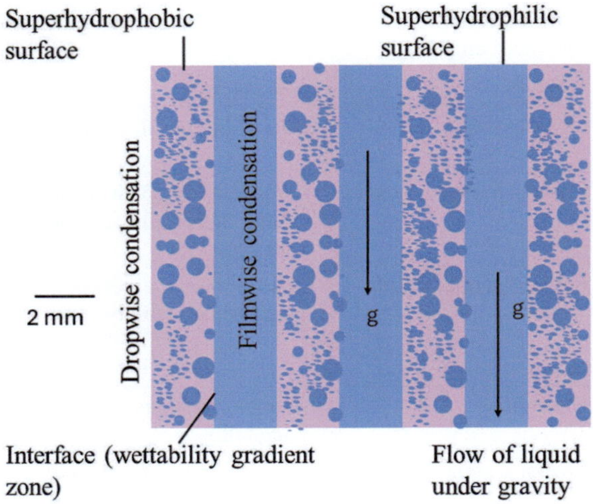

Fig. 9 Condensing surface having alternate zones that are superhydrophobic and superhydrophilic. Condensation over the superhydrophobic part is dropwise, where smaller drops offer smaller thermal resistance. In contrast, a smooth superhydrophilic zone will have micron-order film thickness whose thermal resistance is equally small. Migration of droplets into the liquid film by capillary action will augment condensation rate

Table 3 Flat surface condenser versus coil-type condenser

Flat plate condenser	Coil-type condenser
Consists of smooth superhydrophobic, flat surfaces for condensing surface	Superhydrophobic smooth coiled tubes, offering a larger surface area within a smaller volume
Efficient heat transfer due to uniform surface contact, but limited in surface area	Extended surface to enhance heat dissipation
Easier to clean and ideal for applications with moderate heat loads	Complex to clean but suitable for higher heat loads and space-constrained setups

relationships between the heat transfer rate and parameters such as surface temperature, vapor pressure, and droplet size distribution. This approach is often used for its simplicity and ease of implementation though it lacks accuracy in predicting performance under conditions not considered in the data set. In other approach, the individual subprocesses of the dropwise condensation cycle are modeled and each of these are linked in such that they complete the cycle, Fig. 3. Each subprocess could be modeled by a mechanistic approach though a semi-empirical route may be adopted for computational efficiency. Alternatively, a probability distribution may be used to model the size distribution of droplets while a population balance model will link the subprocesses of the dropwise condensation cycle.

In the semi-empirical approach, a combination of physical principles and experimentally observed parameters are jointly used. Rose and Glicksman (1973) setup one of the earliest semi-empirical models for nucleation site density (i.e., number of droplets per unit area). Later, Rose (2011) used empirical constants for modeling of droplet growth by direct condensation and linked droplet growth to droplet radius, vapor properties, and surface temperature. Sikarwar et al. (2012) described a semi-empirical treatment of coalescence and proposed correlations for wall heat flux and shear stress during droplet sliding and detachment.

In a fully mechanistic model, the subprocesses of dropwise condensation are modeled from first principles, relying on transport phenomena and thermodynamics. Molecular dynamics (MD) simulation offers a detailed view of dropwise condensation at the atomistic level, shedding light on the early phases of droplet formation. This simulation models the interactions between individual molecules, revealing how surface chemistry and molecular forces drive condensation. It enables researchers to explore the intricate dynamics at the molecular scale, enhancing our understanding of how specific surface characteristics influence the onset of dropwise condensation. Lattice Boltzmann method (LBM) is a computational technique for modeling fluid flow and heat transfer during dropwise condensation. By discretizing the fluid domain into a lattice, LBM solves the Boltzmann equation to simulate the interaction of droplets among themselves and with the condensing surface. Phase-field modeling is a computational tool for simulating the evolution of droplet shapes on surfaces with prescribed wettability during dropwise condensation. Along with the governing equations for mass, momentum, and energy, it effectively captures the details of droplet nucleation, growth, and coalescence.

As dropwise condensation is a combination of subprocesses that occur over a wide spectrum of length and timescales, a complete mechanistic model is complex and computationally intensive but offers the potential for greater accuracy and insight. Several challenges persist, including the complexity of droplet interactions, surface heterogeneity, high computational demands, and the seamless integration of multi-scale phenomena. These challenges are outlined and briefly discussed here.

1. Complexity of droplet interactions: The stochastic nature of droplet nucleation, growth, coalescence, and departure is a source of difficulty. Interactions between droplets such as coalescence and jumping are difficult to resolve on the timescale of the condensation cycle.
2. Surface heterogeneity: Real-world surfaces are often heterogeneous, with variations in roughness, chemistry, and surface energy. Modeling these variations to yield a form of a contact angle and hysteresis is a source of difficulty.
3. Computational demand: High-fidelity mechanistic models, such as MD, LBM and phase-field simulations are computationally intensive. Simulating the behavior of millions of droplets over time requires significant computational resources.
4. Integration of multi-scale phenomena: Dropwise condensation involves processes that occur at widely differing scales and integrating them without approximation into a single coherent model remains a challenge.

Three theories of dropwise condensation and the related heat transfer rate have been reported in the literature, as shown in Fig. 10. In the first, vapor condenses first over the free surface of the droplets and forms a thin film. However, vapor transport from the film to droplets by surface diffusion takes place within the adsorbed film. In the second, vapor condenses initially to form a thin film that ruptures due to

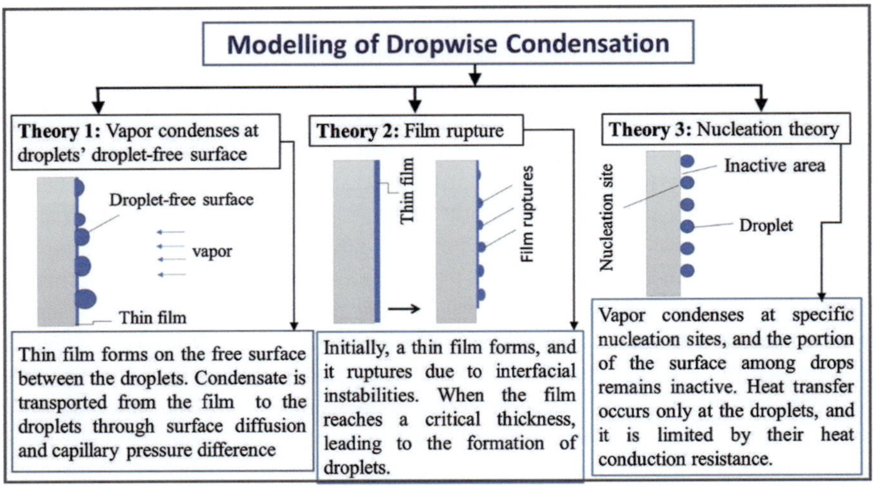

Fig. 10 Various theories of droplet formation over a hydrophobic surface (after Singh et al. 2022, reproduced with permission from Elsevier)

interfacial instability to form droplets. These droplets subsequently grow by direct condensation of vapor. In the third, vapor condenses practically instantaneously at specific nucleation sites and further condensation occurs only at the free surface of droplets appearing at the nucleation sites. The portion of the condensing surface among the droplets remains inactive in terms of heat and mass transfer. According to this theory, thermodynamically stable droplets are assigned at nucleation sites and vapor condenses at the liquid–vapor interface. Experiments support the premise of the third hypothesis and is further described below.

The number of nucleation sites per unit area of surface is known as the nucleation sites density. Although many empirical correlations and theory are available for estimating the initial nucleation site density, its exact number is an area of conjecture. It depends on the thermophysical properties of the condensing vapor as well as surface characteristics, such as wettability, roughness, and the presence of impurities. It is also influenced by surface properties such roughness, chemical composition, and wettability. Rough surfaces with microscale or nanoscale features generally have higher nucleation site densities because they provide more locations where vapor molecules can aggregate to form stable nuclei. Additionally, surfaces treated with hydrophobic coatings tend to exhibit higher nucleation site densities for dropwise condensation, as they reduce the energy barrier for droplet formation and prevent droplets from spreading into a continuous film. Environmental conditions, such as temperature and vapor pressure, also affect nucleation site density. Higher vapor supersaturation levels can increase nucleation site density by providing more vapor molecules that can condense into droplets. Similarly, a lower surface temperature relative to the vapor saturation temperature can increase the likelihood of nucleation by creating a larger driving force for condensation. Numerous experimental studies

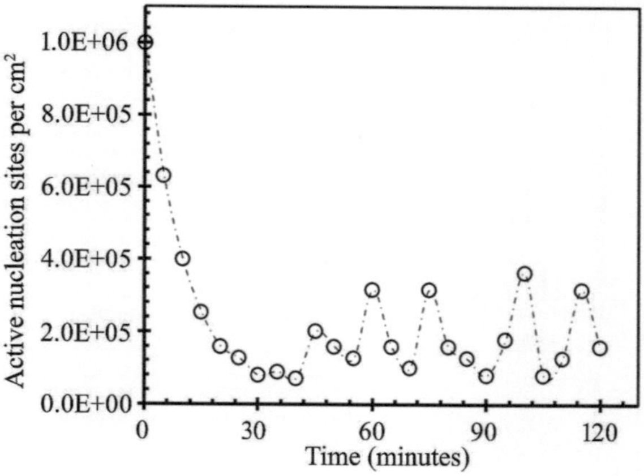

Fig. 11 Active nucleation sites versus time during the condensation process. Initially, active nucleation sites equal the initial nucleation site density that is characteristic of the surface, condensing vapor, and thermodynamic conditions

and semi-empirical models suggest that the initial nucleation site density ranges from 10^5 to 10^7 per cm^2. Some studies, based on various experimental and empirical correlations, have rounded off this number at 10^6 per cm^2. Baghel et al. (2019a; b) reported that droplets grow and coalesce, while hiding some of the nucleation sites. Hence, initial nucleation sites are not activated as, shown in Fig. 11. Active nucleation sites are only equal to initial nucleation sites density at the initiation of condensation and will diminish with the progress of the condensation cycle.

Sikarwar et al. (2012, 2013a; b) developed a multi-scale model of the dropwise condensation cycle using nucleation theory blending theoretical principles with semi-empirical data for the subprocesses. A framework that explains the steps involved in the model is depicted in Fig. 12. The atomistic model, which relies on population balance, is the starting point for determining the size of the smallest thermodynamically stable drop at nucleation sites over the condensing surface. The nucleation site density is kept a constant 10^6 cm^{-2} throughout the study. These sites are randomly distributed on the substrate and all the nucleation sites are initially occupied by liquid drops of the smallest radius (maximum size of stable cluster) as determined from the atomistic model.

Atomistic modeling carried out in these studies showed that the size of the maximum stable cluster was indeed equal to the size of minimum stable droplet obtained from thermodynamic consideration (Van Carey 2020). The nucleation process begins when the surface is cooled below the saturation temperature of the vapor and condensation begins with the nucleation of tiny droplets at specific sites on the cold surface. At this stage, vapor molecules lose energy and tend to cluster together, forming nuclei or embryos of liquid droplets. The size of these embryos plays a crucial role in their stability; smaller clusters have large internal pressure and may evaporate back into the vapor, while larger ones grow into stable droplets. The radius of this minimum stable droplet ($r_{\min}$) is given by the expression (Sikarwar et al. 2013):

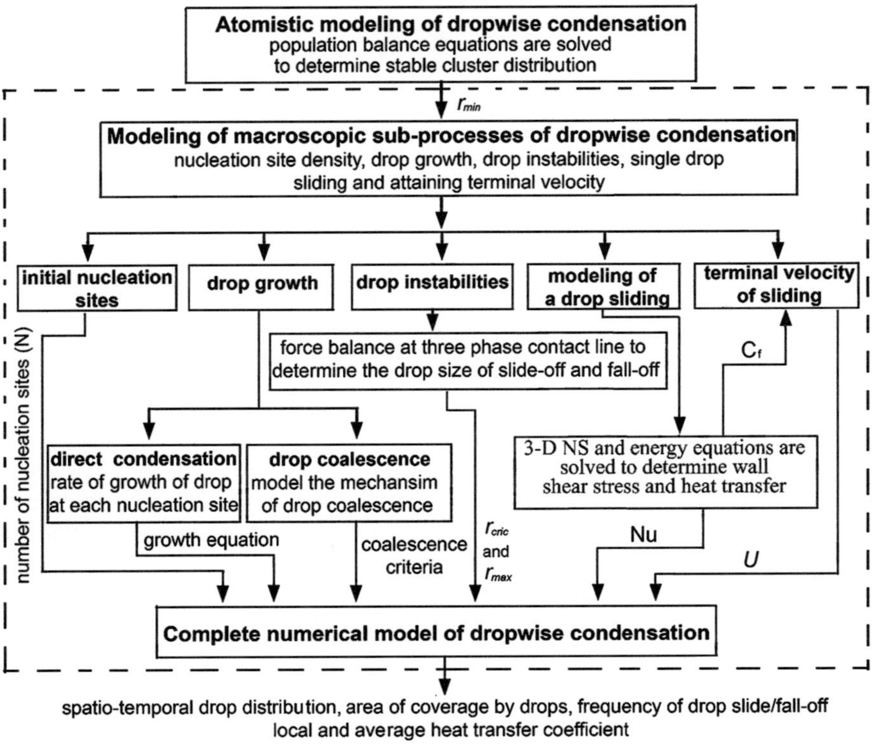

Fig. 12 Flowchart of the complete model representing dropwise condensation, as proposed in the doctoral dissertation of Sikarwar et al. (2013)

$$r_{\min} = \frac{2\sigma v_l T_w}{h_{lv}(T_{\text{sat}} - T_w)} \tag{7}$$

It is found that condensation patterns at longer timescales are not sensitive to the atomic level processes that fix the minimum drop radius. Therefore, in subsequent studies, atomic level modeling of condensation is dispensed with and drops formed at the initial nucleation sites are directly assigned a minimum-possible stable radius at each initial nucleation sites. After nucleation, droplets grow by direct condensation of vapor and by coalescence with neighboring droplets. Initially, tiny droplets form at nucleation sites and grow by direct condensation of vapor onto their surfaces. As these droplets increase in size, they coalesce with neighboring droplets, further accelerating their growth. Equations 8–10 summarizes the growth rate of droplets in the presence of pure vapor and moist air on a nanotextured textured surface coated with a hydrophobic promoter layer. These equations are derived by considering thermal resistances to the transfer of sensible and latent heat at the interface all the way to the cold substrate, as shown in Fig. 13. The subsequent steps that follow nucleation are growth by direct condensation, coalescence, instability, and drop sliding (or,

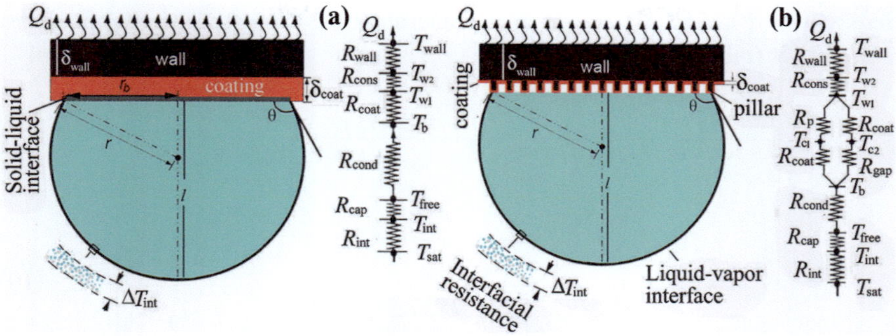

Fig. 13 Mechanism of heat transfer through the droplet: **a** droplet growth on a surface with a hydrophobic coating and **b** droplet grown on a nanotextured pillared surface

shedding) over the surface. The model yields the instantaneous drop size distribution, instantaneous rate of growth of drops, area of coverage by drops, frequency of drop slide/fall-off and local and average heat transfer coefficient over vertical and horizontal surfaces. Details of derivation of these equations are reported elsewhere (Baghel et al. 2020).

The rate of droplet growth is governed by factors such as surface wettability, thermal conductivity of the surface, and the temperature difference between the condensing vapor and the surface. Optimal droplet growth is essential for maintaining a high rate of heat transfer in dropwise condensation systems.

Pure vapor (plain surface followed by a pillared surface)

$$\frac{\mathrm{d}r}{\mathrm{d}t} = \frac{T_{\mathrm{sat}} - T_{\mathrm{wall}}}{\rho_l h_{lv}\left(2 - 3\cos\theta + \cos^3\theta\right)} \times \left(\frac{\overbrace{(1 - r_{\min}/r)}^{\text{Capillary term}}}{\underbrace{\frac{1}{2h_{\mathrm{int}}(1-\cos\theta)}}_{\text{Interfacial term}} + \underbrace{\frac{\delta_{\mathrm{coat}}}{k_{\mathrm{coat}}\sin^2\theta}}_{\text{Coating term}} + \underbrace{\frac{r}{k} f(\theta)}_{\text{Conduction term}}} \right) \quad (8)$$

$$\frac{\mathrm{d}r}{\mathrm{d}t} = \frac{T_{\mathrm{sat}} - T_{\mathrm{wall}}}{\rho_l h_{lv}\left(2 - 3\cos\theta + \cos^3\theta\right)} \times \left(\frac{(1 - r_{\min}/r)}{\frac{1}{2h_{\mathrm{int}}(1-\cos\theta)} + \underbrace{\frac{1}{k_{\mathrm{coat}}\sin^2\theta}\left[\frac{k_p\phi}{\delta_{\mathrm{coat}}k_p + h_p k_{\mathrm{coat}}} + \frac{k(1-\phi)}{\delta_{\mathrm{coat}}k + h_p k_{\mathrm{coat}}}\right]^{-1}}_{\text{Additional term for nano-pillared surface with coating}} + \frac{r}{k} f(\theta)} \right) \quad (9)$$

Here, $f(\theta) = \left(0.255e^{0.3\theta}\right)\frac{1}{4\cos(\theta/2)}$ and contact angle θ is expressed in units of radians.

Moist air (pillared surface)

$$\frac{\mathrm{d}r}{\mathrm{d}t} = \frac{1}{(2 - 3\cos\theta + \cos^3\theta)}\left(\frac{T_{\mathrm{dew}} - T_{\mathrm{wall}}}{\rho_l\left(\frac{1+\omega}{\omega}\right)C_{pm}(T_m - T_{\mathrm{dew}}) + h_{lv} + C_{pl}(T_{\mathrm{dew}} - T_{\mathrm{wall}})}\right) \times \left(\frac{(1 - r_{\min}/r)}{\frac{1}{2h_{\mathrm{int}}(1-\cos\theta)} + \underbrace{\frac{1}{k_{\mathrm{coat}}\sin^2\theta}\left[\frac{k_p\phi}{\delta_{\mathrm{coat}}k_p + h_p k_{\mathrm{coat}}} + \frac{k(1-\phi)}{\delta_{\mathrm{coat}}k + h_p k_{\mathrm{coat}}}\right]^{-1}}_{\text{Additional term for nano-pillared surface with coating}} + \frac{r}{k}f(\theta)}\right) \tag{10}$$

Droplet coalescence in dropwise condensation is a key step where adjacent droplets merge to form larger ones. As droplets grow by condensing vapor, they eventually come into contact with neighboring droplets, leading to coalescence. This merging process reduces the surface area available for condensation but simultaneously facilitates the shedding of larger droplets due to gravity, clearing the surface for new nucleation sites. Efficient droplet coalescence is crucial in maintaining the cycle of droplet formation, growth, and removal, which underpins the high heat transfer rates characteristic of dropwise condensation. Despite the abundance of literature on coalescence, a simple model adopted in most studies is the following. As soon as two droplets contact each other (or three droplets, and rarely, four contact each other simultaneously), they are substituted with an equivalent single drop while maintaining constancy of total volume located at the weighted center of mass of the individual coalescing drops. Figure 14 shows the two droplets coalescence process leading to the formation of a single droplet. The volume and position of the combined droplet are

$$V_c = \left(V_i + V_j\right)$$

$$x_c = \frac{\left(V_i x_j + V_j x_i\right)}{V_c} \quad y_c = \frac{\left(V_i y_j + V_j y_i\right)}{V_c}$$

As droplets grow and coalesce, they reach a size where gravity overcomes surface tension holding them in place, causing them to slide-off the surface. This movement removes larger droplets and exposes fresh surface area for new droplets to form. The ease and speed of droplet sliding depend on factors such as surface wettability, inclination, and the size of the droplets. Effective droplet sliding ensures the continuous renewal of the condensing surface, maximizing the condensation rate and heat transfer efficiency. The terminal velocity of the droplet sliding is estimated by balancing forces at the inception of droplet sliding, Fig. 15.

For an inclined substrate, the drop achieves terminal velocity when the component of weight parallel to the surface, the retention force of the deformed drop at the

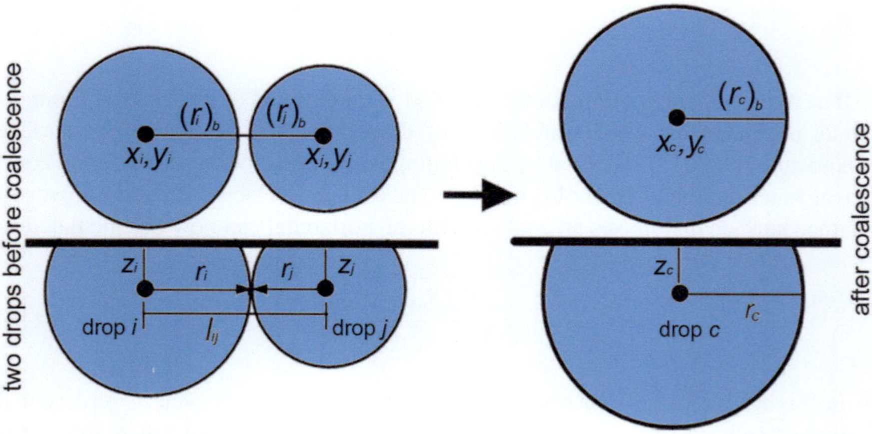

Fig. 14 Schematic diagram of the footprint and front view before and after coalescence of two micron-sized droplets. Two droplets *i* and *j* coalesce to form drop*c*

Fig. 15 Representation of forces on a sliding drop underneath a substrate moving with terminal velocity. See the text for the meaning of the symbols

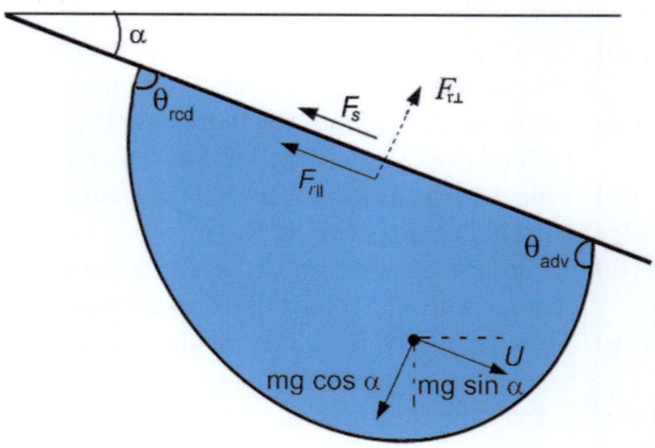

three-phase contact line owing to surface tension and wall friction are in balance. Hence,

$$F_{g\|} + F_{r\|} + F_s = 0 \tag{11}$$

where $F_{g\|}$ is the component of weight parallel to the inclined substrate, $F_{r\|}$ is the retention force opposing drop motion, and F_s is the wall shear associated with the relative velocity between the fluid and the substrate. An estimate of the retention force underneath a drop sliding on an inclined substrate is given by Eq. (12). Hence

$$F_{r\|} = -(4/\pi)\sigma r_b(\cos\theta_{\text{rcd}} - \cos\theta_{\text{adv}}) \tag{12}$$

$$F_s = \frac{1}{2} C_f A_{sl} \rho_l U^2 \tag{13}$$

The average skin friction coefficient (C_f) is determined at the scale of the individual droplet using CFD simulations of droplet sliding on an inclined surface (Sikarwar et al. 2013) followed by developing corrections for the skin friction coefficient as a function of Reynolds number. The terminal velocity of the drop over an inclined surface that makes an angle α with the horizontal can now be calculated as

$$U = \sqrt{\frac{2(F_{g\parallel} - F_{r\parallel})}{C_f \rho_l A_{sl}}} \tag{14}$$

In this way, each subprocess of the dropwise condensation cycle is represented mathematically to form a full cycle. Arising from the model described above, the important steps in the numerical algorithm can be stated as follows:

(i) Initialize all thermophysical properties, physico-chemical properties of the substrate, wettability, orientation of substrate, nucleation site density, time step, and total time of simulation.
(ii) Distribute the nucleation sites on the substrate using a random number generator and place the drop of minimum radius at all the nucleation sites.
(iii) Solve the growth rate equation numerically, say the 4th-order Runge–Kutta method over a time step, and find the new radius.
(iv) Calculate the distance between nucleation sites.
(v) Check for the coalescence.
(vi) Identify nucleation sites covered by the resulting coalesced drops and keep them deactivated till the drop covers them.
(vii) Simultaneously, search for newly exposed sites created by drop coalescence and provide a minimum radius on such newly exposed sites.
(viii) For all the drops, check for the critical radius of slide-off and the sliding velocity.
(ix) Re-activate the exposed sites created by drop slide-off and provide a minimum radius drop on newly exposed sites.
(x) Check for the possibility of drop-off on inclined and horizontal surfaces.
(xi) Re-activate the exposed sites created by drop-off and provide a drop of a minimum radius on the newly exposed sites.
(xii) Repeat steps (iii)–(xi) till a dynamic steady state is reached.

It may be that the simulation needs to track multiple generations of droplets—nucleating, growing by direct condensation, by coalescence and some slide or falling-off, when the virgin surface thus exposed, is re-nucleated.

In view of the severity in computational effort, efforts are ongoing to simplify the intermediate steps and reduce CPU time. Parallel computing is one such strategy. On the other hand, advancements in modeling techniques by implementing machine learning and data-driven models will improve the accuracy and applicability of dropwise condensation models. By training models on large datasets, researchers can develop predictive tools that capture complex patterns in condensation behavior, potentially reducing the need for extensive computational simulations. In hybrid modeling, distinct modeling techniques are combined, for example, coupling MD simulations with continuum models, allowing for accurate and efficient simulations. These hybrid approaches leverage the strengths of each method to overcome individual limitations. For validation of simulation studies, advanced experimental techniques, such as high-speed imaging, atomic force microscopy, and observations of droplet behavior during condensation are carried out to capture real-world phenomena. Future research in modeling of dropwise condensation is likely to develop sophisticated multi-scale models that integrate molecular dynamics, fluid dynamics, and heat transfer principles. In addition, the development of adaptive models that can adjust to changing conditions in real-time could enable more responsive and efficient thermal management systems while incorporating dropwise condensation models in smart systems, leading to intelligent and autonomous and reliable thermal management solutions.

4 Parametric Studies

Parametric studies on dropwise condensation of pure vapor and model validation against experimental data is available in the literature (Sikarwar et al. 2012, 2013a; b). Simulations can be used to record condensation patterns and match with the visualization of dropwise condensation in the experiment. Figure 16 shows such a comparison of simulation data with respect to the experimental, and a good agreement is to be seen between them. Additional parameters under investigation include surface wettability and the degree of subcooling. These parameters are critical in understanding the effectiveness of the condensation processes in heat rejection. The condensation cycle is analyzed to determine its impact on the rate of droplet growth and detachment. Surface wettability is varied to assess how surface properties influence droplet behavior and the overall heat transfer rate. Finally, the degree of subcooling is adjusted to explore its effect on the nucleation rate and average heat flux during condensation. This research is expected to provide valuable insights for enhancing the performance of heat exchangers and other systems where efficient condensation is crucial.

4.1 Dropwise Condensation Cycle

Figure 3 illustrates the cycle of dropwise condensation of pure vapor over a subcooled substrate schematically. The condensation patterns are derived from simulations where the initial nucleation site density is 10^6 per cm^2. Drops grow on a hydrophobic coated surface of contact angle 145° and a degree of subcooling is 2 K. The figure clearly shows that semi-empirical models, which blend theoretical principles with

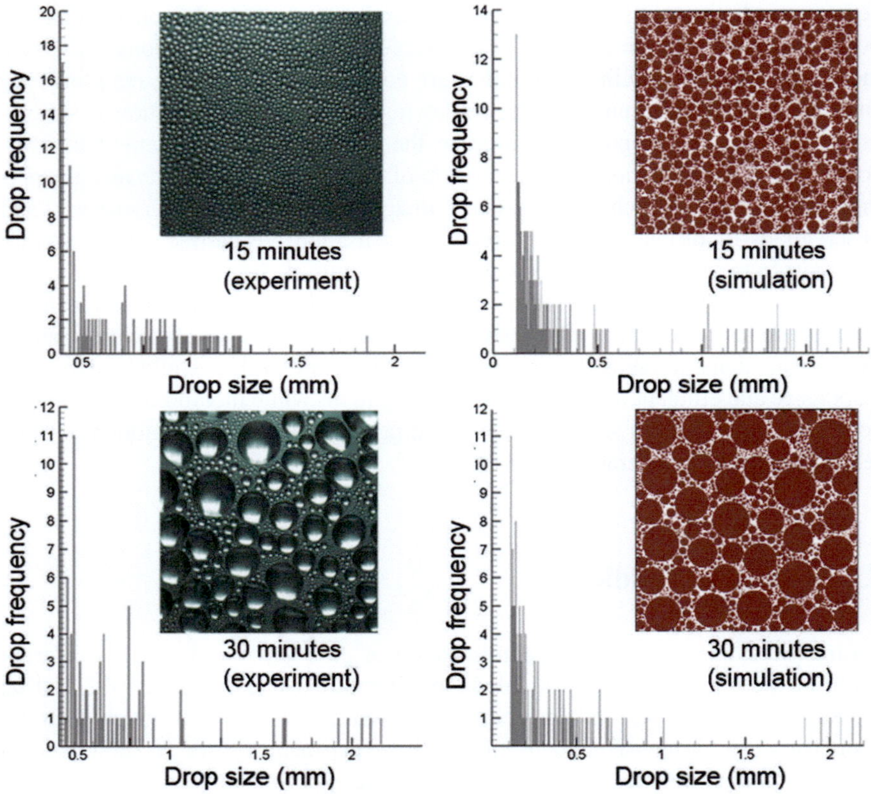

Fig. 16 Visual and statistical comparison of experimental and simulated spatial drop distribution patterns and the corresponding histograms of droplet size after dynamic steady state is reached

empirical data, realistically capture the complete physics of dropwise condensation (Fig. 17).

4.2 *Effect of Surface Wettability*

For equilibrium contact angles in the range of 120°–160°, condensing patterns formed on a colder surface from pure water vapor are shown in Fig. 18. These simulation results show that the critical droplet size for gravitational instability decreases as the contact angle increases. Consequently, smaller droplets slide down more easily from superhydrophobic surfaces when compared to hydrophobic ones. For the latter, detachment is delayed on hydrophilic and hydrophobic surfaces. Drop shedding owing to excess weight of the drop on a vertical surface begins at 45 min when $\theta = 120°$, at 27 min when $\theta = 140°$, and at 20 min when $\theta = 160°$. This is a direct result

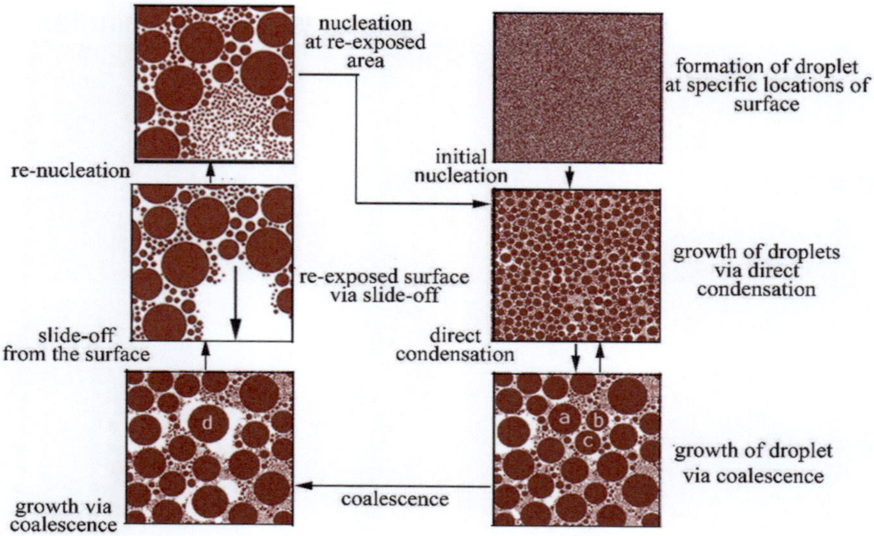

Fig. 17 Snapshots of individual stages of a dropwise condensation cycle on a hydrophobic surface showing the formation of initial droplets, growth due to direct condensation and coalescence, slide-off from the substrate, and re-nucleation. The condensation patterns shown are recorded from simulations. Slide-off of the drops is in the vertically downward direction (first column, second row)

of decreasing critical drop size with increasing hydrophobicity. Figure 19 shows how the critical radius varies with contact angle and hysteresis. It is clear that the critical droplet radius decreases significantly as the surface contact angle increases and contact angle hysteresis decreases, causing early detachment on a condensing surface with a high contact angle and low hysteresis. The gradual decrease in average droplet size indicates an increase in heat transfer rate and, so, the condensation rate on surfaces with higher hydrophobicity.

4.3 *Effect of Degree of Subcooling*

Figure 20 shows the effect of saturation temperature and degree of subcooling on heat transfer coefficient during dropwise condensation underneath a horizontal and over a vertical substrate. All the data show a tendency of increasing heat transfer coefficient with increasing saturation temperature and degree of subcooling. Apart from higher availability of water vapor at higher saturation temperatures, the increase in heat transfer coefficient is also caused by the reduction in the interfacial resistance. In addition, a high degree of subcooling increases the number density of active nucleation sites on the condensing substrate, increasing the wall heat flux.

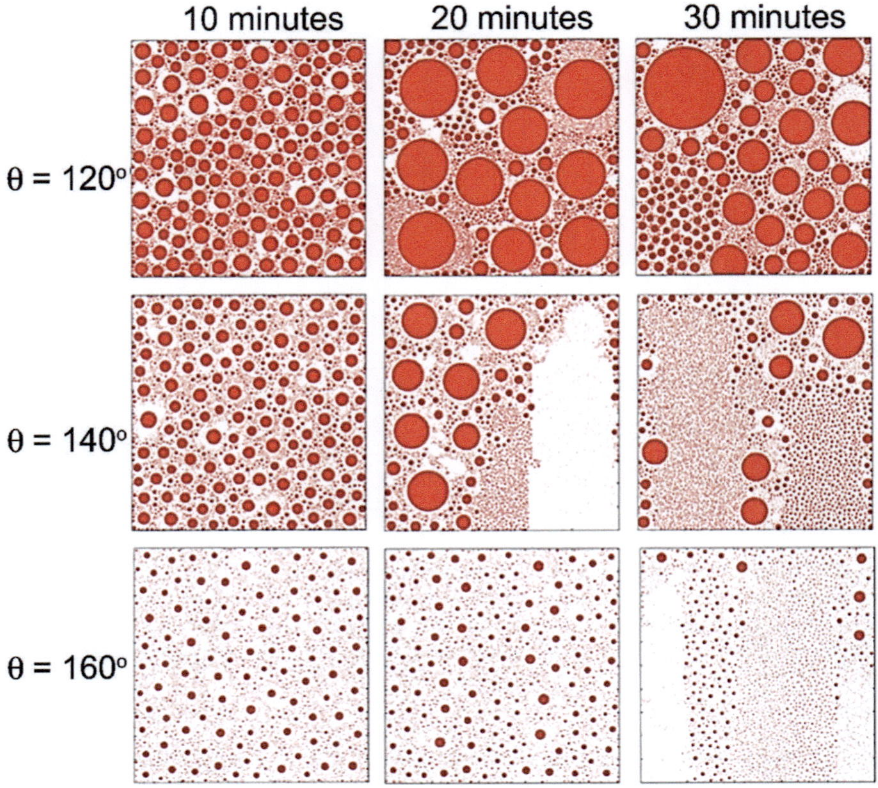

Fig. 18 Spatio-temporal droplet distribution on a cold condensing substrate with varying contact angles. Simulation is carried out considering contact angle hysteresis = 20°, pure vapor temperature = 303 K, degree of subcooling = 5 K (Baghel et al. 2020; reproduced with permission form Elsevier)

Based on numerical data, the heat transfer coefficient (kW/m^2-K) is empirically correlated with the critical radius of drop (mm) at instability, the degree of subcooling ΔT and saturation temperature (both in units of °C) for water vapor condensation underneath an inclined substrate with a nucleation site density of 10^6 per cm^{-2} as follows:

$$h = (0.42\Delta T + 6.4)T_{\text{sat}}^{0.75} r_{\text{crit}}^{-1.18} \tag{15}$$

Here, the critical radius depends on contact angle, contact angle hysteresis and angle of inclination of the substrate with respect to the horizontal. It is preferable to cast this correlation in dimensionless form, applicable for all the inclinations, saturation temperature and degree of subcooling. The dimensionless heat transfer correlation for condensation of water vapor underneath an inclined substrate is obtained as

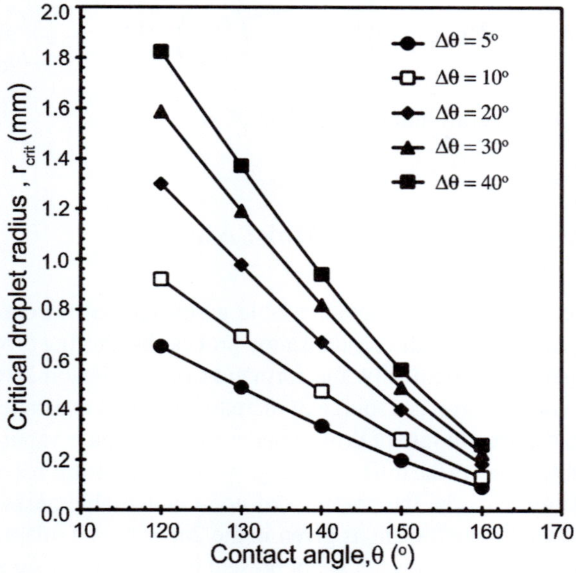

Fig. 19 Variation of the critical droplet radius for a vertical surface as a function of equilibrium contact angle for various values of contact angle hysteresis. Simulations were carried out considering pure vapor at a temperature = 303 K, degree of subcooling = 5 K

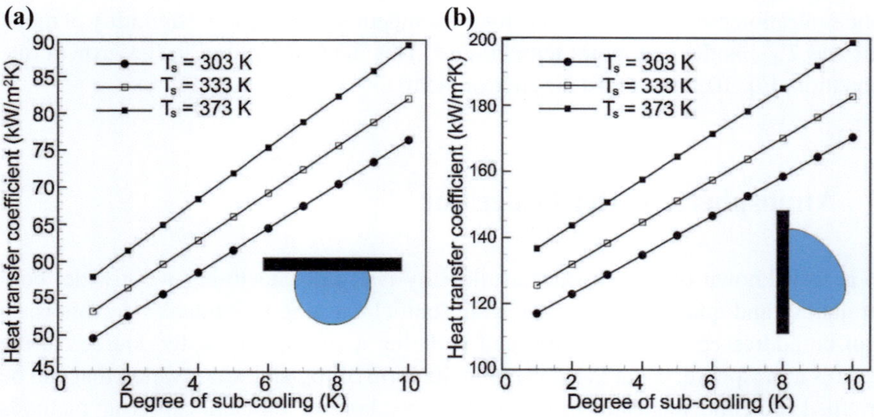

Fig. 20 Dependence of the heat transfer coefficient on saturation temperature and degree of subcooling on **a** a horizontal substrate with contact angle = 110° and **b** vertical substrate. In (**b**), the condensing water droplet has advancing and receding contact angles of 117° and 103°, respectively. The nucleation site density is 10^6 cm^{-2}

$$\mathrm{Nu} = \left(8.54 \times 10^3\ \mathrm{Ja} + 240\right)\left(\frac{T_{\mathrm{sat}}}{T_{\mathrm{ref}}}\right)^{0.75}\left(\frac{r_{\mathrm{crit}}}{r_{\mathrm{cap}}}\right)^{-1.18}; \quad \mathrm{Ja} = \frac{C_p}{h_{fg}}\Delta T \tag{16}$$

Equation (16) has a correlation coefficient of 98.5%.

5 Dropwise Condensation in Moist Air

In moist air, the presence of non-condensable gases such as nitrogen and oxygen (air, in general) can influence the condensation process by altering droplet formation and growth dynamics. Specifically, the formation of a diffusion layer of the non-condensable around the drop is a source of increased thermal resistance. Equations 9 and 10 compare the growth rate of drops condensing from pure vapor and moist air, respectively. Table 4 shows the difference in growth rates of drops forming from pure vapor and moist air. Clearly, the presence of non-condensable gases decreases the interfacial heat transfer coefficient as given in the 2nd row of Table 4.

Here χ_v is the percentage of water vapor present in moist air at a given temperature and pressure. Further, M_v and M_a are the molecular masses of water vapor and dry air (non-condensable gases), respectively. The revised formula for the interfacial heat transfer coefficient that includes non-condensable gases is given in the 2nd row and second column of Table 4. In this expression, T_v and P_v are the wet bulb temperature and saturation pressure appropriate for the temperature and relative humidity of moist air and T_{dew} is the dew point temperature. This data is required in the growth rate equation, Eq. 10, for a moist air environment.

6 Atmospheric Water Generator

It is well known that freshwater availability is rapidly declining worldwide, both in quality and quantity, presenting a substantial warning to humankind. This issue can be addressed via developing and realizing a sustainable water source. In the earth's atmosphere, water is available in the form of fog and water vapor, that can be trapped and collected using various techniques. For fog, efficient capturing methods are available as discussed in Chap. 7. However, water vapor needs to be transformed into a liquid phase, typically by the process of condensation. An Atmospheric Water Generator (AWG) is a device that extracts water from humid air via condensation of moist air, as shown in Fig. 21. Water vapor present in moist air alone is condensed into water when the air is cooled below the dew point. Hence, the condensed water does not have impurities and is suitable for consumption. On the other hand, AWG techniques are energy-intensive and require more attention if low-cost and sustainable power sources are to be adopted. It is clear that to maximize water production, a

Table 4 Comparison of subprocesses of dropwise condensation for pure vapor and moist air

	Pure vapor condensation	Moist air condensation
Minimum size of droplet at the nucleation sites	$\frac{2\sigma T_w}{h_{lv}\rho_l(T_{sat}-T_w)}$	$\frac{2\sigma T_w}{h_{lv}\rho_l(T_{dew}-T_w)}$
Growth rate	Equation 9	Equation 10
Interfacial heat transfer coefficient	$h_{int} = \left(\frac{2\hat{\sigma}}{2-\hat{\sigma}}\right)\frac{h_{lv}^2}{T_{sat}\upsilon_{lv}}\left(\frac{\overline{M}_v}{2\pi\overline{R}T_{sat}}\right)^{1/2} \times \left(1-\frac{P_v\upsilon_{lv}}{2h_{lv}}\right)$	$h_{int(NCG)} = \chi_v\left(\frac{2\hat{\sigma}}{2-\hat{\sigma}}\right)\left(\frac{h_{lv}^2}{T_v\upsilon_{lv}}\right)\left(\frac{\overline{M}_v}{2\pi\overline{R}T_v}\right)^{1/2} \times \left(1-\frac{P_v\upsilon_{lv}}{2h_{lv}}\right)$
Coalescence and slide-off	Instantaneous	Instantaneous

Here, σ is the surface tension at the air–water interface, T_w is the condensing wall temperature, T_{sat} is the saturation temperature of vapor, T_{dew} is dew point temperature of water vapor in moist air, ρ_l is the density of water, h_{lv} is the latent heat of water vapor. Further, h_{int} is the interfacial heat transfer coefficient for the droplet placed in pure vapor. In addition, $\hat{\sigma}$ is the accommodation coefficient, v_{lv} is the difference in specific volume between the liquid and vapor, $\overline{m}_v$ is the molecular mass of water, $\overline{R}$ is the universal gas constant and p_v is the vapor pressure.

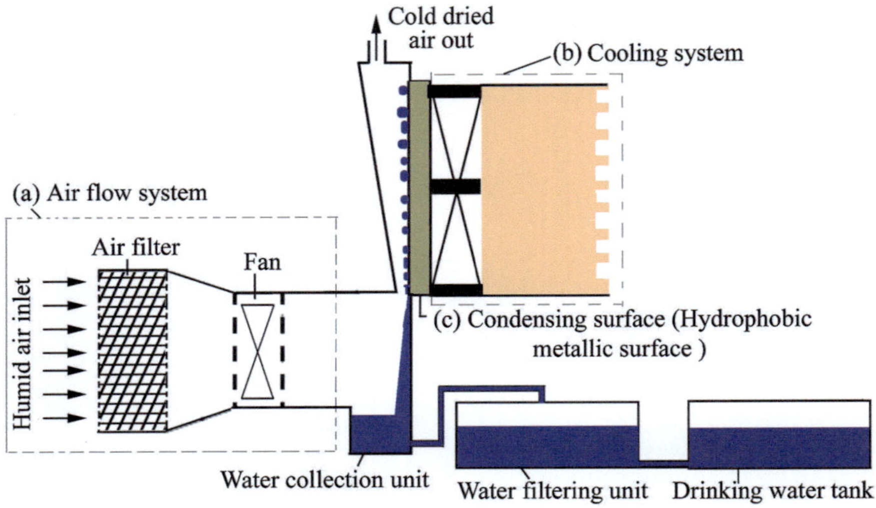

Fig. 21 Schematic diagram of an atmospheric water generator that flows moist air over a cold surface, condenses water vapor, and releases practically dry air

dropwise mode of condensation should be preferred over the filmwise mode, calling for texturing of the cold substrate wall.

Despite being in the initial stages of its research, AWG holds potential as a future solution to water scarcity because the atmosphere contains around 1.29×10^{13} m^3 of extractable water. It is a renewable process and does not negatively impact the environment. Other existing techniques, such as water desalination and membrane-based water purification units, require access to conventional water sources and generate brine that needs to be carefully disposed. However, AWG requires relative humidity above 40% and temperatures between 25 and 40 °C to economically produce water. Additionally, it needs a power source to cool the condensing surfaces. Research is in progress to make AWG suitable for locations where there is a need for pure water and renewable energy is abundant to produce cold surfaces.

A schematic diagram of the AWG system is shown in Fig. 21. It has (a) an airflow section that flows ambient air toward the condensing surface using fans, (b) a cooling system that cools the condensing surface via various a refrigerator, thermoelectric cooler, pumped single-phase cooler, pumped two-phase cooler, or simply a passive cooler, (c) cold surface that condenses water vapor, and (d) collecting unit for the condensate. The cooling technique lowers air temperature near the surface to fall below the dew point, thus enabling condensation. In addition, the device contains a water treatment system for adding minerals and UV treatment, thus ensuring that it meets drinking water standards. A storage tank is needed to hold the purified water until it is ready for use.

The choice of the cooling technique is important from the viewpoint of operating cost while the characteristics of the condensing surface determine the reliability and lifetime of the AWG apparatus. Thus, research is in progress on improving the cooling technique to reduce grid power consumption via the use of renewables on one hand and surface modification to enhance wettability of the condensing surface. Sustainable hydrophobic metallic surfaces are preferred because of long life and lower manufacturing costs with the possibility of recycling.

The water generation rate measures the performance of AWG systems (WGR: the amount of water produced per unit time, sometimes scaled by the surface area of condensation) and specific energy consumption (SEC: the energy consumed per unit mass of water generated). Factors such as ambient temperature, relative humidity (RH), and dew point temperature significantly impact WGR and SEC, making system performance dependent on the local climatic and seasonal conditions.

6.1 Classification AWG

Based on the choice of the cooling technology, atmospheric water generator systems (AWGs) are classified as follows:

(a) *Active AWGs*: In this category, vapor compression refrigeration (VCR) or thermoelectric cooling (TEC) is used to cool air below the dew point. These techniques can easily produce the required cooling for condensation of water vapor in moist air, but are energy-intensive with a requirement of high-grade electrical energy. Figure 22a shows the schematic diagram of a VCR cooling-based AWG; Fig. 22b shows the schematic diagram of a thermoelectric cooling-based configuration. In these systems, ambient air flows toward the condensing surface by fans through filters that remove particulates. Compressor, condenser, expansion valve, and evaporator are the main components of the refrigeration cycle. Compressor will compress refrigerant gas, raising its pressure and temperature, condenser dissipates heat, converting the refrigerant into a high-pressure liquid, expansion valve reduces pressure by throttling, causing cooling of the gas–liquid mixture, and the evaporator cools the hydrophobic condensing surface to induce water vapor condensation. In contrast, a Peltier module with heat sink cools the condensing surface below the dewpoint temperature using a thermoelectric cooling system.

Although active cooling systems are energy-intensive, they are widely used due to their ability to generate water in areas without groundwater sources. These systems offer an environmentally friendly alternative for dependable production of water from atmospheric moisture. As a result, they are ideal for emergency water supplies during natural disasters and military applications where reliable water sources are unavailable. However, low water productivity and high energy consumption are twin factors that call for extensive research.

(b) *Passive AWGs:* In this category, cooling is achieved by natural processes of heat exchange to extract water from humid air without relying on energy-intensive systems such as the vapor compression cycle or a Peltier module. Cooling effect

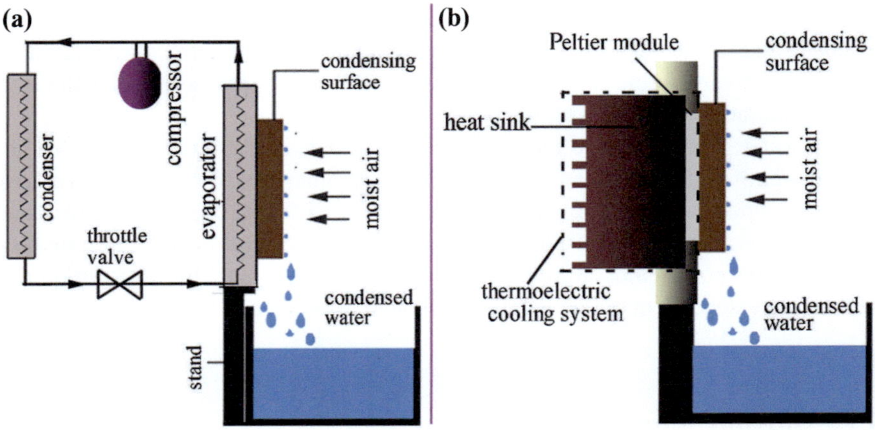

Fig. 22 Schematic diagram of AWGs based on active cooling. **a** Vapor compression system-based AWG and **b** thermoelectric cooling-based AWG

in passively operated AWGs rely on naturally occurring temperature gradients, such as nocturnal radiation, evaporative cooling, and conductive cooling from the ground. In nocturnal radiative cooling, the condensing surface exposed to clear sky radiates heat into the atmosphere facilitating condensation conditions of moist air. Additionally, coatings with high emissivity and materials with moderate thermal conductivity will enhance this effect. In evaporative cooling, a naturally porous substrate is wetted with water. As water evaporates, the porous substrate absorbs heat from the surroundings. This mechanism creates a temperature difference below the ambient, aiding condensation. For conductive cooling, the condensing surface is in direct contact with the cooler part of the earth to dissipate the latent heat released. These cooling techniques emphasize sustainability and simplicity, leveraging ambient environmental conditions and materials, yielding high device productivity of potable water. Table 5 compares actively and passively cooled AWGs in terms of their characteristics.

Dew harvesting, fog harvesting, and desiccant-assisted AWGs are further examples of passive systems used to extract water from moist air using renewable energy sources. Dew water harvesting is a passive technique that utilizes the concept of radiative sky cooling (see Fig. 23). This phenomenon enables a surface to achieve temperatures below the ambient environment by exchanging radiation with the cold temperature of outer space through selective emissivity in the atmospheric transparency window, specifically in the infrared wavelength range of 8–13 μm. Dew formation occurs at night and is influenced by various factors such as wind velocity, air temperature, relative humidity (RH), cloud cover, and the structure and properties of the dew-harvesting material. The cooling power from natural radiation is limited to <100 W/m^2 (Beysens 2022). As a result, this method requires extensive surface areas in favorable locations with high nocturnal RH for effective performance. The water yield from passive systems is generally low and is measured in terms of daily

Table 5 Comparison between active and passive atmospheric water generators

Characteristic	Active Cooling-based AWG	Passive Cooling-based AWG
Cooling techniques	Utilizes vapor compression cycle to cool air below the dew point	Relies on natural processes such as nocturnal radiative cooling, evaporative cooling, and cold earth
Energy consumption	High energy consumption due to compressors and fans	No external energy required
Water production	High productivity. These systems can be efficient in extracting water even in low humidity conditions	Low water production because effectiveness depends on ambient conditions and sky conditions
Operational costs	High due to intense energy consumption and maintenance	Low because there are no mechanical or energy-intensive components
Environmental impact	Moderate to high owing to high grade energy usage and refrigerant emissions	These are environmentally friendly and sustainable
Dependency on environmental condition	They can operate efficiently over a wide range of humidity conditions	They are highly dependent on high humidity and conducive environmental conditions
Maintenance	Regular maintenance needed for compressors, filters, and refrigerant levels	Minimal maintenance, limited to cleaning and occasional replacement of materials
Initial cost	High due to advanced components and technology	Low to moderate, as the design is simpler and uses fewer materials
Applications	Used in homes, disaster relief, military operations, and industrial settings	Ideal for remote, off-grid areas or regions with limited energy infrastructure
Scalability	Can be scaled for both domestic and industrial applications	Suitable for small-scale applications, with limited potential for large-scale deployment
Examples	Refrigeration-based AWGs, TEC, HVAC-integrated systems	Dew collectors, fog nets, desiccant-assisted condensation systems

productivity per unit area. Interestingly, radiative sky cooling can also function during the day if solar inputs are blocked and specialized materials with selective optical properties are employed (Hu et al., 2022, Raman et al.2014). Despite its energy efficiency and potential to operate under low RH and daytime conditions, this passive cooling approach has limitations. It is geographically restricted and often associated with high costs due to the need for large surface areas. The primary challenge lies in developing cost-effective and scalable metamaterials optimized for efficient radiative cooling.

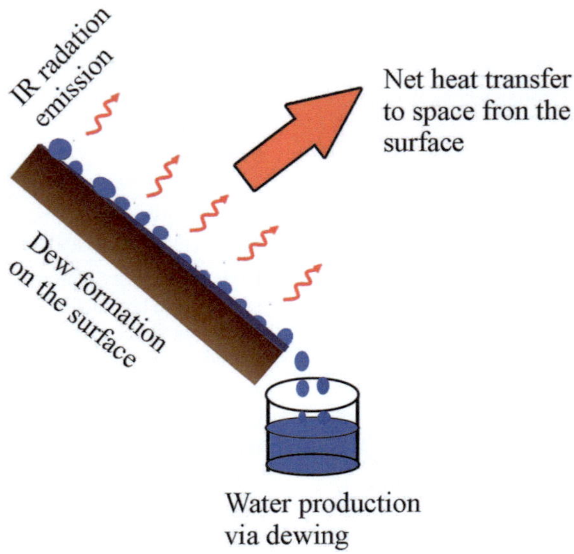

Fig. 23 Schematic diagram of water extraction from atmospheric air via dew harvesting

Fog harvesting is optimal in areas with persistent fog cover and requires minimal infrastructure (Fig. 24). Fog consists of visible aerosols made up of tiny water droplets formed by condensation in the atmosphere. These droplets are captured by mesh-like structures on which they are allowed to grow by coalescence and then collected by gravity. These techniques are simple, inexpensive, and operate without additional energy requirements. The method is not popular because water generation is slow and its performance is assessed based on the amount of water generated per unit area of the collector per unit time.

Desiccant-assisted AWGs are popular for regions with moderate humidity and access to renewable or low-cost energy sources. It is based on vapor concentration technology in which sorbent materials can extract water vapor from atmospheric air even at low relative humidity conditions. In this technique, sorbent materials capture moisture, and low-grade heat releases water vapor from the sorbent materials. Finally, water vapor condenses on a cooler surface to yield potable water. Figure 25 shows the schematic diagram of desiccant-assisted atmospheric water harvesting. The moisture-absorbing sorbent, possessing a strong attraction to water molecules, captures and accumulates atmospheric moisture on its surface or within its internal pores. Upon heating, the accumulated water vapor is released from the sorbent and then condensed into liquid form. Following this release, the sorbent can be regenerated and is ready for reuse in subsequent cycles. In this device, the sorbent material plays a crucial role in defining the harvester efficiency and suitability for various environmental conditions. Ideal sorbents possess the following characteristics: (i) a high water-sorption capacity under operating conditions, (ii) rapid kinetics to maximize the number of sorption–desorption cycles within a given timeframe, (iii) the ability to be easily regenerated using low-grade energy, and (iv) consistent, long-term stability in performance. Dew harvesting is passive and best suited for humid regions with

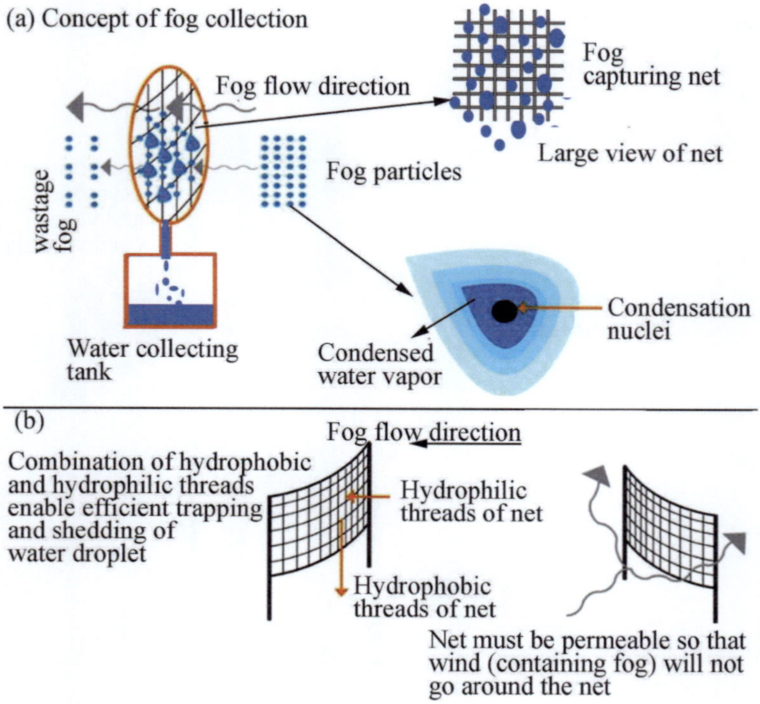

Fig. 24 Schematic diagram of water extraction from atmospheric air via fog harvesting (also see Chap. 7)

cool night-time temperatures, and desiccant-assisted techniques are versatile but energy-intensive, making them ideal for regions with moderate humidity and access to renewable or low-cost energy sources. Table 6 compares AWGs based on dew harvesting, fog harvesting, and desiccant-adsorption. This comparison illustrates that fog harvesting is ideal for regions with consistent fog cover, while dew harvesting works best in areas with high humidity and cool night-time conditions. These are passive, sustainable water collection methods, but tailored to specific environments.

Table 6 summarizes issues and research opportunities for developing AWG technologies further.

AWG technologies are at the crossroads of innovation and practical challenges. To address issues like high energy consumption, scalability, and cost, interdisciplinary research involving material science, energy systems, and policy is needed. The integration of renewable energy sources, advanced materials, and AI for optimization offers promising pathways for the future, enabling AWG to become a cornerstone in global water sustainability efforts, as shown in Table 7.

(c) *Hybrid AWGs*: Recently, many attempts have been made to optimize cooling power and enhance efficacy by developing hybrid atmospheric water generators. In hybrid AWGs, passive and active cooling techniques are integrated to

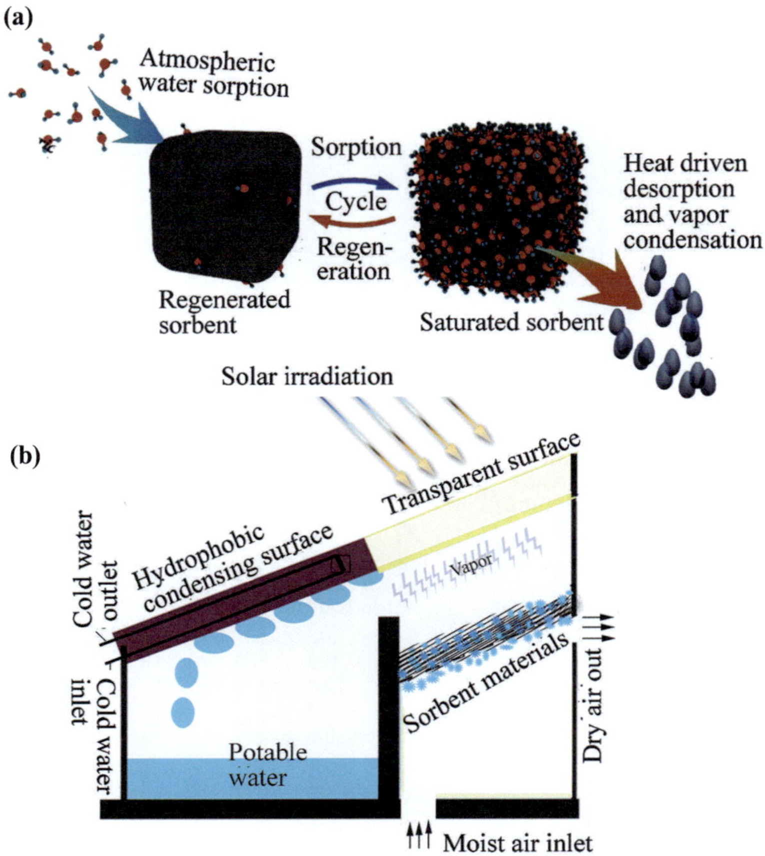

Fig. 25 Schematic diagram of (**a**) the mechanism of water recovery by a desiccant and (**b**) desiccant-assisted atmospheric water harvesting

optimize water production while balancing energy consumption and yield. In the literature, desiccant-based moisture capture with active cooling has been used for efficient water production in low humidity conditions. Fog harvesting has been integrated with dew collection to optimize atmospheric water production in regions with appropriate weather patterns. Figure 26 shows the schematic diagram of a hybrid AWG incorporating a pulsating heat pipe (passive cooling) with vapor compression refrigeration (VCR) and pumped two-phase cooling to extract water from moist air. The working fluid for VCR and pulsating heat pipe is refrigerants such as R134a and R22, respectively. The refrigerant of the pulsating heat pipe takes heat from moist air that flows in the lower chamber and releases it to the top chamber, exchanging heat through refrigerant flow in the vapor compression cycle, as shown in Fig. 26(a). This technique is less effective at low relative humidity (RH < 40%) and low-temperature conditions ($T_a < 15$ °C).

Table 6 Comparison of fog harvesting, dew harvesting, and desiccant-assisted AWGs

Aspect	Fog harvesting	Dew harvesting	Desiccant-assisted AWG
Working principle	Captures water droplets suspended in fog using fine mesh nets	Collects water vapor as it condenses onto cooled surfaces below the dew point	Absorbs water vapor from air using hygroscopic materials and extracts moisture via heating first followed by condensation
Cooling mechanism	No active cooling; relies on natural fog movement	Passive cooling through nocturnal radiation or conductive cooling	Requires heating (for desiccant regeneration) and cooling to release and condense water
Energy requirement	Minimal; may require small fans for air circulation	Minimal; no external energy needed for cooling	Moderate to high; requires energy for heating or cooling during desiccant regeneration
Water yield	Moderate to high in foggy regions	Low; yield depends on night-time cooling and ambient humidity	Moderate; effective in both low and high humidity, depending on desiccant properties
Dependence on weather	Requires frequent and dense fog for efficient operation	Requires clear skies, high humidity, and low night temperatures	Less dependent on weather; works in a wide range of humidity conditions
Optimal conditions	Coastal, mountainous, or desert regions with regular fog	Clear nights with high humidity and significant temperature drops	Works in both humid and semi-arid environments, where active desiccant regeneration is feasible
Material requirements	Fine mesh nets made of hydrophilic or hydrophobic materials	Smooth, thermally efficient surfaces with hydrophobic coatings	Hygroscopic materials such as silica gel, zeolites, or lithium chloride
Maintenance	Periodic cleaning of nets to prevent clogging and degradation	Occasional cleaning of condensation surfaces	Regular replacement or regeneration of desiccants; maintenance of heating or cooling systems
Cost	Low; simple design and inexpensive materials	Low; basic setup with minimal components	High; involves advanced desiccants and energy-consuming systems for regeneration
Environmental impact	Negligible; fully passive and sustainable	Negligible; sustainable and ecofriendly	Moderate; depends on an energy source for regeneration (e.g., solar or grid energy)

(continued)

Table 6 (continued)

Aspect	Fog harvesting	Dew harvesting	Desiccant-assisted AWG
Applications	Used in foggy regions for drinking water, agriculture, and reforestation	Ideal for rural or off-grid areas with favorable night-time conditions	Suitable for areas with moderate humidity and reliable energy sources; industrial and domestic uses
Examples	Fog nets in the Atacama desert (Chile) and Morocco	Radiative cooling panels in arid regions	Solar-powered desiccant systems or hybrid AWGs for desert or semi-arid zones

The principle underlying a pumped two-phase (PTP) cooling loop is predicated on the efficient transfer and removal of heat via the latent heat of vaporization of a working fluid. This system is particularly advantageous for cooling components with high heat flux, such as those found in electronic devices or laser systems. In the present context, it is employed to cool tubes exposed to moist air, as illustrated in Fig. 6(b). A pumped flow of saturated fluid traverses the loop, absorbing heat at the evaporator section, which results in the boiling and partial vaporization of the working fluid (e.g., R134a). Subsequently, it proceeds to a condenser, where heat is dissipated to the environment or to a secondary coolant, causing the vapor to condense into liquid form. The cooled liquid is then circulated to the evaporator by a pump, thereby completing the cycle.

Renewable energy sources can make active AWG systems more sustainable by reducing the demand for grid electricity. Theoretically, energy demand is estimated based on the proximity of inlet air conditions to the saturation curve in a psychrometric chart. In this respect, inlet air must be sensibly cooled first to the dew point temperature, followed by the removal of latent heat at the wall.

Active refrigeration techniques enable controlled condensation of water vapor but require significantly high-grade energy, such as grid electricity, to cool air below its dew point, typically using conventional refrigeration methods. As a result, this approach is less effective in low relative humidity (RH) and low-temperature conditions. Alternatively, vapor concentration methods employing sorbent materials may be adopted, capable of operating at low RH levels, where heating is an additional step to release the trapped water vapor. Solar radiation can assist this process, depending on the sorbent properties. Vapor concentration is also possible using selective membranes to separate water vapor from other non-condensing atmospheric gases. Passive AWG techniques, such as fog and dew harvesting, are effective only in high RH conditions. Consequently, no single AWG technology is universally effective across all climatic conditions. The choice of AWG technology depends on the psychrometric properties of atmospheric air at a specific location and time, such as RH, dry bulb temperature, and dew point temperature. There are ongoing efforts to develop integrated AWG systems with other technologies, but currently, no

Table 7 Issues and prospects with research opportunities in building AWG technologies

Technology	Prospects	Issues	Future research opportunities
Vapor compression refrigeration	Refrigerant-based AWG systems are built on a well-established working principle and are commercially available in various water generation capacities. They require minimal infrastructure but primarily rely on high-grade electrical energy, making them most suitable for hot and humid climates	Refrigerant-based AWG systems are considered environmentally unfriendly due to their high specific energy consumption, which leads to an increased carbon footprint from the use of high-grade energy	Focus on optimizing and modifying evaporator coil designs, including studies on hydrophobic material thickness, coating, and thermal conductivity to enhance water generation. Furthermore, new methods for selective water vapor separation and system integration with other techniques are being explored
Peltier cooling	This technology is compact, portable, and easily powered by solar photovoltaic systems Require minimal infrastructure and investment	Low COP (<1), economic concerns when scaling up, thermal management issues, very low capacity, limited water handling capability, and overall low performance	Improved thermal management by recirculating cold exit air to the hot side of the TEC module enhances its efficacy TEC modules powered by renewable energy and implementing new configurations can further enhance water generation rates and thermal efficiency Additionally, applying hydrophobic coatings to the condensing surface contributes to more efficient water production
Passive radiative cooling	It requires no external power source Easy to install and deploy Highly effective under conditions of high ambient humidity and clear skies	Need a large collector area for significant water generation Output is inconsistent and yielding less than 0.8 L/m^2/day Daytime operation with enhanced deep radiative cooling	Research and development in nano photonics and metamaterials focused on coupling with active cooling systems, exploring bio-inspired designs, and enhancing deep and daytime radiative cooling. Additionally, the decoupling of dew collectors and emitters is being investigated to improve efficiency

(continued)

Table 7 (continued)

Technology	Prospects	Issues	Future research opportunities
Fog harvesting	Fog collectors are most suitable for highly humid locations with frequent fog occurrence. They require less space than passive dew collectors due to their vertical placement of fog mesh	Fog mesh systems face challenges such as water droplet re-entrainment and clogging, necessitating highly sophisticated fabrication methods for advanced designs	Advanced fog mesh designs aim to prevent issues such as water droplet re-entrainment and clogging of the mesh's void spaces. These improvements include more bio-inspired designs, various geometric and surface modifications, and techniques like electrostatic fog harvesting. Additionally, fog harvesting from cooling tower plumes of power plants is being explored as a viable method
Sorption	This system can function as both a passive (solar-assisted) and active system, operating effectively even at low ambient relative humidity levels of 15–20%. It has significant potential for scalability and sustainability	The system faces challenges such as a low sorption cycle, contamination of the sorbent bed by airborne pollutants, and the high cost of advanced sorbents. However, it can achieve multiple sorption cycles in a day, producing between 1 and 2.8 kg of water per day per square meter under low relative humidity conditions with solar-assisted desorption	Key areas of focus include the recovery and efficient utilization of latent heat from sorption beds, addressing the trapping of airborne contaminants, and enhancing the sorption kinetics of materials. Innovations such as night-time sorption cycles with thermal storage, the concept of continuous sorption, and high-frequency sorption cycles are also being explored. Additionally, techniques like interfacial solar heating combined with sorption, along with the development of novel sorbents, are contributing to advancements in this field

single technique has proven to be energy-efficient for large-scale freshwater production (Bayer 2020; Zhang et al. 2022). Also see the work of Naveira-Cotta (Moreira et al., 2025) and Cameron and Ready (Cameron et al. 2022) who describe a sustainable polycogeneration prototype (TRL-6) designed and built for the simultaneous and decentralized generation of electricity, distilled water, biofuel, and other possible outputs.

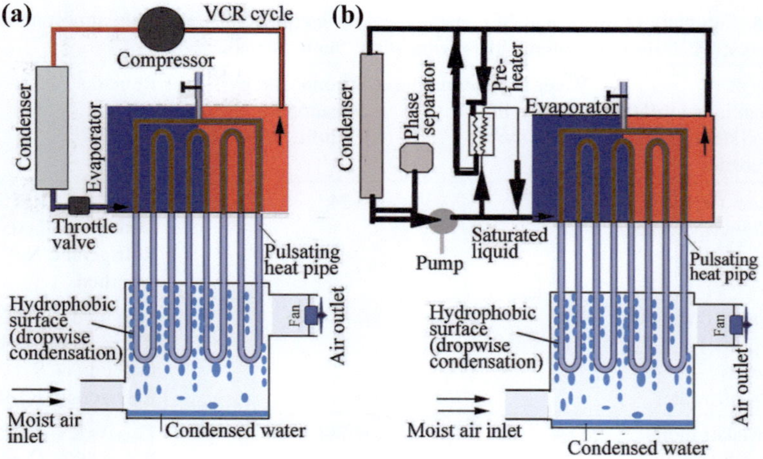

Fig. 26 Schematic diagram of a hybrid AWG apparatus. (**a**) A vapor compression refrigeration (VCR) cycle is integrated with a pulsating heat pipe. (**b**) A pumped two-phase (PTP) cooling arrangement is integrated with a pulsating heat pipe. A pulsating heat pipe extracts the required heat from moist air, leading to condensation. In a pulsating heat pipe, the outer tubes of the evaporator zone are superhydrophobic, allowing water vapor to condense in the dropwise mode

6.2 Performance Evaluation and Effect of Humidity Parameters

Environmental conditions play a pivotal role in the performance of an atmospheric water generator. Systems may also be designed to adapt to the prevailing relative humidity, either by using specialized materials or adopting hybrid technologies to enhance water yield and lower high-grade energy consumption. Performance evaluation using laboratory-scale tests as well as mathematical modeling helps identify the most appropriate solutions for each environment.

The performance parameters used for evaluating AWG systems are water generation rate (WGR) and specific energy consumption (SEC). Here, WGR is expressed in units of lit/h and the SEC is expressed in units of kWh/lit. These two parameters vary drastically with the change in ambient conditions. Thus, the feasibility of the AWG system is questionable if daily productivity and water generation costs are unfavorable. Table 8 summarizes the WGR and SEC parameters for commercially available atmospheric water generators.

Table 8 Summary of WGR and SEC parameters for commercially available atmospheric water generators; VCR—vapor compression refrigeration; 'amb' is ambient

Details of commercially available AWG systems (model and website)	Water generation rate (WGR), lit/h	Specific energy consumption (SEC), kWh/lit	Remarks
Vayuneer (VN30) https://www.vayuneer.com/	1.25	0.24	30 °C, 80% RH Cooling method: VCR Refrigerant: Not specified
Airowater (dew point smart prime) Dewpoint prime—airowater	1.04	0.48	T_{amb} = 30–32 °C and RH = 70%. Cooling method: VCR Refrigerant: R134a
WaterMaker India (magic), https://watermakerindia.com	1.04	0.384	T_{amb} = 25–32 °C and RH = 80%. Cooling method: VCR Refrigerant: not specified
GenaQ (Stratus S50) Solutions—GENAQ	2.16	0.39	T_{amb} = 30 °C and RH = 80% Cooling method: VCR Refrigerant: R290 (propane)
Maithri aquatech (signature 2 K) https://www.maithriaqua.com/	1.66	0.42	T_{amb} = 30 °C and RH = 80% Cooling method: VCR Refrigerant: R450
Watergen (GEN-M pro) watergen|water from Air	23	0.35	T_{amb} = 26 °C and RH = 60% Cooling method: VCR Refrigerant: not specified
Clean water generator (morning dew AWG-30) https://www.cleanwaterproblem.com/	1.25	0.36	T_{amb} = 30 °C, RH = 65% Cooling method: VCR Refrigerant: R-22 and R134a
Accairwater (HR-77L) https://www.accairwater.com/	1.25	0.36	T_{amb} = 30 °C and RH = 80% Cooling method: VCR Refrigerant: not specified

(continued)

Table 8 (continued)

Details of commercially available AWG systems (model and website)	Water generation rate (WGR), lit/h	Specific energy consumption (SEC), kWh/lit	Remarks
Akro (Pod): Home—Akvo	2.08	0.312	$T_{amb} = 30$ °C and RH = 80%. Cooling method: VCRs Refrigerant: not specified
Aguaer (AGR 20L-RIVA) https://www.aguaer.com/	0.83	0.564	$T_{amb} = 30$ °C and RH = 80%. Cooling method: VCRs Refrigerant: not specified

7 Closure

The present chapter lays out the fundamentals of dropwise condensation and describes a space- and time-resolved mathematical model that can trace all important steps of the condensation cycle. Examples of condensation patterns formed over vertical and horizontal surfaces are provided. The second part of the chapter is concerned with an essential application of dropwise condensation related to atmospheric water generators. The basic operating principles, configurations, design considerations, and the possibility of integrating renewable energy with conventional techniques for creating a cold substrate are described.

References

Ahlers M, Buck-Emden A, Bart H-J (2019) Is dropwise condensation feasible? A review on surface modifications for continuous dropwise condensation and a profitability analysis. J Adv Res 16:1–13. https://doi.org/10.1016/j.jare.2018.11.004

Ahmad D et al (2018) Hydrophilic and hydrophobic materials and their applications. Energy Sour A Recov Utilizat Environ Effects 40(22):2686–2725. https://doi.org/10.1080/15567036.2018.1511642

Baba S et al (2021) Condensation behavior of hierarchical nano/microstructured surfaces inspired by *Euphorbia myrsinites*. ACS Appl Mater Interf 13(27):32332–32342. https://doi.org/10.1021/acsami.1c01400

Baghel V et al (2019a) Tailoring the hydrophobicity of copper surface with ion beam irradiation. Radiat Effects Defects Solids 174(3–4):307–319. https://doi.org/10.1080/10420150.2018.1564149

Baghel V, Sikarwar BS, Muralidhar K (2019b) Modeling of heat transfer through a liquid droplet. Heat Mass Transf 55(5):1371–1385. https://doi.org/10.1007/s00231-018-2520-2

Baghel V, Sikarwar BS, Muralidhar K (2020) Dropwise condensation from moist air over a hydrophobic metallic substrate. Appl Therm Eng 181:115733. https://doi.org/10.1016/j.applthermaleng.2020.115733

Bayer IS (2020) Superhydrophobic coatings from ecofriendly materials and processes: a review. Adv Mater Interf 7(13):2000095. https://doi.org/10.1002/admi.202000095

Beysens D (2022) The physics of dew, breath figures and dropwise condensation. Springer, Cham

Bonner RW (2013) Correlation for dropwise condensation heat transfer: water, organic fluids, and inclination. Int J Heat Mass Transf 61:245–253. https://doi.org/10.1016/j.ijheatmasstransfer.2012.12.045

Boreyko JB, Chen C-H (2009) Self-propelled dropwise condensate on superhydrophobic surfaces. Phys Rev Lett 103(18):184501. https://doi.org/10.1103/PhysRevLett.103.184501

Cameron WJ, Reddy KS, Mallick TK, (2022) Review of high concentration photovoltaic thermal hybrid systems for highly efficient energy cogeneration. Renew Sust Energy Rev 163:112512 https://doi.org/10.1016/j.rser.2022.112512

Carey VP (2018) Liquid vapor phase change phenomena: an introduction to the thermophysics of vaporization and condensation processes in heat transfer equipment, 2nd edn. CRC Press, New York

Carey VP (2020) Liquid-vapor phase-change phenomena: an introduction to the thermophysics of vaporization and condensation processes in heat transfer equipment. Third Edition, CRC Press, Boca Raton. https://doi.org/10.1201/9780429082221

Chu F et al (2017) Relationship between condensed droplet coalescence and surface wettability. Int J Heat Mass Transf 111:836–841. https://doi.org/10.1016/j.ijheatmasstransfer.2017.04.052

Das SK, Chatterjee D (2023) Introduction to two-phase flow—flow patterns and maps. In: Das SK, Chatterjee D (eds) Vapor liquid two phase flow and phase change. Springer, Cham, pp 1–22

de Gennes P-G, Brochard-Wyart F, Quéré D (2010) Capillarity and wetting phenomena: drops, bubbles, pearls, waves. Translation of the French ed. Springer, New York

Drelich J et al (2011) Hydrophilic and superhydrophilic surfaces and materials. Soft Matter 7(21):9804. https://doi.org/10.1039/c1sm05849e

El-Fil B, Kini G, Garimella S (2020) A review of dropwise condensation: theory, modeling, experiments, and applications. Int J Heat Mass Transf 160:120172. https://doi.org/10.1016/j.ijheatmasstransfer.2020.120172

Fu J et al (2024) Research progress on preparation of superhydrophobic surface and its application in the field of marine engineering. J Mar Sci Eng 12(10):1741. https://doi.org/10.3390/jmse12101741

Ho JY, Leong KC (2021) A critical review of filmwise natural and forced convection condensation on enhanced surfaces. Appl Therm Eng 186:116437. https://doi.org/10.1016/j.applthermaleng.2020.116437

Hu M, Zhao B, Suhendri Ao X, Cao J, Wang Q, Riffat S, Su Y, Pei G (2022) Applications of radiative sky cooling in solar energy systems: Progress, challenges, and prospects. Renew Sust Energy Rev 160:112304 https://doi.org/10.1016/j.rser.2022.112304

Incropera FP et al (2007) Fundamentals of heat and mass transfer, 6th edn. Wiley, Hoboken

Jain S, Sikarwar BS, Muralidhar K (2023) Improved heat transfer modeling of moist air condensation on hydrophobic metallic surface. Heat Transf Eng 54:1–19. https://doi.org/10.1080/01457632.2023.2275237

Jeevahan J et al (2018) Superhydrophobic surfaces: a review on fundamentals, applications, and challenges. J Coat Technol Res 15(2):231–250. https://doi.org/10.1007/s11998-017-0011-x

Kataria S et al (2022) Plasma techniques for the fabrication of hydrophobic substrates. Biennial international conference on future learning aspects of mechanical engineering. Springer, Singapore, pp 831–846

Kataria S et al (2024) Metallic hydrophobic surfaces: fabrication methods and applications in water vapor condensation. Int Commun Heat Mass Transf 159:108306. https://doi.org/10.1016/j.icheatmasstransfer.2024.108306

Kekaula K, Chen Y (2020) Theoretical analysis of filmwise condensation in inclined tubes with nondivergent and irrotational flow components. Int J Heat Mass Transf 151:119444. https://doi.org/10.1016/j.ijheatmasstransfer.2020.119444

Khandekar S, Muralidhar K (2020) Introduction to condensation. In: Khandekar S, Muralidhar K (eds) Drop dynamics and dropwise condensation on textured surfaces. Springer, Cham, pp 149–164

Lee A et al (2020) Hydrophilic/hydrophobic characteristics on the carbon nanotube buckypapers with various mechanical and chemical manufacture process. Diam Relat Mater 110:108152. https://doi.org/10.1016/j.diamond.2020.108152

Leipertz A (2010) J3 dropwise condensation. In: VDI heat atlas. Springer, Berlin, pp 933–938. https://doi.org/10.1007/978-3-540-77877-6_61

Leipertz A, Fröba AP (2008) 'Improvement of condensation heat transfer by surface modifications. Heat Transf Eng 29(4):343–356. https://doi.org/10.1080/01457630701821563

Lo C-W et al (2019) Enhancing condensation heat transfer on three-dimensional hybrid surfaces. Joule 3(11):2806–2823. https://doi.org/10.1016/j.joule.2019.08.005

Ma C et al (2022) Condensation droplet sieve. Nat Commun 13(1):5381. https://doi.org/10.1038/s41467-022-32873-1

Miljkovic N, Wang EN (2013) Condensation heat transfer on superhydrophobic surfaces. MRS Bull 38(5):397–406. https://doi.org/10.1557/mrs.2013.103

Miljkovic N, Enright R, Wang EN (2012) Effect of droplet morphology on growth dynamics and heat transfer during condensation on superhydrophobic nanostructured surfaces. ACS Nano 6(2):1776–1785. https://doi.org/10.1021/nn205052a

Miljkovic N, Rong X et al (2013) Condensation on hydrophilic, hydrophobic, nanostructured superhydrophobic and oil-infused surfaces. Nenad Miljkovic. https://dspace.mit.edu/handle/1721.1/84996. Accessed 24 Oct 2024

Moreira RSM, Peñaranda E, dos Reis BS, de Moraes DB, Rodrigues G, Naveira Cotta CP (2025) Cogeneration in a high concentration photovoltaic (hcpv) system using microchannel heat exchangers for heat recovery. Proceedings of the 27th international congress of mechanical engineering. https://doi.org/10.26678/ABCM.COBEM2023.COB2023-2037

Nenad M, Preston DJ, Enright R, Solomon A, Youngsuk N, Wang EN (2013a) Jumping droplet dynamics on scalable nanostructured superhydrophobic surfaces. J Heat Transf 135:080907.https://doi.org/10.1115/1.4024189

Nenad M, Rong X, Preston DJ, Enright R, McKay I, Wang EN (2013b) Condensation on hydrophilic, hydrophobic, nanostructured superhydrophobic and oil-infused surfaces. Nenad Miljkovic

Parvate S, Dixit P, Chattopadhyay S (2020) Superhydrophobic surfaces: insights from theory and experiment. J Phys Chem B 124(8):1323–1360. https://doi.org/10.1021/acs.jpcb.9b08567

Raman AP, Anoma MA, Zhu L, Rephaeli E, Fan S (2014) Passive radiative cooling below ambient air temperature under direct sunlight. Nature 515: 540–544. https://doi.org/10.1038/nature13883

Rezaee B et al (2023) Enhancement of dropwise condensation heat transfer through a sprayable superhydrophobic coating. Langmuir 39(23):8354–8366. https://doi.org/10.1021/acs.langmuir.3c00947

Rohsenow WM (2022) Heat transfer and temperature distribution in laminar-film condensation. Trans Am Soc Mech Eng 78(8):1645–1647. https://doi.org/10.1115/1.4014125

Rose JW, Glicksman LR (1973) Dropwise condensation—the distribution of drop sizes. Int J Heat Mass Transf 16:411–425. https://doi.org/10.1016/0017-9310(73)90068-9

Rose JW (1998) Condensation heat transfer fundamentals. Chem Eng Res des 76(2):143–152. https://doi.org/10.1205/026387698524712

Rose JW (2011) Condensation, overview. In: Thermopedia. Begel House Inc

Shadmani S et al (2020) Superhydrophobicity through coatings prepared by chemical methods. In: Khodaei M, Chen X, Li H (eds) Superhydrophobic surfaces: fabrications to practical applications. IntechOpen

Sharma DK et al (2020) Self assembly of super-hydrophobic nanotextured methyl functionalized silica on copper and aluminium surfaces for moist air condensation. Colloids Surf A Physicochem Eng Aspects 605:125379. https://doi.org/10.1016/j.colsurfa.2020.125379

Sharma DK, Pachchigar V et al (2022a) Plasma nano-patterning for altering hydrophobicity of copper substrate for moist air condensation. Appl Surf Sci Adv 11:100281. https://doi.org/10.1016/j.apsadv.2022.100281
Sharma DK, Sikarwar BS et al (2022b) Super-hydrophobic nanostructured silica coating on aluminum substrate for moist air condensation. J Mater Eng Perform 31(2):1266–1276
Sheng Q et al (2016) On the onset of surface condensation: formation and transition mechanisms of condensation mode. Sci Rep 6(1):30764. https://doi.org/10.1038/srep30764
Sikarwar B (2023) Visualization of dropwise condensation of water vapor process from experiments and simulation. J Flow vis Image Process 30(4):25–45. https://doi.org/10.1615/JFlowVisImageProc.2022045741
Sikarwar BS et al (2012) Dropwise condensation studies on multiple scales. Heat Transf Eng 33(4–5):301–341. https://doi.org/10.1080/01457632.2012.611463
Sikarwar BS, Khandekar S, Muralidhar K (2013a) Mathematical modelling of dropwise condensation on textured surfaces. Sadhana 38(6):1135–1171. https://doi.org/10.1007/s12046-013-0190-9
Sikarwar BS, Khandekar S, Muralidhar K (2013b) Simulation of flow and heat transfer in a liquid drop sliding underneath a hydrophobic surface. Int J Heat Mass Transf 57(2):786–811. https://doi.org/10.1016/j.ijheatmasstransfer.2012.10.066
Sikarwar BS, Khandekar S, Muralidhar K (2015) Dropwise condensation of metal vapors underneath inclined substrates. Interf Phenom Heat Transf 3(1):5425
Sikarwar BS et al (2020) Simulation: dropwise condensation of water. In: Khandekar S, Muralidhar K (eds) Drop dynamics and dropwise condensation on textured surfaces. Springer, Berlin, pp 259–281
Sikarwar B (2013) Modelling of dropwise condensation underneath inclined textured surfaces. PhD thesis, Indian Institute of Technology Kanpur, India
Singh NS et al (2021) Implementing superhydrophobic surfaces within various condensation environments: a review. Adv Mater Interf 8(2):2001442. https://doi.org/10.1002/admi.202001442
Singh PL et al (2022) Enhancing dropwise condensation of vapor from moist air over a copper substrate by temperature-controlled chemical etching. Therm Sci Eng Progr 34:101403. https://doi.org/10.1016/j.tsep.2022.101403
Sotoudeh F et al (2023) Natural and synthetic superhydrophobic surfaces: a review of the fundamentals, structures, and applications. Alexand Eng J 68:587–609. https://doi.org/10.1016/j.aej.2023.01.058
Tang Y et al (2021) Robust micro-nanostructured superhydrophobic surfaces for long-term dropwise condensation. Nano Lett 21(22):9824–9833. https://doi.org/10.1021/acs.nanolett.1c01584
Thomas TM et al (2023) Preferred mode of atmospheric water vapor condensation on nanoengineered surfaces: dropwise or filmwise? Langmuir 39(15):5396–5407. https://doi.org/10.1021/acs.langmuir.3c00022
Thome JR (2015) Encyclopedia of two-phase heat transfer and flow I. World Sci 15:9310
Vemuri S, Kim KJ (2006) An experimental and theoretical study on the concept of dropwise condensation. Int J Heat Mass Transf 49(3):649–657. https://doi.org/10.1016/j.ijheatmasstransfer.2005.08.016
Vemuri S et al (2006) Long term testing for dropwise condensation using self-assembled monolayer coatings of n-octadecyl mercaptan. Appl Therm Eng 26(4):421–429. https://doi.org/10.1016/j.applthermaleng.2005.05.022
Veronesi F et al (2021) Evaluation of the durability of slippery, liquid-infused porous surfaces in different aggressive environments: influence of the chemical-physical properties of lubricants. Coatings 11(10):1170. https://doi.org/10.3390/coatings11101170
Wang J et al (2020) Improved modeling of heat transfer in dropwise condensation. Int J Heat Mass Transf 155:119719. https://doi.org/10.1016/j.ijheatmasstransfer.2020.119719

Wang X et al (2022) Review of droplet dynamics and dropwise condensation enhancement: theory, experiments and applications. Adv Colloid Interf Sci 305:102684. https://doi.org/10.1016/j.cis.2022.102684

Yilbas BS, Al-Sharafi A, Ali H (2019) Water-droplet dynamics and heat transfer. In: Self-cleaning of surfaces and water droplet mobility. Elsevier, Amsterdam, pp 133–284. https://doi.org/10.1016/B978-0-12-814776-4.00005-7

Zhang T-Y et al (2022) Advances in modeling investigations of multimode dropwise condensation heat transfer on smooth and textured surfaces: a review. Int J Therm Sci 172:107309. https://doi.org/10.1016/j.ijthermalsci.2021.107309

Zheng S-F, Gross U, Wang X-D (2021) Dropwise condensation: from fundamentals of wetting, nucleation, and droplet mobility to performance improvement by advanced functional surfaces. Adv Colloid Interf Sci 295:102503. https://doi.org/10.1016/j.cis.2021.102503

Zhu J et al (2015) Clustered ribbed-nanoneedle structured copper surfaces with high-efficiency dropwise condensation heat transfer performance. ACS Appl Mater Interf 7(20):10660–10665. https://doi.org/10.1021/acsami.5b02376

Fog Harvesting

Sohom Goswami and Malay K. Das

1 Introduction

Fog harvesting has emerged as an attractive unconventional source of water to supplement conventional water sources like groundwater, rivers, and lakes to meet growing water demands. Fog is usually available in high altitudes and coastal desert regions, which are mostly remote, making conventional sources relatively inaccessible to them. The first documented historical evidence of fog harvesting originates from El Hierro in the Canary Islands in the period of fifteenth to seventeenth century (Marzol-Jaen 2011). Several American and European travellers passing through El Hierro have described the "fountain tree" which intercepted the dense fog passing over the island and "sweated" water during the nighttime, providing sufficient water for the survival of the island people, even in absence of any other freshwater source like lakes, streams, rivers, etc. In nature, several organisms have developed adaptations capable of harvesting freshwater from fog and utilising it for their physiological needs. Perhaps the most widely studied among them is the Namib desert beetle (*Stenocara c.*), which captures water by exposing its back to the foggy winds (Parker and Lawrence 2001), and guiding the water collected to their mouths through special mechanisms. Another famous example is that of the spines of the redwood trees, *Sequoia sempervirens*, which capture droplets on parallel needle arrays, imitating the musical instrument harp (Dawson 1998; Shi et al. 2018). Other examples include desert lizard (Philipp et al. 2015), cactus spines (Ju et al. 2012), spider webs (Zheng et al. 2010) and so on. These examples have served as inspiration for researchers, who have mimicked these mechanisms to develop efficient fog harvesting strategies and devices, which will be discussed in detail in subsequent sections in this chapter.

Artificial fog harvesting involves trapping suspended water droplets present in fog on a substrate, while the air is filtered out. Typically, porous substrates normal to

S. Goswami · M. K. Das (✉)
IIT Kanpur, Kanpur, India
e-mail: mkdas@iitk.ac.in

R. Narayanaswamy et al. (eds.), *Sustainable Water Harvesting*, Springer Tracts in Mechanical Engineering, https://doi.org/10.1007/978-981-95-3373-2_7

the flow direction are preferred to allow maximum interaction between fog droplets and the substrate. One must note here the distinction between fog harvesting and dew harvesting as two different methods of atmospheric water harvesting (Nioras et al. 2021; Yu et al. 2015). While dew harvesting consists of essentially condensing water from the water vapor contained in the atmosphere, fog harvesting consists of capturing pre-existing liquid droplets suspended in air, formed as a result of mixing of hot and cold air layers present in the atmosphere. A typical fog harvesting system is illustrated in Fig. 1a. It consists of a porous medium like screen mesh, nets, textile fabrics, metal foams, etc. stretched in path of the fog flow, which captures fog droplets on its surface, effectively separating out liquid water from the air–water suspension.

In the current discussion, we consider a screen mesh as a representative example, although all discussions throughout the chapter are also valid for other kinds of porous medium mentioned above, unless otherwise mentioned. The fog droplets strike the fibres of the mesh, and get captured, eventually falling into the sloping gutter, and into the reservoir as shown in Fig. 1a. This water can then be utilised for domestic, commercial as well as agricultural use. This is a multiscale phenomenon, ranging from several meters at the device scale (i.e., the scale of the entire mesh), to few hundreds of microns at the individual fibre or pore scale. It is largely governed by aerodynamic and surface tension forces. At device scale, the mesh obstructs the fog, such that a part of the incoming fog bypasses the fog collecting medium, as shown in

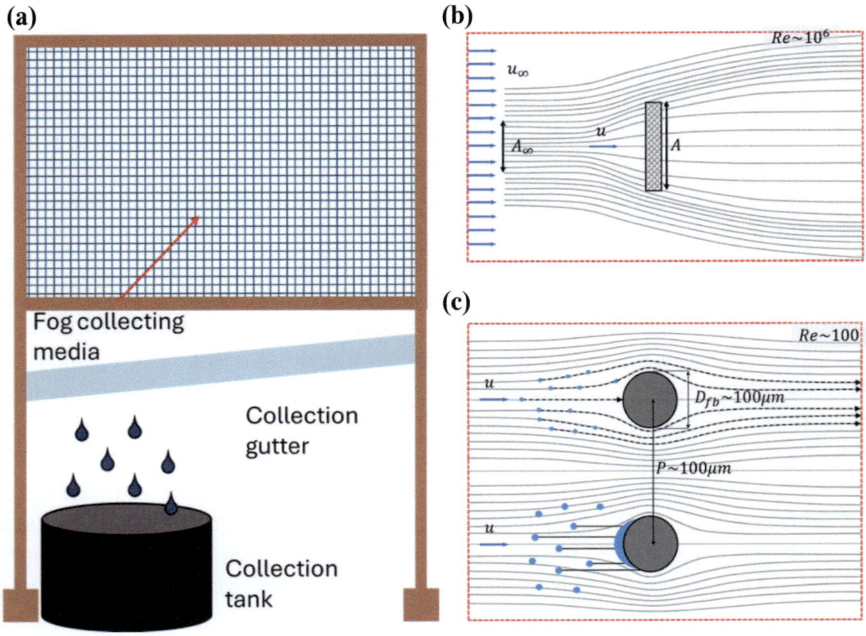

Fig. 1 **a** A typical fog collection system consisting of fog collecting mesh, water collection gutter, and water reservoir **b** flow past fog collecting mesh at system scale, with length scales ~ 1 m and $Re \sim 10^6$, **c** fog flow past mesh fibres at pore scale, with length scales ~ 100 μm

Fig. 1b. The water droplets present in this part of the fog can never interact with the mesh fibres and are therefore not collected. The remaining fog droplets are incident on the mesh, and may strike the fibres, or flow around them, passing through the pores of the mesh. Thus, we can define a first stage of efficiency in the fog collection process, called the aerodynamic efficiency (η_a), which is defined as the fraction of total fog mass flux which may strike the fibres of the mesh. This is governed by aerodynamics at the device length scale (O(~1 m)) and depends on geometric properties like shade coefficient (SC), fibre diameter (d_f), as well as aerodynamic properties like drag coefficient (C_d).

At the pore scale, as mentioned earlier, the fog droplets can either strike the mesh fibres or pass through the pores. This is governed by the inertia of the incoming fog droplets. Larger droplets with higher inertia tend to continue along their initial straight path, along the direction of the undisturbed flow and strike the fibres to get captured. The smaller droplets tend to follow the air streamlines, flow around the cylindrical fibres and pass through the mesh pores, as shown in Fig. 1c. This is then taken into account by the second stage efficiency, known as the capture efficiency or the deposition efficiency. In the present discussion, we will use the term capture efficiency, denoted by η_c. Quantitatively, capture efficiency can be defined as the ratio between the mass of fog droplets striking the mesh fibres and getting captured to the total mass of fog droplets incident on the mesh. It is governed by aerodynamics on pore scale and depends on the response time of the air flow to the disturbance caused by the presence of mesh fibres and the response time of the fog droplets to the same disturbance. As several droplets strike and get captured on the mesh fibre surface, they start growing. Initially the growth of these droplets is due to the vertical coalescence with droplets impinging from fog stream. As the droplets start growing, adjacent droplets tend to undergo lateral coalescence, as shown in Fig. 2a. Eventually, the volume of the captured droplet grows to a critical value, called critical volume of droplet shedding ($\overline{V}_{cr}$), where the gravitational force overcomes the surface tension force which tends to hold the droplets to the fibre surface, and the droplets are shed into the collector placed below the mesh. Often the droplets growing on adjacent fibres tend to coalesce. These droplets do not shed due to strong adhesion to the mesh fibres, thus contributing to losses in the collection process. Further, this leads to an increase in local aerodynamic resistance due to clogging of the mesh pores (as shown in Fig. 2b), which in turn reduces η_a. Another form of losses encountered at this stage includes re- entrainment losses, which occurs due to droplets getting detached from the fibre surface due to strong drag forces of wind flowing over them, which overcomes the adhesion force between the droplet and the fibre surface, see Fig. 2b. These losses, along with leakage and evaporation losses, are taken into account by a third stage efficiency, called as the drainage efficiency (η_d). Only a part of the total amount of fog captured on the fibre surface is shed into the collector placed below it and is available for use. This introduces the final efficiency term, called the drainage efficiency. Thus, the overall fog collection efficiency is given by product of the three efficiencies defined above (since the denominator of each stage efficiency becomes the numerator of the next),

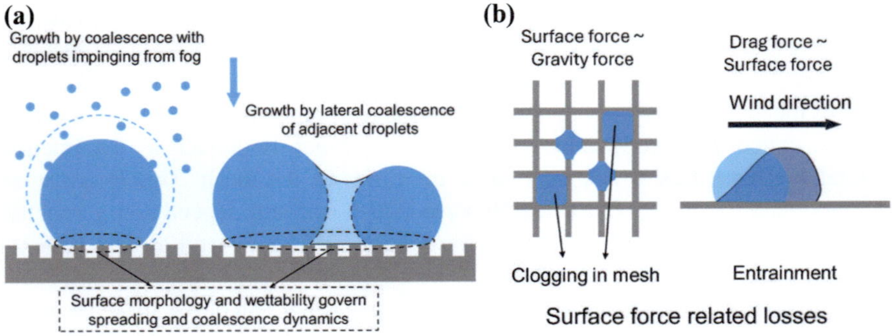

Fig. 2 Fog collection at droplet scale. **a** Droplet growth and coalescence mechanisms, governed by surface morphology and wettability, **b** surface force-related losses: clogging, entrainment

$$\eta = \frac{\dot{m}}{LWC \times \dot{Q}} = \eta_a \eta_c \eta_d \tag{1}$$

where, $\dot{m}$ is the rate of water collected, $LWC \times \dot{Q}$ gives the total mass flow rate of water through the collector, given by the product of liquid water contained in the fog per unit volume of dry air (in g/m^3), *LWC,* and the fog flow rate $\dot{Q}$.

2 Aerodynamics of Fog Collection

2.1 Aerodynamic Efficiency

As discussed in the previous section, aerodynamic efficiency is the first stage efficiency of fog harvesting process. By definition, it is the fraction of fog droplets which is incident on the solid mesh area and can strike the mesh fibres to get captured. It can therefore be also interpreted as the maximum fraction of fog droplets that may be captured by the mesh. This efficiency is governed by the aerodynamics of the fog flow at the device scale. It primarily depends on the 'bluffness' of the fog collecting mesh, which in turn depends upon the solidity of the mesh, called shade coefficient (SC). It is the ratio of area occupied by the mesh fibres and the total area of the mesh. Thus, for a completely solid plate, the SC = 1, while for a completely hollow frame, SC = 0. A typical flow field for fog flow through and around porous mesh is shown in Fig. 3, top. Here, the undisturbed fog flowing through the area A_∞ spreads out to cover the area A of the mesh. Thus, the amount of fog flowing through the mesh is $\rho_{\text{air}} A_\infty U_\infty = \rho_{\text{air}} A u$, by mass conservation. In absence of the mesh, this would have been, $\rho_{\text{air}} A U_\infty$. Out of the total area of the mesh, only the solid fraction, SC, is capable of capturing fog droplets. Thus, the fraction of the total fog flow that may be captured by the mesh is given by the ratio of the amount of fog incident on the mesh to the maximum possible fog flow through the mesh area, multiplied by the fraction of area occupied by the mesh fibres,

$$\eta_a = \frac{\rho_{\text{air}} A_\infty U_\infty}{\rho_{\text{air}} A U_\infty} \times SC = \frac{A_\infty}{A} SC = \frac{u}{U_\infty} SC \tag{2}$$

where u is the velocity of fog just upstream of the mesh, which is slower than the undisturbed flow, as mass remains conserved. Equation 2 gives the most general form for aerodynamic efficiency of fog collection. As one might readily observe, the problem of finding aerodynamic efficiency now reduces to essentially finding A_∞ or u, since A and U_∞ can be readily measured. This problem is solved by the principle of superposition, a common methodology often adopted in fluid mechanics to solve irrotational flows. The flow past the mesh is assumed to be a superposition of irrotational flow of velocity u, forced to flow through a mesh and viscous flow of velocity u′ past a solid plate, as shown in Fig. 3. Pressure drop across the mesh in potential flow forced through a mesh can be expressed in terms of mesh resistance coefficient, k (Taylor and Davies 1944) and pressure drop across a plate placed in a viscous flow can be expressed in terms of drag coefficient C_d (White 2001). Since the viscous flow through and around the mesh is the superposition of the two, the pressure-drop across it should be unique. Thus, equating the pressure drop in the two cases and solving for velocities, we have an expression for aerodynamic efficiency, η_a (Rivera 2011):

$$\eta_a = \frac{SC}{1+\sqrt{\frac{k}{C_D}}} \tag{3}$$

η_a shows a non-monotonic variation with SC, governed by the competition between increase in available fibres per unit area with increasing SC which increases the probability of fog droplets being captured by the mesh fibres, and the increase in aerodynamic resistance to the fog flow as it tends more towards a solid plate with the increase in SC, leading to larger fractions of fog circumventing the mesh, thus reducing η_a. At lower values of SC, the increase in availability of fibres per unit area dominates, while at higher values of SC, the increase in aerodynamic resistance increases. The optimum value of SC for maximum aerodynamic efficiency lies between 0.5 and 0.55 depending on the mesh fibre materials (Park et al. 2013; Rivera 2011) with a maximum theoretical efficiency of 24%. The effect of mesh material and fibre cross section also affect the aerodynamic efficiency, primarily through the parameter k, which is discussed in more detail in the next section (Fig. 4).

The aerodynamic efficiency is also affected by the shape of the mesh, i.e., concave meshes tend to have higher aerodynamic efficiency as compared to flat meshes on account of their higher drag coefficients (Carvajal et al. 2020). Mesh with larger curvature cause lower pressure drop, such that the momentum downstream of the mesh is sufficient to counter the adverse pressure gradient in the wake leading to a higher aerodynamic efficiency. From a practical perspective, the atmospheric boundary layer formation on the ground surface also affects the performance, since in such a case, the freestream velocity of the unperturbed fog is no longer constant but depends on the velocity profile inside the boundary layer. This is not considered

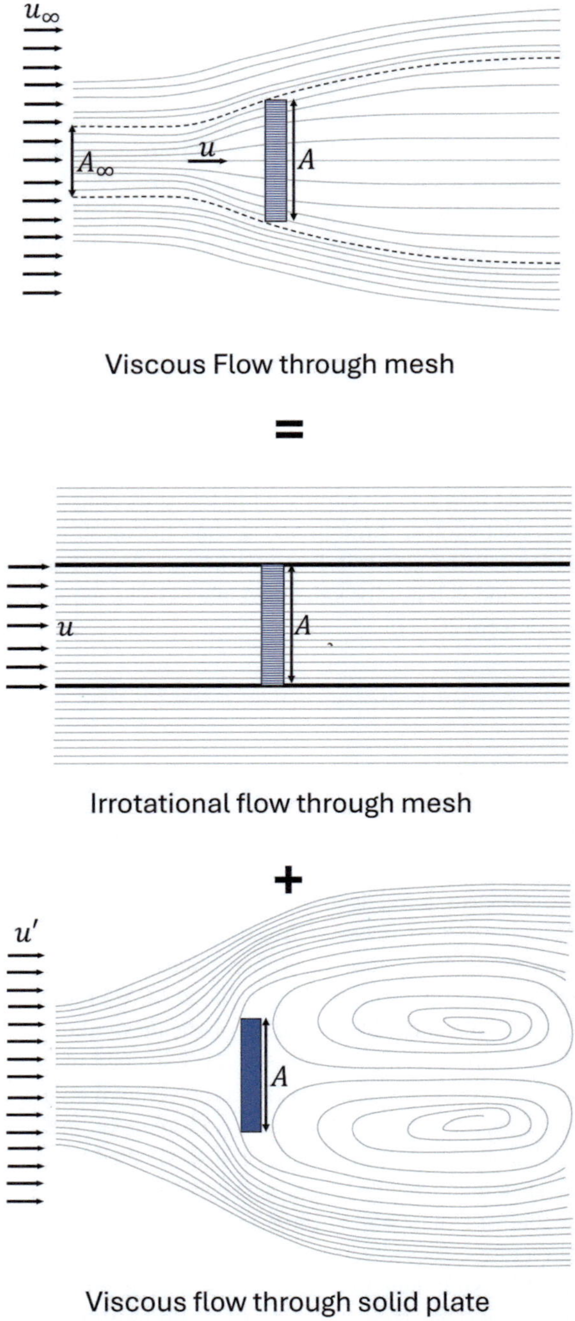

Fig. 3 Decomposition of flow past mesh into linear superposition of potential flow of velocity u, forced through the mesh and viscous flow of velocity u′ around a solid plate

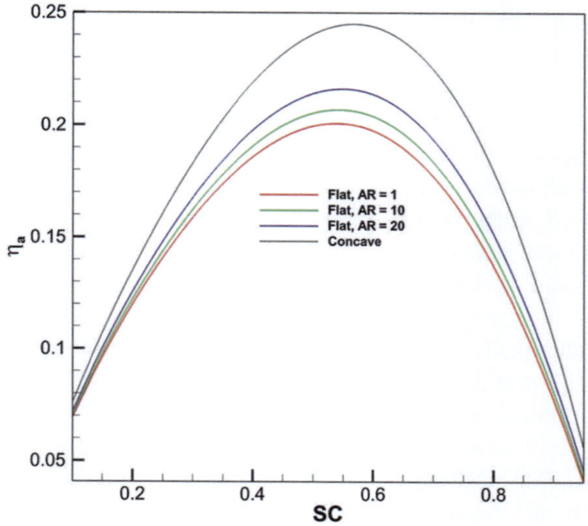

Fig. 4 Plot showing variation of aerodynamic efficiency with SC as predicted by Rivera's model, with a maximum value of 24% at $SC = 0.55$ (Rivera 2011)

in the model discussed above, which is reasonable if the mesh is mounted at a height greater than the atmospheric boundary layer thickness. Considering the Reynolds number based on mesh characteristic length (O (1 m)) which is of the order of 10^6, the boundary layer thickness is found to be of the order of ~ 1 m (Carvajal et al. 2020).

Recent studies on air flow past porous structures have shown that the parameters C_d and k are indeed coupled with the porous media parameters of the mesh, i.e., the shade coefficient, pore size, and pore structure (Ledda et al. 2021; Steiros and Hultmark 2018). To understand these in detail, we briefly discuss the general problem of flow past a porous structure, and the theoretical framework developed to model the flow field, which allows us to have greater physical insight into the parameters governing aerodynamic efficiency.

2.2 General Problem of Flow Past a Porous Mesh

The fundamental flow physics related to aerodynamics of fog harvesting at the device scale can be described by the general problem of flow past a porous media placed in an infinite flow. This can be considered as a special case of flow past a solid plate kept normal to the flow. In case of a solid plate, as expected for flow past bluff bodies, boundary layer separation occurs at the edges of the plate, and a wake is formed behind it. As a result, there is a substantial pressure drop across the plate, leading to a drag force commonly known as form drag. Similarly, when fluid flows past a porous plate, similar flow physics is observed, but with certain differences due to the ability of fluid to flow through the mesh as opposed to the solid plate. From fog harvesting perspective, it is important to know how the mesh parameters like porosity, fibre diameter, etc. affect the term u/U_∞ which is the most prominent term governing aerodynamic efficiency. The modelling and measurement of pressures and velocities for such a flow pose significant challenges. The difficulties primarily arise due to the presence of small pores, with a dimension much lesser than the device scale, such that conventional pressure taps cannot be used due to local disturbances associated with intrusive measurements. In computational studies, resolving the pore scale as well as the device scale becomes prohibitively computationally costly. Therefore, several quasi-analytical approaches have been explored to solve this problem using potential flow theory, as discussed below.

This problem was first studied as early as 1944, when Taylor and Davies (1944) attempted to model the flow of air through porous meshes spanning an entire channel. This was done to study flow through screen meshes installed in wind tunnels. Their approach was to replace the mesh with a distribution of "centres of resistance." The flow outside the wake was then modelled as potential flow occurring due to said distribution. Typically, for flow past such screen meshes, the pressure drop is characterised by a pressure drop coefficient k, defined as $\Delta p = 1/2k\rho u^2$ where ρ is the density of the fluid, u is the velocity just upstream of the mesh, and k is the coefficient of resistance of the mesh. Using their model, Taylor and Davies found a relation between this resistance coefficient, and the drag coefficient (C_d) of the mesh. However, their formulation did not model the entire flow field, especially in the wake, where the flow is intuitively expected to be rotational due to viscous effects. To include these effects and to solve the entire flow field, more comprehensive models were subsequently developed (Cumberbatch 1982; Koo and James 1973). These methods are similar to panel methods (see Katz and Plotkin 2001) commonly used in aerospace industry to solve aerodynamics problems at low computational cost. In this method, the mesh is modelled as a continuous distribution of sources. One of such methods is discussed here. In this method, the flow domain is divided into two regions as shown in Fig. 5. The first region is upstream of the mesh, where the flow can be modelled as potential flow due to superposition of rectilinear, undisturbed flow and flow due to the continuous distribution of sources. The velocities in this region can be obtained from the gradient of potential function of the above superposed flow. The flow in the wake near the mesh is rotational and cannot be modelled as a potential flow. However, it can be assumed that the streamlines follow a similar pattern as they would have done in case of a purely potential flow (Koo and James 1973). This enables the evaluation of stream functions in region II as a function of region I, i.e.,

$$\psi_{II}(x, y, z) = Ef(\psi_I) \tag{4}$$

where E is an attenuation function, depending on the bluffness of the mesh (Steiros and Hutmark 2018). This is a free parameter and is not readily known from physical arguments. Thus, each velocity in region II can be expressed as functions of velocity components in region I and the attenuation function, E. Specifically, we notice that the normal component immediately upstream and downstream (i.e., $x \to 0^-$ and $x \to 0^+$) of the mesh should remain constant, i.e., $u_I = u_{II}$. The pressure just

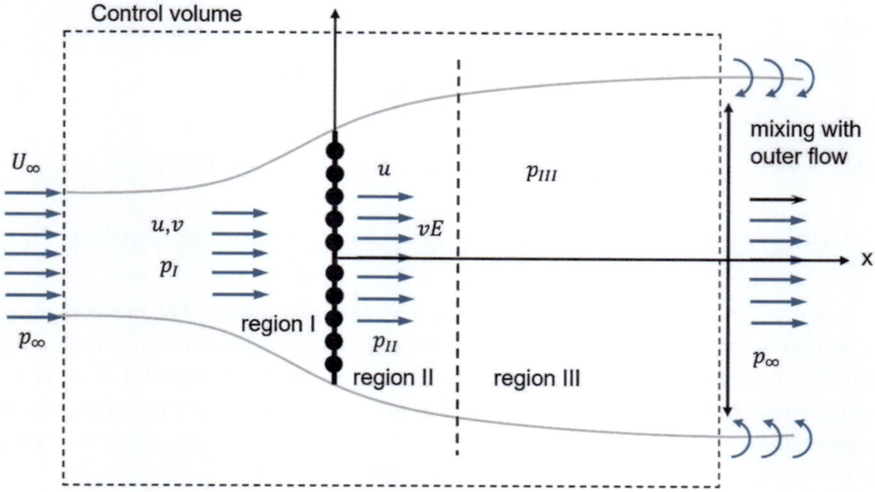

Fig. 5 Control volume analysis for solving flow field for flow past a mesh with arbitrary shape and porosity by superposition of rectilinear flow and flow due to continuous distribution of sources

downstream of the mesh decreases to p_{II}, then increases to p_{III} in the near wake, before being fully recovered at the far wake to p_∞. This pressure p_{III} is also a free parameter, resulting from the mixing of outer flow with the wake flow and is termed as "base suction" (Roshko 1954). Thus, we have, at this point, three free parameters, i.e., the source strength (Ω), attenuation function E, and the base suction pressure, p_{III}. To solve the complete velocity field, we need to evaluate these free parameters. This is done using energy, mass and momentum conservation across the mesh, as well as in a control volume as shown in Fig. 5. The energy balance across the mesh requires expressing the pressure jump across the mesh in terms of the mesh parameters,

$$p_I - p_{II} = f(u_\infty, E) + \tfrac{1}{2} k \rho u_I^2 \tag{5}$$

where k depends on mesh properties. The function $f(u_\infty, E)$ appears due to the ability of the flow to circumvent the mesh. When viscous forces are negligible, k can be expressed in terms of SC only. Several authors have found functional forms of k based on theoretical arguments (Wierghardt 1953) and empirical correlations (Brundrett 1993; Idelch'k 1986; Xu et al. 2020). Following Wierghardt, the mean velocity of flow in the holes can be expressed as $u_I/(1 - SC)$ and subsequent energy conservation gives $k = \left[1 - 1/(1 - SC)^2\right]$. For flows with low Reynolds number, viscous effects become important, such that permeability also affects the pressure drop coefficient (Pezzulla et al. 2020; Ledda et al. 2021). Further, angle of attack, β has an effect on the pressured drop, since it reduces the effective pore area accessible to the flow. Marchand et al. (2024), proposed that for a thin, porous media like a screen mesh, Re_{d_f}, the Re based on local fibre diameter is much more meaningful in capturing viscous effects than permeability, and expressed k in terms of SC, Re_{d_f}

and β,

$$k = \left[1 - \frac{1}{(1-SC)^2}\right] g\left(Re_{d_f}, \beta\right) \tag{6}$$

where $g\left(Re_{d_f}, \beta\right)$ is an empirical relation, given by (Brundrett 1993),

$$f(\mathrm{Re}_{d_f}, \beta) = \sin^2(\beta)\left(\frac{c_1}{\mathrm{Re}_{d_f}\sin(\beta)} + \frac{c_2}{\ln(\mathrm{Re}_{d_f}\sin(\beta)) + 1.25} + c_3 \ln(\mathrm{Re}_{d_f}\sin(\beta))\right) \tag{7}$$

where c_1, c_2 and c_3 are constants obtained by empirical fitting. The pressure jump condition is then applied to the mass and momentum conservations to estimate u_1. The above methodology leads to solution of complete velocity field for flow past a porous mesh and allows evaluation of u_I/U_∞ as a function of mesh properties, which is much more generalized, and can be applied to evaluate aerodynamic efficiencies of a broad range of fog collector designs and flow conditions.

2.3 *Aerodynamics of Droplet Capture*

Fog is an aerosol, and there are several mechanisms by which droplets can get captured on a surface. These include diffusion, interception, inertial impaction and electrostatic precipitation.

The mode of collection in absence of electric field is governed primarily by the velocity of flow and particle size distribution. For smaller particles and lower velocities, diffusion is the dominant mechanism of droplet deposition, while for larger particles and higher velocities, inertial impaction dominates. In case of fog, the droplet diameters range from 1–40 μm, while the velocity ranges from 0.5 to 8 m/s. With this range of droplet sizes and velocities, the droplets are captured by inertial impaction alone (Hinds and Zhu 2020). Since the droplets are immersed in the air flow, the drag force exerted by the air on the droplets tends to make them follow the streamlines of the air. On the other hand, the inertia of the droplets tends to allow them to maintain their original trajectory as the undisturbed flow. The trajectory of droplet motion decides whether the droplet hits the mesh fibres or flows around them through the mesh pores and hence determines the deposition efficiency. The problem of finding deposition efficiency of a droplet thus reduces to calculating the trajectories of large number of droplets and finding what fraction would actually strike the mesh. These droplets initially start moving from different spanwise (y) locations, at a fixed distance upstream of the mesh fibre as shown in Fig. 6a. Thus, it is very important to calculate the droplet trajectories in the presence of a cylinder. The trajectory of the droplets can be found by a simple force balance of these two forces. This problem can be solved by solving the trajectories of individual droplets for given initial conditions in Lagrangian frame of reference. The governing equations for droplet motion is then given in the non-dimensional form (Langmuir and Blodgett 1946; Brun et al. 1955),

$$\frac{d\boldsymbol{u}_p^*}{dt^*} = \frac{C_d \mathrm{Re}_{d_p}}{24} \frac{1}{St} \left(\boldsymbol{u}_p^* - u_f^* \right) \tag{8}$$

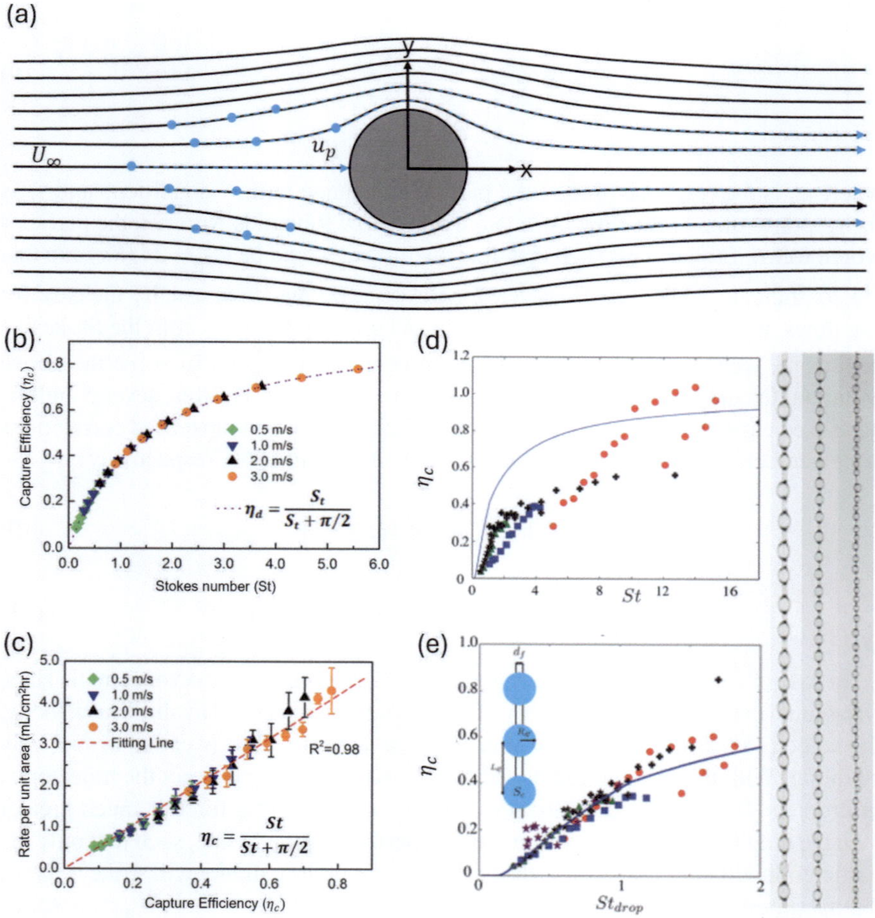

Fig. 6 **a** Schematic representation of flow domain for solving droplet trajectories. **b** Experimentally measured fog capture efficiency plotted as a function of Stokes number. **c** Linear correlation for rate of fog collection per unit area on a superwetting fibre, and theoretical curve obtained from Langmuir and Blodgett model (1946) (Jiang et al. 2019), reproduced, with permission, from American Institute of Physics©. **d** Capture efficiency on vertical fibre array with partial wettability as a function of Stokes number. **e** linear fit for theoretical curve of capture efficiency on partially wetting fibers with Stokes number based on droplet length scale rather than fibre diameter. *Inset* shows pattern of droplets formed on vertical fibre arrays (Labbe and Duprat 2019), reproduced, with permission, from Royal Society of Chemistry©

where $\boldsymbol{u_p}$ is the particle velocity, $\boldsymbol{u_f}$ is the flow velocity, C_d is the drag coefficient, Re_{d_p}. is the Reynolds number based on particle diameter, μ. is the viscosity of air d_p is the particle diameter. U_∞, the undisturbed fog velocity far away from the fibres is used as a velocity scale, diameter of the fibre is the length scale, and the time is scaled by d_f/U_∞. Here, the left-hand side is the inertial term, while the right hand side is the drag term. Non-dimensionalisation leads to a new term, called the Stokes number, St, given by,

$$St = \frac{\frac{\rho_p d_p^2}{18\mu_{air}}}{\frac{U_\infty}{d_f}} = \frac{\tau_p}{\tau_{flow}} \tag{9}$$

where τ_p is the response time of the particle to change in flow conditions and τ_f is the response time of the flow itself. Note that at low Reynolds Number, the particles would follow Stokes' law, such that drag coefficient would be $C_d = 24/Re_{d_p}$, so that the coefficient $\frac{C_d \mathrm{Re}_{d_p}}{24} = 1$. However, for $\mathrm{Re}_{d_p} > 100$, which is usually the case for fog flows, the behaviour of such particles can considerably vary from the Stokesian behaviour, which is then quantified by the coefficient $\frac{C_d \mathrm{Re}_{d_p}}{24}$. To solve the above equation, the drag coefficient needs to be given as an input. For this, several authors have used experimental semi-empirical correlations. The two most used correlations are given below (Beard and Pruppacher 1969; Goldstein 1965, respectively).

$$\frac{C_d \mathrm{Re}_{d_p}}{24} = \left(1 + 0.158\mathrm{Re}_{d_p}^{\frac{2}{3}}\right) \tag{10}$$

$$\frac{C_d \mathrm{Re}_{d_p}}{24} = 1 + 0.197\mathrm{Re}_{d_p}^{0.63} + 2.63 \times 10^{-4}\mathrm{Re}_{d_p}^{1.38} \tag{11}$$

In case of fog flow, the concentration of fog droplets in the air is very small. Thus, we assume that although the velocity of the droplets is influenced by the flow velocity, the velocity of droplets does not alter the flow velocity. This is called as one way coupling. With this assumption, the above Eq. 8 can be solved to get the trajectories of droplets. If the droplets are heavy enough, i.e., the inertia force is much greater than the drag force, then they will travel in a perfectly straight line, such that only the particles directly headed towards the projected area of the cylinder will strike it. Thus, the maximum water that can be collected in the limit of $\mathrm{St} \to \infty$ is $U_\infty d_f \times LWC$. Thus, for a finite Stokes number, the capture efficiency is defined as the mass of particles which hit the fibre as a fraction of $U_\infty d_f \times LWC$, denoted by η_c. Over the years, several approaches have been used to solve the above equation, primarily using numerical approaches, beginning with the first computer ever invented. The solution for a circular cylinder, which is most commonly the shape of fibre cross section, is given as (Langmuir and Blodgett 1946),

$$\eta_c = \begin{cases} \frac{St}{St+\frac{\pi}{2}} forSt > 1.1 \\ 0.466 \log(8St)^2 forSt < 1.1 \end{cases} \tag{12}$$

this relation for capture efficiency has also been experimentally validated for fog harvesting by measuring water collection on a single superhydrophilic wire placed in fog flow inside a wind tunnel (Jiang et al. 2019). Using a single wire makes sure that there is no circumvention of the fog at device scale, i.e., $\eta_a \rightarrow 1$, while the superhydrophilicity makes sure there is negligible hold-up on the wire, leading to $\eta_d \rightarrow 1$. Under such circumstances, the volume of water collected per unit area per unit time is proportional to the capture efficiency. Plotting the experimental data for capture efficiency as a function of St gives a good agreement with Eq. 12 as shown in Fig. 6b. Further, a linear correlation is obtained when the rate of collection per unit area is plotted with the capture efficiency (see Fig. 6c), confirming that this theory holds accurately for a single superhydrophilic fibre placed in fog flow. However, when partially wetting fibres are used as opposed to superwetting fibres in the previous case, the experimental data obtained for capture efficiency seem to deviate from the theoretical curves (Labbé and Duprat 2019), as shown in Fig. 6d. This is due to the presence of droplets on the fibres. As fog harvesting progresses, droplets get formed and shed under the effect of gravity. Thus, at any time, the surface of the fibre is covered by droplets of a mean size of d_{drop} (see Fig. 6 inset). Thus, the length scale corresponding to response time of the flow should be d_{drop} instead of the fibre diameter d_f. Thus, the Stokes number should be based on droplet diameter, denoted by $St_{drop} = St \times d_f / d_{drop}$. Plotting the capture efficiency as a function of this St_{drop} shows a good agreement with experiments, validating this hypothesis, see Fig. 6e.

3 Role of Wettability and Surface Morphology in Fog Harvesting

3.1 Need for Tuning Surface Properties

As mentioned earlier, the shedding of droplets from mesh fibres are dependent on gravity. As the droplet grows, two effects can arise which adversely affect the fog collection efficiency. Firstly, droplet detachment and loss from fibre surface due to drag force of air flow, and secondly, the droplets on adjacent fibres may coalesce, leading to clogging of mesh pores, increasing its shade coefficient. This in turn degrades the aerodynamic performance of the mesh as it tends towards a solid plate as more and more pores get clogged. The first of these effects, i.e., re-entrainment occurs due to the competing effects of drag force and surface forces. Thus, a critical radius for the limiting case when these two forces balance can be defined, given by,

$$r_e = \frac{4\pi\gamma_{lv}\sin^2\theta(1+\cos\theta_r)}{\rho_a U_\infty^2 C_d(\theta-\sin\theta\cos\theta)} \tag{13}$$

where θ is the apparent contact angle, θ_r is the receding contact angle, and γ_{lv} is the surface tension. Here, r_e is called the critical radius of entrainment, i.e., once the droplet grows to a size larger than this, the drag force overcomes the adhesion force, thus entraining the drop into the fog stream. Next, we look at the problem of clogging of mesh pores. As the droplet grows, its weight increases, and eventually, the gravitational force overcomes the surface tension force, leading to droplet shedding. Again, we want to estimate a critical radius, beyond which the droplet will be shed, called the critical radius of droplet shedding. Equating these two forces, we get the critical radius of droplet shedding,

$$r_c = \sqrt{\frac{(12\gamma_{lv}(\cos\theta_r-\cos\theta_a)}{\pi\rho_w gB(3+B^2)}} \tag{14}$$

where $B = \left(1 - \frac{\sin\theta}{\cos\theta}\right)$. As before, here r_c is the critical radius beyond which gravitational force overcomes the force due to contact angle hysteresis, leading to droplet shedding. Now, if this $r_c \geq p/2$, where p is the centre-to-centre distance between adjacent fibres, then droplets will coalesce, leading to clogging of the mesh pore. Thus, for mesh of a given material with optimal *SC*, Eq. 14 sets a limit on the minimum pore size, i.e., p_{cr}. If p_{cr} is less than 2 r_c, drops on adjacent fibres will coalesce, leading to clogging. To understand the significance of this limit, we must recall that while maintaining the optimal SC ~ 0.55 for maximum aerodynamic efficiency, typically, the fibre diameter has to be minimized in order to maximize capture efficiency, which would mean reducing the pore diameters along with fibre diameter, eventually reaching p_{cr}. Thus, to overcome this limitation, the value of r_c needs to be minimized. This can be achieved by reducing the contact angle hysteresis (CAH) given by $(\cos\theta_r - \cos\theta_a)$, the key parameter which determines the value of r_c. A small CAH ($< 5°$) is typically observed for substrates with very high or very low contact angles, i.e., the mesh fibres should be made superhydrophobic or superhydrophilic. This can be done by tuning the wettability and surface morphology of the fibres/substrate. This is an area of very intense research activity, especially since the advent of advanced technologies for micro and nanofabrication, as well as additive manufacturing.

3.2 Drop Dynamics on Single Fibre

Droplets on smooth, chemically homogenous surfaces exhibit a spherical cap shape depending on the contact angle, with a circular triple contact line. The shape of any droplet can be found by solving the Young–Laplace equation given by,

$$\Delta p = \gamma\left(\frac{1}{R_1} + \frac{1}{R_2}\right) \tag{15}$$

where R_1 and R_2 are two principal radii of curvature. In order to solve for the droplet profile, these have to be expressed as a function of the drop surface coordinates. Considering a case where the droplet has an axisymmetric contact line such that the shape can be approximated to be two dimensional, we have,

$$\frac{1}{R_1} + \frac{1}{R_2} = \frac{y''}{(1+y')^{\frac{3}{2}}} \tag{16}$$

where $y\ (x, z)$ gives the coordinate of drop surface. The nonlinear nature of the terms in this equation poses significant challenges to find analytical solutions. Thus, only few simple shapes, like the one described above can be obtained analytically. More involved shapes usually have to be solved using some approximate methods. Drops on inclined and vertical surfaces under the influence of gravity are usually asymmetric due to contact angle hysteresis mentioned earlier. While several approximate methods have been utilized in solving drop shapes on flat surface, like perturbation methods (De Coninck et al. 2017), two-circle method (ElSherbini and Jacobi 2004), surface energy minimization etc., droplet morphologies on cylindrical fibres, which play a significant role in droplet shedding and dynamics are not very well understood.

Droplets on fibre have been found to exhibit two stable morphologies (Adam 1937): the barrel shape, which is symmetric in shape, and engulfs the entire fibre, and the clamshell shape, which is asymmetric in nature and has a continuous triple contact line, lying on the fibre surface at some preferred orientation. Topologically speaking, the clamshell shape with its simply connected triple contact line is distinct from the barrel shape, which has two circular contact lines where the droplet engulfs the fibre. The stability of each of these morphologies is critical in designing an efficient droplet transport pathway in fog harvesting. The preferred morphology of drops on fibre is governed by its volume, contact angle, fibre diameter and gravity. The quantitative effects of these parameters on stability of barrel and clamshell morphologies were investigated experimentally by Eral et al. (2011). For a given volume at small contact angle, a barrel shape is observed, characterized by the centroid of the droplet coinciding with that of the fibre when viewed from top, as shown in Fig. 7a. As the contact angle is increased, at a volume dependent critical contact angle, θ_{c1}, there is a sudden movement of the centroid of the droplet radially away from the fibre centroid, marking the transition from barrel to clamshell shape. The clamshell shape is asymmetric in nature and lies on a preferred orientation depending on local surface heterogeneities. As the contact angle is decreased, at a second distinct critical contact angle, clamshell-to-barrel transition occurs. Repeating this experiment with different volumes of droplet results in a morphological phase diagram illustrated in Fig. 7b. This diagram reveals that droplets with large volume and small contact angle prefer a barrel configuration while those with small volumes and large contact angle prefer a clamshell configuration, with a bistable region at intermediate values, where both morphologies coexist. The stability criterion for such morphologies can be determined by evaluating the free energy of a given set of volume and contact angle. A morphology is found to be stable if the second derivative of the free energy is positive for small arbitrary perturbations of the interface. Such stability analysis also reveals that the transition from barrel to clamshell morphologies is more susceptible to local heterogeneities and small external forces. The clamshell to barrel configuration is unaffected by such perturbations and is physically achievable only when the two opposite sections of the clamshell contact line merge and coalesce.

From the above discussion, it is clear that gravity may influence the barrel to clamshell transition and hence alter the morphological phase diagram significantly. In presence of gravity, drops on a horizontal fibre (i.e., with its axis perpendicular to gravity) exhibit an asymmetric barrel shape and a "downward clamshell" shape, both lying on the bottom face of the fibre (Chou et al. 2011) as shown in Fig. 7c. The morphological phase diagram for such droplets shows a finite bistable regime at low contact angles, as shown in Fig. 7d, and there is no "purely barrel" regime. The presence of gravity hinders the merging of the opposing sections of TCL of clamshell droplets, leading to downward clamshell shape for most contact angles and volumes. For very large volumes, the surface tension is unable to hold the droplet on the fibre surface, marking a "falling off" regime. In fog harvesting, we encounter both horizontal as well as vertical fibre for fabrics or meshes. On a vertical fibre, presence of gravity leads to asymmetry in the vertical direction, which can be approximated as a cone-on-hemisphere geometry (Shi et al. 2020a, b). More importantly, when gravity overcomes surface forces, the droplets slide along the fibre rather than falling off.

For a droplet sliding on a fibre, in addition to surface and gravity forces, viscous effects become important. For barrel droplets, a steady state is achieved shortly after the onset of motion, as viscous forces balance the gravitational forces (Gilet et al. 2010). This steady motion is however short lived, as the droplet starts leaving a thin film of liquid on the fibre. This process is similar to pulling a plate out of a liquid bath, called dip coating and has been extensively studied in the literature (Que're' 1999). The thickness of this film relative to the fibre diameter varies as $Ca^{2/3}$, where $Ca = \mu U_{drop}/\gamma_{lv}$ quantifies the relative magnitudes of viscous and surface tension forces. Considering the loss in volume due to this film as a function of distance z travelled along the wire, combined by momentum conservation of the droplet yields an ordinary differential equation in z, such that the long-term trajectory of the droplet can be solved.

From practical perspectives, in case of real fog harvesting devices on wire meshes, clamshell shapes are more prevalent (except for some special cases like spider web). Clamshell drops on vertical wires assume a tear drop like shape, which can be approximated as an ellipsoid with a wedge, as shown in Fig. 7e Just like the barrel droplets, the clamshell drops also leave a thin film along the fibre, which later breaks into small droplets due to Rayleigh-Plateau instability. This leads to the formation of a wedge, where strong viscous dissipation occurs (see Fig. 7e). Assuming a uniform fog flow over a finite length of fibre, several droplets are formed on the fibre adjacent to each other leading to coalescence. This coalescence has a predominant effect on onset of droplet sliding as well as the subsequent sliding motion (Kowalski and Boreyko 2022). For droplets near critical volume of shedding, coalescence leads to increase in volume as well as release of energy due to reduction of surface area, which overcomes force of gravity, starts sliding and sweeps all the droplets in its

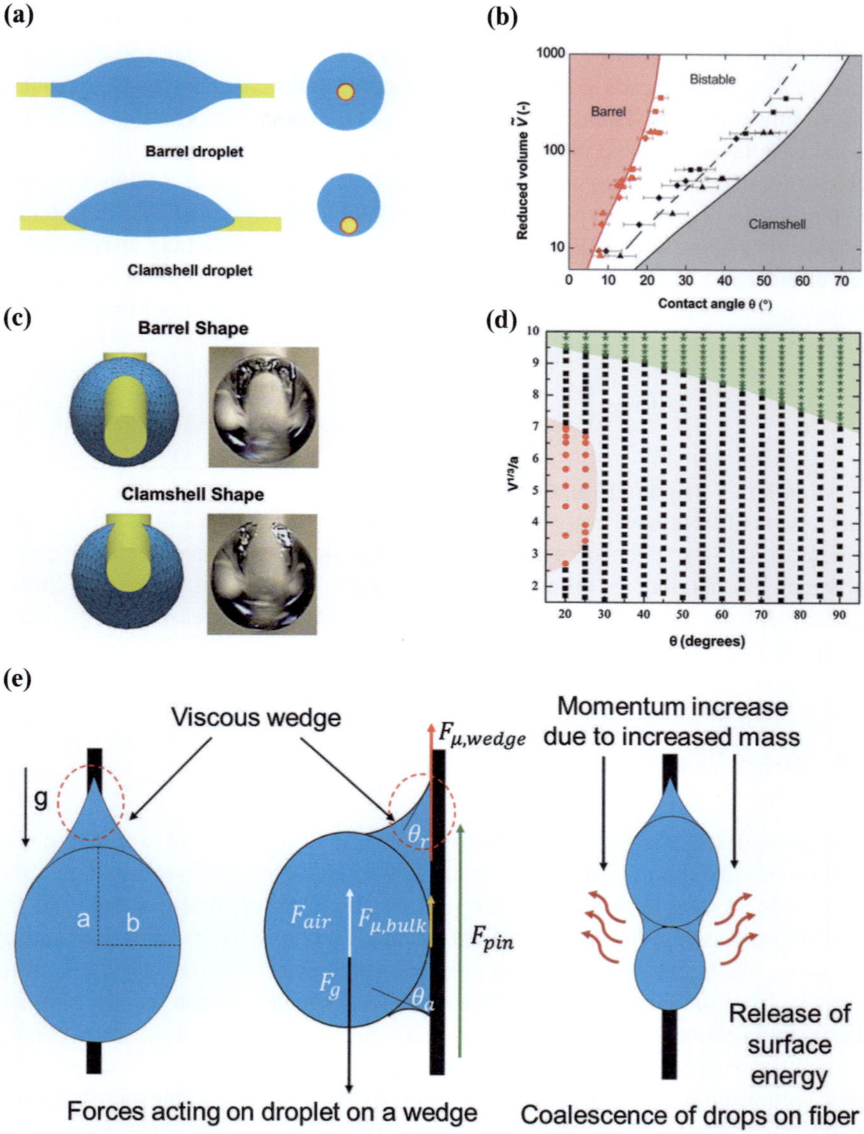

Fig. 7 **a** Barrel and clamshell configuration without gravity effects **b** phase diagram for barrel, clamshell and bistable droplet shapes in absence of gravity (Eral et al. 2011), reproduced, with permission, from Royal Society of Chemistry© **c** asymmetric barrel shape and downward clamshell shapes in presence of gravity **d** morphological phase diagram in presence of gravity (Chou et al. 2011), reproduced, with permission, from American Chemical Society© **e** fog droplet on fibre showing viscous wedge and different forces governing its onset of sliding and subsequent motion; coalescence of droplets leads to increase in momentum and release of surface energy which gets converted into kinetic energy

path, like droplet sweeping observed in case of dropwise condensation. For droplets smaller than critical volume, the surface energy released due to coalescence causes a small sliding, which leads to further coalescence, such that critical volume is reached and sliding continues. Thus, coalescence leads to increase in gravitational force due to increase in volume and also increase in energy due to release of surface energy which is converted into kinetic energy of the droplet. The onset of motion is resisted by pinning force, which is proportional to contact angle hysteresis, air drag on the droplet and viscous forces. Viscous forces in this context include bulk viscous forces due to diffusion of momentum from the fibre surface into the bulk of the droplet, as well as the viscous dissipation at the wedge. Thus, the equation of motion of the droplet can be written as,

$$m\frac{dU_{\text{drop}}}{dt} = F_g - F_{pin} - F_{air} - F_{\mu,bulk} - F_{\mu,wedge} \tag{17}$$

An order of magnitude analysis of all the forces shows that the drag forces and bulk viscous forces have a negligible contribution to the acceleration of the droplet as compared to the other forces. Once a droplet coalesces with another, the increase in velocity is obtained by momentum and energy conservation, where the former accounts for the increased mass and the latter accounts for increased energy due to release of free energy upon coalescence, as illustrated in Fig. 7e.

3.3 Drop Dynamics on Fibre Intersections

With the understanding of droplet motion on single fibres, we now explore the next level of complication in a fog harvesting device, i.e., fibres intersecting at an arbitrary angle. As we have seen for droplets on single fibres, first we look into the stable morphological states of droplets at fibre intersections, followed by their sliding dynamics. We begin by considering droplets which exist as barrels on single fibre, far away from the intersection, i.e., for fibres with smaller contact angles. Droplets at fiber intersection can either remain as quasi-spherical droplets or assume a conical column shape due to the Laplace pressure gradient originating from varying area of cross section along the intersection. Accordingly, three possible stable morphological shapes arise: column on both side of the intersection, drop engulfing the intersection, and column on one side with droplet on the other (Sauret et al. 2014), as shown in Fig. 8a. The effect of increasing the angle of intersection of the fibre (α) on morphological states can be characterised using the length of wetting normalized by the fibre diameter (see Fig. 8a). Experimental investigations reveal that for small angles, column shapes are preferred, followed by the mixed morphology having column on one side and droplet on the other, and finally only droplet morphology dominates at higher angles. This can be explained by calculating the free surface energy of each configuration as a function of angle of intersection and normalized droplet volume. At smaller angles, column and mixed morphologies have lower surface energy, while at higher angles, droplets have lower surface energy. Theoretically, the length of wetting can be predicted for a pure column shape, by considering the energy balance for a small movement dl of the meniscus with an area *A(y)*, where y represents the vertical length along the wire, as shown in Fig. 8a. Here, the work done by the forces acting on the liquid vapor and liquid vapor interfaces are utilized to overcome the Laplace pressure gradient due to variation of curvature of the area of cross section of the column along the length of the fibre. For purely drop morphology, the shape is approximated as a sphere and the deviation from sphere is accounted for by a fitting parameter evaluated from experiments.

A special case of fibre intersection arises when the angle of intersection between the fibres, $\alpha = 0°$, i.e., the fibres are parallel. In such cases, only pure columns with constant area of cross section and droplets are observed to be stable morphological states, the transition between the two governed by the inter fibre distance (Protiere et al. 2013), as shown in Fig. 8b. As in previous cases, a morphological phase diagram can be constructed by experimentally observed morphological state as a function of droplet volume and inter-fibre distance. At smaller inter fibre distances, columns are stable having lower surface energy, while at larger distances, droplet morphology is

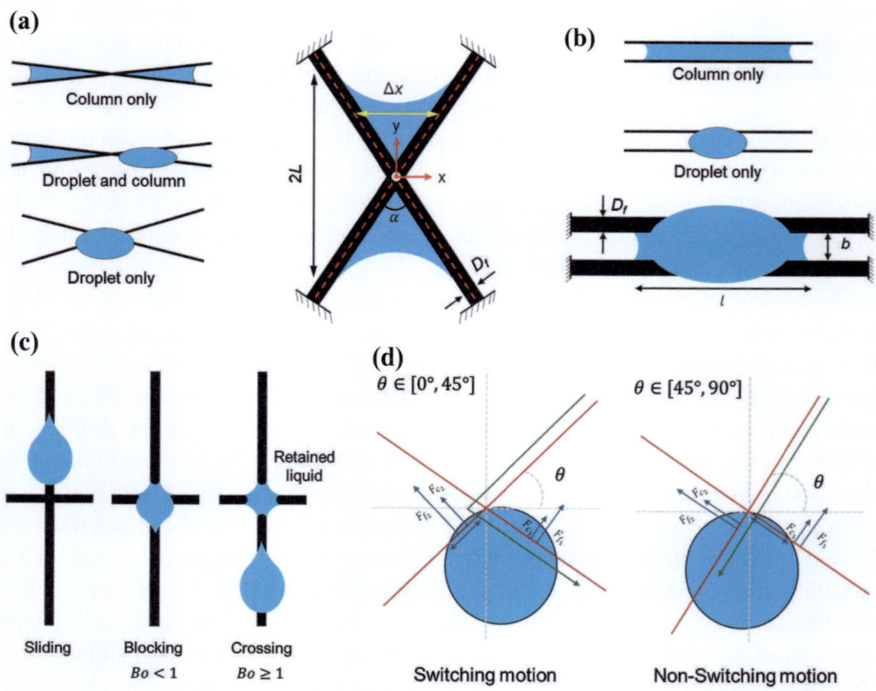

Fig. 8 **a** Stable morphological states of droplets at fibre intersection and schematic representation of purely column state at fibre intersection **b** stable droplet and column morphologies on parallel wire and schematic representation of columns on parallel fibres **c** blocking and crossing of droplet at perpendicular fibre intersection. **d** Forces acting on a droplet at fibre intersection, leading to switching and non-switching droplet behaviour

found to be stable, crossing over at $\frac{b}{a} = 1.4$, where b is the inter-fibre distance and a is the fibre diameter. This value, obtained from theoretical energy calculations, is well matched by experimental results.

We now look at the dynamics of droplets sliding on fibre networks and crossing an intersection. We begin with fibres intersecting at right angle, which is the most common configuration observed in case of woven meshes. The stable morphology at this angle is a droplet morphology as shown in Fig. 8c. A droplet approaching a fibre intersection can either be blocked by the intersection or it may cross the intersection, governed by the relative influence of surface tension and gravity forces, quantified by bond number $Bo = \rho_w \overline{V} g / 2\pi \gamma_{lv} d_f$, where $\overline{V}$ is the droplet volume and d_f is the fibre diameter. For $Bo < 1$ droplets are blocked by the intersection, while for $Bo \geq 1$ droplets are observed to cross the intersection, with a small amount of liquid being retained at the intersection (Gilet et al. 2010). For droplets with large viscosity, there is a significant time delay in crossing the node when the Bo criteria is met, while for small viscosity, no delay is observed. Note that these studies were performed for droplets which were initially in barrel state on single fibres, but the results can be extended to clamshell drops also, by slightly modifying the definition of Bo to $Bo = \lambda_v / \lambda_c$ (Semprebon and Brinkmann 2014), where $\lambda_v = \overline{V}^{1/3}$ corresponds to a length scale based on droplet volume and $\lambda_c = \gamma / \rho g$ is the capillary length scale. Prompted by these experimental observations, Gilet et al. modelled the dynamics of droplets at fibre intersection as a damped harmonic oscillator where gravity appears as the forcing term, viscous forces appear as damping terms and surface tension forces appear as restoring terms of the equation. The damping factor has the form of Ohnesorge number, $Oh = \mu_w / \sqrt{\rho_w \gamma_{lv} g}$. When $Oh > 1$, viscous forces dominate, the system becomes overdamped, exhibiting blocking and crossing with delay, while for $Oh < 1$, surface tension forces dominate, and the system becomes underdamped, resulting in blocking with smooth crossing.

Once a droplet crosses such an intersection, it may either remain on the same fibre, or cross over to the other fibre forming the intersection, depending upon the fibre diameters and angle of inclination of the fibres. This switching behaviour can therefore determine the droplet pathway on a fibre network. This phenomenon was studied extensively using fibres of different diameters, angles of inclination and droplets of different surface tensions (Weyer et al. 2017). Accordingly, fibre intersections can be classified as symmetric nodes or asymmetric nodes formed by fibres with equal and unequal diameters respectively. The droplet switching behaviour is characterized by the maximum volume of liquid held at the intersection before crossing happens. This volume increases as angle of inclination is increased, reaches a maximum at 45° and then decreases symmetrically for symmetric nodes. Droplets on the rising half switch from their initial fibre, while droplets on the falling half remain on the fibre. The forces acting on the droplets is shown in Fig. 8d and theoretical modelling of such forces enable prediction of the maximum volume for given set of fibre diameter, surface tension and angle of inclination. The results suggest that droplets at symmetric nodes select steeper fibres. However, on unsymmetric nodes, droplets switch not to the steeper, but to the thicker fibres. Thus, by selecting fibres of appropriate diameter, droplets can be forced to pick a preferential path along the thicker fibre. Changing the surface tension on symmetrical nodes show similar trends, with an increase or decrease in maximum volume at a given angle depending on whether surface tension is increased or decreased. For fibres of equal diameter intersecting at angles other than at right angles, droplets still select the steeper fibre, just the angle for the maxima achieved is different, and it is always at an angle where both the fibres have equal slopes. This happens to be 45° for intersection of 90°, but changes to 75° for intersection of 30°. We can appreciate that clamshell droplets with smaller contact area with the fibre, facilitated by larger contact angles, are easier to detach from fibres and are favourable for enhanced collection. The droplet interaction with fibre intersection enables us to select weaving or knitting patterns optimized for high fog collection performance. One important effect of coupling between surface tension and aerodynamic force driven phenomena occurs as fog collection proceeds on fibre assemblies. As fog collection proceeds on mesh fibres, droplet laden fibres occlude more area than dry meshes, altering the *SC* at device length scale. Long term experiments performed on nylon harps (Goswami et al. 2023) show that the *SC* grows as droplets start growing on harp fibres, eventually reaching a quasi-steady value of *SC* as droplets start shedding, and increase in global projected area due to droplet growth is balanced by droplet shedding from different locations of the collector. When this quasi-steady value of *SC*, distinguished as SC_{act} from the initial geometrical SC_{geo} is considered for calculation of η_a, the experimental results show a better match with theoretical results. This increase is a function of droplet shapes, which are in turn governed by surface properties of the fibre. In this context, two relevant non-dimensional groups can be defined, namely $SC^* = SC_{\text{act}}/SC_{\text{geo}}$ and $\lambda_v^* = \overline{V}_{cr}^{\frac{1}{3}}/d_f$. SC^* captures the increase in *SC*, while λ_v^* captures the length scale associated to critical volume. The SC^* is high for higher λ_v^*, which corresponds to higher critical volume relative to thinner fibres (see Fig. 9a,b). The regular increase of SC^* with $\lambda_v^* = 1.38,\ 1.94,\ 2.38$ correspond to $d_f = 0.8, 0.5$, and 0.35 mm respectively illustrates this fact. Comparing hydrophobic fibres with uncoated nylon fibres, it can be observed that the λ_v^* corresponding to hydrophobic fibres is 2.08, as opposed to 2.36 of uncoated wires on account smaller CAH resulting in a smaller $\overline{V}_{cr}$ at the same d_f, resulting in smaller SC^*.

These physical insights allow researchers to effectively design fog collection surfaces with optimal combinations of properties in terms of fibre diameter, wettability, surface roughness to prevent clogging and facilitate fast and efficient water collection. Such surface design has been a topic of intense research, and we explore some of such strategies in the later sections.

4 Enhancing Fog Collection by Tuning Mesh Geometry

We know that for aerodynamic efficiency, there is an optimum value of SC, governed by the competition between increasing surface area for droplet collision and increased aerodynamic resistance leading to bypass of fog. One approach to enhance this surface area for collision, without drastically increasing the aerodynamic resistance, is to stack multiple meshes one after another, with some definite gap in between these layers (Azeem et al. 2020). Such an arrangement increases the surface area available for droplet collision, but in a three-dimensional manner, rather than the in-plane increase in surface area typically achieved by increasing the SC as shown in Fig. 10a In such a case, the fog harvesting efficiency increases with addition of successive layers as shown in Fig. 10b. Obviously, as the number of meshes increase, the pressure drop increases, but the probability of droplet collision with one of the fibres increases, as droplets not colliding with the first can collide with the second and third and so on. It can be shown, assuming that the mesh fibres are sufficiently small such that the capture efficiency of each individual stage $\eta_{c,i} \to 1$, and assuming no clogging in the mesh pores such that $\eta_d \to 1$, the total efficiency of fog collection is given by (Azeem et al. 2020)

$$\eta = \tfrac{A_\infty}{A}\left[(1 - (1 - SC)^n\right] \tag{18}$$

where A_∞ and A have the usual definition as in case of aerodynamic efficiency, and n is the number of layers of mesh used. The term outside square braces is called the *filtered fraction* denoted by ϕ, which signifies the amount of fog intercepted by the mesh, while the term within the square braces is called the *incident fraction* denote by χ, which represents the probability of fog droplets striking the mesh fibres. As the number of meshes increases, the amount of fog intercepted decreases, as each mesh contributes to the overall pressure-drop across the collector which, in the limit of the interlayer spacing being much greater than the fibre diameter, becomes additive, i.e., the overall pressure drop of the device is simply the addition of pressure drop across each individual mesh (idel'chk 1986). On the other hand, χ increases rapidly with increase in number of layers before reaching a plateau. Thus initially, the total efficiency, which is a product of χ and ϕ, increases rapidly with increase in number of mesh layers because the increase in χ with n dominates. Later, with further increase in n, the value of ϕ as well as χ saturate, leading to a constant value of efficiency. Thus, beyond a certain number of layer n, there is no increase in collection efficiency. Azeem et al. theoretically predicted this critical value of n to be 10, which was also corroborated by their experimental results presented in Fig. 10c. The theoretical maximum efficiency achievable was also predicted to be 49%, significantly higher than that obtained for single layered mesh, 24% as mentioned earlier.

Next, we look at several approaches used in the literature to increase the capture efficiency. Shahrokhian et al. (2020) demonstrated that millimetric structures like spherical cap shaped dimples, spherical bumps and grooves on spherical surfaces increase their capture efficiency by more than 300% as compared to a smooth sphere.

Although a solid spherical surface is not ideal for fog harvesting and has a very poor efficiency as compared to meshes, the authors studied these structures to simulate fog collection by Namib beetles, who are well known for their 'fog basking' behaviour, through which they supply their daily requirement of water. Nevertheless, it was shown that these millimetric structures enhanced the capture efficiency by manipulating flow around them. A comparison of capture efficiency of different shaped millimetric structures revealed that spherical bumps performed best among structures at millimetric scale, giving a deposition efficiency of 0.35% as compared to 0.05% observed on a smooth sphere. While a specific mechanism was not proposed, the authors experimentally showed that this was not merely due to the increase in surface area, neither was it due to boundary layer transition delaying separation, a common phenomenon observed for all the different shapes tested. Apart from the shape of the structures, diameter of the millimetric bump (D_{bump}) and the distance between them (S) are important parameters governing the capture efficiency. The capture efficiency increased almost linearly with increase in D_{bump}/S till the value of 0.6 after which it decreased. Initially, as the diameter-to-separation ratio increases, there is an increase in droplet capture due to larger number of bumps per unit area. However, when such bumps are close to each other, the boundary layers of adjacent bumps interfere, leading to reduction in collection. Following this, another approach was adopted which combined the advantage of bumped structures with the advantage provided by a mesh rather than a solid surface. To achieve this, fibres with regularly spaced spherical bumps were fabricated by microfluidic process (Zhu et al. 2021) and woven together to form a mesh with optimal shade coefficient ~ 0.5. Previously, we have discussed the capture efficiency of a circular cylindrical fibre and its dependence on the St. similar relations have also been found for spherical structures (Langmuir and Blodgett 1946) given by,

$$\eta_c = \frac{f(st)}{1+f(st)} \tag{19}$$

where

$$f(St) = 0.82\left(St - \frac{1}{12}\right)^{1.04} \tag{20}$$

the capture efficiency of such spherical structure is greater than that of a circular cylinder for a given value of St. Experimentally, it was found that a circular cylindrical fibre with spherical bumps at regular intervals has a capture efficiency lying between a capture efficiency values obtained for a smooth cylindrical fibre and a sphere for the same St. Note that the increase in efficiency in this case is not as large as that for smooth sphere and sphere with bumps, which suggests that the two mechanisms of droplet capture are not quite similar, and further investigations are required to understand the latter.

A passive method of enhancing fog capture efficiency utilizing centrifugal forces was proposed by Li et al. (2021) simply by using folded structures placed in the path of fog flow. These folded structures are made of aluminium coated polyethylene terephthalate (PET) sheets as shown in Fig. 11a. These folds cause the incoming fog flow to form stationary vortices which undergo axial stretching. Fog particles rotating in these vortices are acted upon by two forces, drag force due to the rotating airflow in the vortex and centrifugal force due to the rotational motion acting against each other. Thus, there exists a critical size of droplets (and hence a critical stokes number, $St_{critical}$ for which these forces balance each other, given by (Marcu et al. 1995)

$$d_{p,cr} = \frac{12\pi \mu_{air}\delta}{\Gamma_0\sqrt{\frac{\rho_{air}\rho_w}{2}}} \tag{21}$$

where δ is the size of the vortex, and Γ_0 is the strength of the vortex. These two parameters depend on the angle of the folds. Depending on the size of incoming droplets, one of the two following outcomes are possible: for a large droplet with $St > St_{critical}$, the droplets are ejected out of the vortices, strike the sides of the fold, and get collected. For $St < St_{critical}$, the droplets rotate in the vortex, accumulate towards its centre, and grow in size by collision leading to coalescence. As they grow in size, the centrifugal force grows much rapidly than the drag force, as the former is proportional to the cube of droplet diameter, while the latter is a linear

function of the droplet diameter. Once the centrifugal force is large enough to overcome the drag force, the droplets escape the vortex and impinge on the fold surface, thus getting collected. Thus, this approach is much less dependent on wind velocity as compared to collection on flat meshes. The angle of the fold controls the two flow parameters δ and Γ_0, and has a significant effect on collection giving a maximum at 120° and lower collection rates at 90° and 150°. This technique yields an overall efficiency of ~ 17–21% which is among the highest reported fog harvesting efficiencies in existing literature.

As discussed in Sect. 1, coalescence of droplets on adjacent fibres clogs the mesh pores, reducing aerodynamic efficiency, which necessitates the use of surface engineering techniques. However, such surface coatings require sophisticated methods of fabrication and are expensive to implement, limiting scalability. A simple yet novel design was proposed by Shi et al. (2018), which used vertical wire arrays called harps, which perform better than woven/knitted meshes due to absence of cross fibres impeding the droplet motion due to gravity. In such meshes, even if coalescence occurs, liquid bridges are formed as discussed in Sect. 3.3, and these bridges slide off under self-weight (see Fig. 11b). Fog harps with optimal SC of 0.5 obtain an efficiency of ~ 16%, which is an order of magnitude higher than meshes having similar geometrical parameters (Shi et al. 2020a, b; Kowalski et al. 2021). Surface treatment of harps with hydrophilic materials shows only slight increase in efficiency, thus confirming the redundance of such approach for this configuration. Harps however suffer from one limitation, where the coalescence of droplets on adjacent meshes may lead to their sticking or tangling (Shi et al. 2020a, b) leading to opening of large gaps between fibres which in turn reduces aerodynamic efficiency, as shown in Fig. 11c. This is an electrocapillary phenomenon, and depends on the length of harps, which puts a restriction on use of very large harp-like structures.

Apart from geometric parameters, fog capture efficiency can also be enhanced using active techniques, like using an electric field. Following the principle of electrostatic precipitators, a corona discharge is used to charge fog particles, which are then collected by electrode wires with opposite charge. This technique can therefore alter the streamlines of droplets around a fibre, as shown in Fig. 12, enhancing the deposition efficiency beyond 1 (Damak and Varanasi, 2018). The trajectory of droplets is now governed by the ratio of electric force to drag force quantified by electrical number, $Ke = \frac{d_p \in_0 V_e^2}{\mu_{air} L R_c U_\infty}$, where d_p is the diameter of droplet, $\in_0$ is the permeability of free space and V_e is the applied voltage, L is the length of collector, R_c is the radius of collecting wire. This technique when applied to meshes yielded a performance efficiency of 36%, much higher than passive techniques, with an energy requirement of 2 kWh/m^3 of water Zeng et al. (2023a), developed a wire to plate type electrostatic fog collector where wires acted as corona discharge and plates were collectors which gave 60% efficiency at a 0.8 kWh/m^3. They also optimized the geometry of such collectors for maximizing performance using numerical simulations (Zeng et al. 2023b). However, these methods require very high voltages ~ 10 kV, which is difficult to be supplied in resource constrained settings. Further, high wind speeds significantly decrease the efficiency of electrostatic fog collection due to low residence time of the fog droplets near the corona.

5 Surface Engineering for Enhancement of Fog Collection Efficiency

In the previous sections, we have seen that clogging and entrainment losses lead to serious reduction in fog harvesting efficiency and are mitigated primarily by tuning the surface properties. Several authors have also suggested that novel materials and surface design also accelerate droplet motion. From perspective of fog harvesting applications, small contact angle hysteresis and small receding contact angles are of primary interest, as discussed in Sect. 3.1. Typically, surfaces having very high, or very low intrinsic contact angles exhibit such small contact angle hysteresis. Such substrates are termed as superhydrophobic ($\theta > 150°$) and superhydrophilic ($\theta < 10°$) surfaces respectively. However, simply coating surfaces with low and high surface energy chemicals usually do not achieve such extreme contact angles. We know that surface morphology plays an important role in determining the contact angle of a droplet on a substrate. In fact, droplets exist in two different wetting states on textured surfaces, depending whether the liquid penetrates the gaps between asperities on the substrate. When the liquid penetrates the gaps between asperities, it is said to be in Wenzel state. In such a state, the triple contact line follows the contour of all the roughness elements, and the gaps between them, thus changing the apparent contact angle according to the following law,

$$\cos\theta_w = \in \cos\theta \tag{22}$$

where ε quantifies the roughness, i.e. the ratio of the actual distance on the surface to the shortest distance. Thus, a rough surface makes a hydrophilic surface more hydrophilic and a hydrophobic surface more hydrophobic. On the other hand, if a droplet on rough or textured surface sits on the surface asperities such that air is trapped in the gaps between the roughness elements, it is said to be in Cassie Baxter state. The apparent contact angle in this state is given by

$$\cos\theta_{cb} = f\cos\theta - (1 - f) \tag{23}$$

where f is the fraction of the total contact area occupied by the solid–liquid interface. It can be observed that decreasing f increases contact angle θ_{cb} thus increasing water repellence. A droplet in Cassie Baxter state has much lower contact area and thus experiences less surface force at its contact line, enabling easy removal. Thus, using surface textures along with high or low energy coating can result in development of superhydrophilic or superhydrophobic

surfaces. Showket et al. (2024) studied chemically etched metal meshes with superhydrophobic and superhydrophilic coatings in the context of water recovery in cooling towers. They used fluoropolymers and titanium dioxide to achieve superhydrophobicity and superhydrophilicity respectively. Their results indicated that textured superhydrophilic metal meshes performed best, followed by textured superhydrophobic meshes, while uncoated smooth meshes exhibited lowest fog collection rate. This is because in the presence of spherical droplets on superhydrophobic meshes, the effective SC and St increases, leading to lower collection rates, as explained earlier (see Fig. 9). Several such surface engineering approaches have been tried for fog harvesting surfaces, which can be categorized as shown in Fig. 13. Often, multiple approaches are combined to further enhance the performance. In the following sections, we discuss several devices fabricated utilizing such approaches, their advantages and limitations.

5.1 Wettability Gradient Driven Fog Collection

A droplet, when placed on a surface with a wettability gradient, tends to move from hydrophobic regions to hydrophilic regions due to unbalanced forces acting on its triple contact line (Chaudhury and Whiteside 1992). This principle has been observed often in nature, most notably in case of Namib beetle backs, which consists of hydrophilic patches on hydrophobic surface. A comprehensive review on Namib beetle inspired fog harvesting surfaces can be found in (Lei and Guo 2020). Several works on surfaces with hybrid wettability, with bumps of different shape, and spacing were reviewed. It was concluded that hybrid wettability surfaces performed better than purely hydrophobic, hydrophilic, superhydrophobic or super hydrophilic surfaces, and that rolling off angle played a significant role in determining the water collection rate than the contact angle itself. It was found that composite surfaces performed best followed by star shaped, bulged, circular dimpled and flat surface

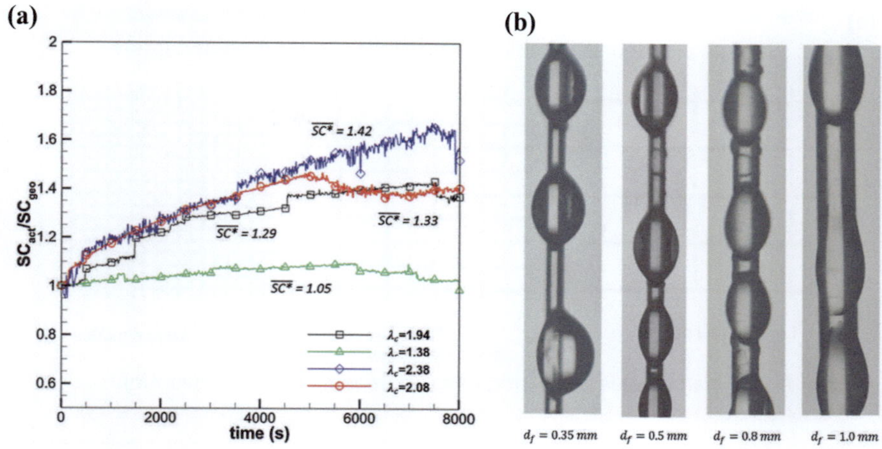

Fig. 9 **a** Effect of fibre geometry and surface properties on increase in SC* **b** droplet shapes on fibres of increasing diameter (Goswami et al. 2023), reproduced, with permission, from American Chemical Society©

respectively. Enhancement of fog collection using heterogeneous wettability has been explored in the textile community as well, for example by making fabrics composed of alternate superhydrophobic (SHB) and superhydrophilic (SHL) fibre woven together (Yu et al. 2021). Intrinsically superhydrophilic yarns were woven with nanoparticle coated superhydrophobic yarns to achieve hybrid structures with alternating wettability. This hybrid surface exhibited better fog collection efficiency as compared to purely SHB and SHL surfaces, due to directional droplet transport on account of wettability driven droplet motion. Surfaces consisting of equal fractions of SHL and SHB coating were found to perform better than surfaces with higher SHL and lower SHB fractions, and vice versa. A similar attempt was made using electrospun hierarchical fabrics (Knapczyk-Korczak et al. 2020). Composite fabric consisting of hydrophobic Polystyrene (PS) microfibers and hydrophilic polyamide (PA6) nanofibers were made by electrospinning. Again, it was found that the water harvesting performance was best when the hydrophilic fraction was 0.5. This increase was explained on the basis of reducing contact angle hysteresis, i.e., with increasing hydrophilic fraction the contact angle hysteresis on the fabric decreases.

5.2 *Fog Collection on Janus Membranes*

As opposed to the Namib beetle like heterogenous wettability, which consists of hydrophilic and hydrophobic patches in the same plane, membranes having a front-to-back wettability gradient can also be used for fog harvesting. In this context, "front" refers to the upstream side of the collector, while the downstream face is called "back" side. The basic idea in such a case is to decouple the surfaces used for

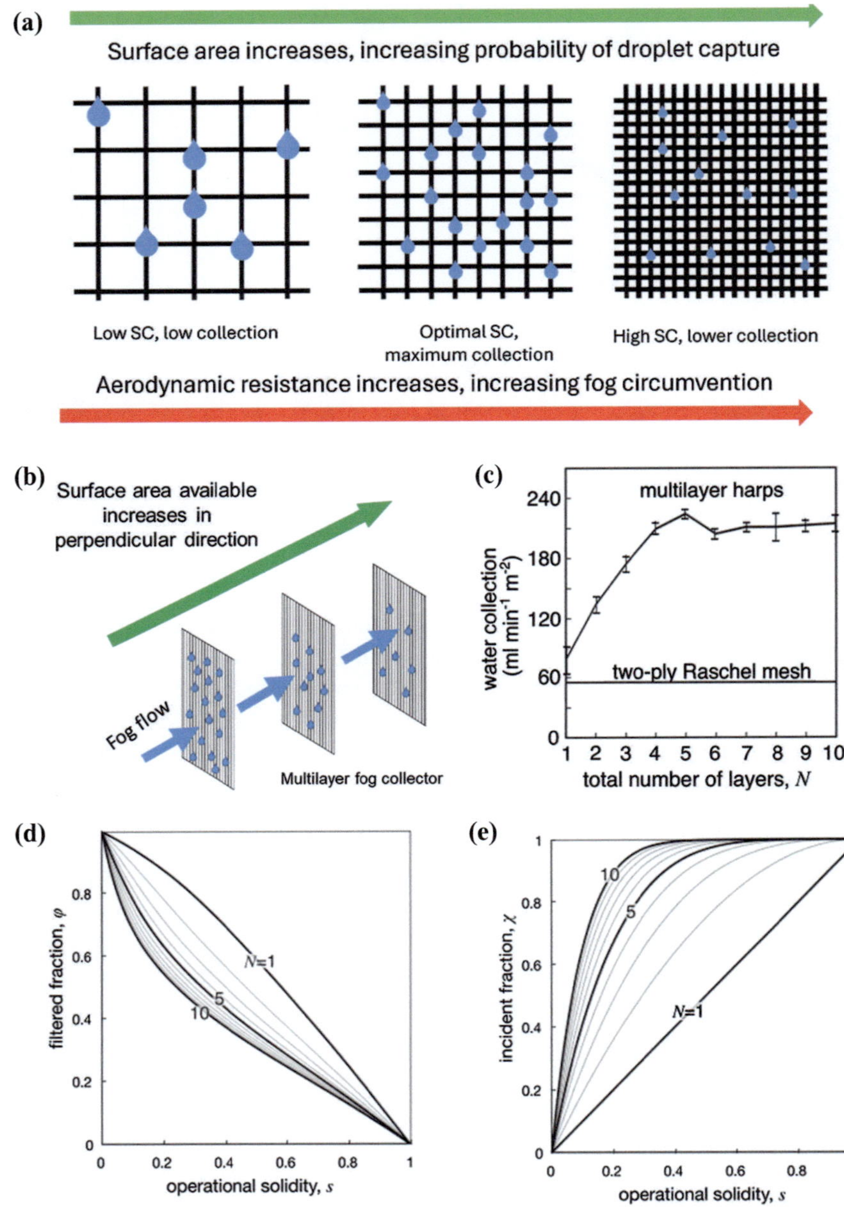

Fig. 10 **a** Schematic representation of in-plane increase in surface area and aerodynamic resistance with increase in SC **b** increase in available fog collection fibres for droplet collision perpendicular to the plane of each mesh **c** experimental data for water collection of multilayer vertical wire arrays or harps **d** filtered fraction plotted as a function of SC for different number of layers, n **e** plot showing variation of incident fraction with increasing SC for different number of layers, n (Azeem et al. 2020), reproduced, with permission, from American Chemical Society©

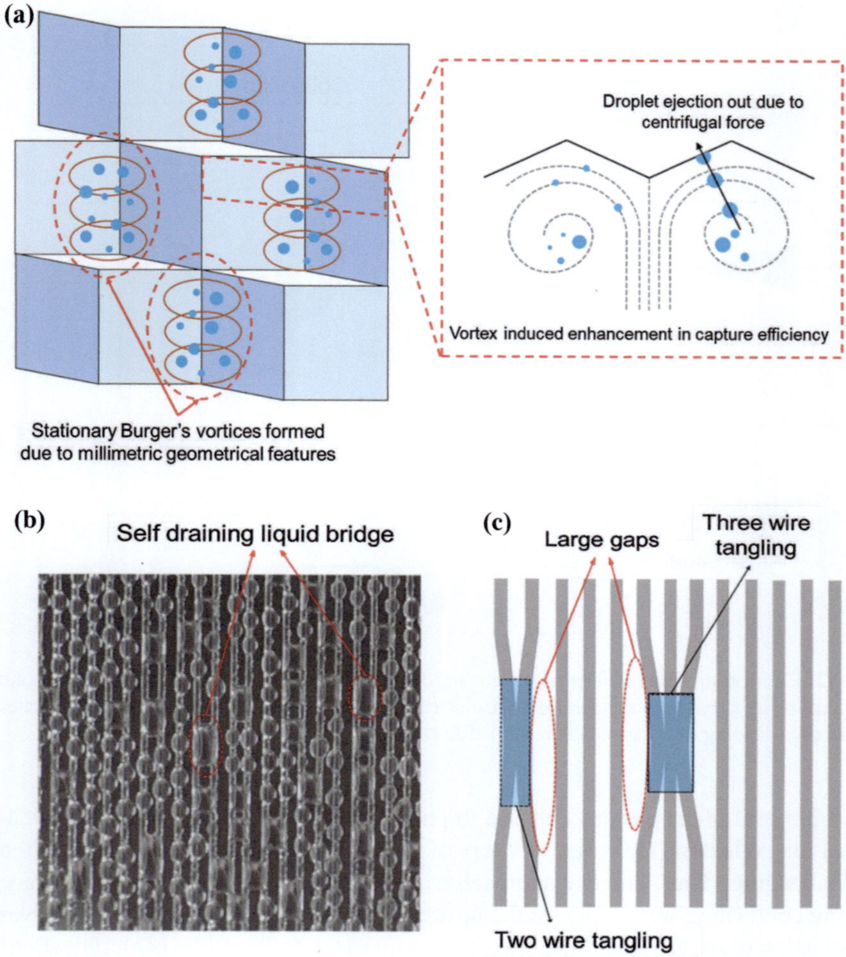

Fig. 11 **a** Millimetric folded structure of aluminium sheet promotes formation of stationary counter rotating vortices (see inset). The droplets rotate due to drag force, collide among themselves, grow in size and eject out of the vortex due to centrifugal forces. **b** Fog collection on vertical fibre arrays called fog harp **c** elastocapillary tangling of fog harp fibres due to coalescence of droplets on adjacent fibres, opening large gaps and reducing fog collection

droplet capture and droplet drainage. Such membranes, called Janus membranes can be easily fabricated using two different layers with contrasting wettability and binding them together. One simple Janus membrane can be made with silanized hydrophobic copper mesh bounded to commercial cotton absorbent (Cao et al. 2015). On such a membrane, the droplets grow on the hydrophobic copper surface, and as soon as they come in contact with the cotton absorbent, they are absorbed readily, regenerating the collection surface. It was found that these membranes gave better collection than the individual hydrophobic mesh or the cotton, owing to the wettability gradient driven

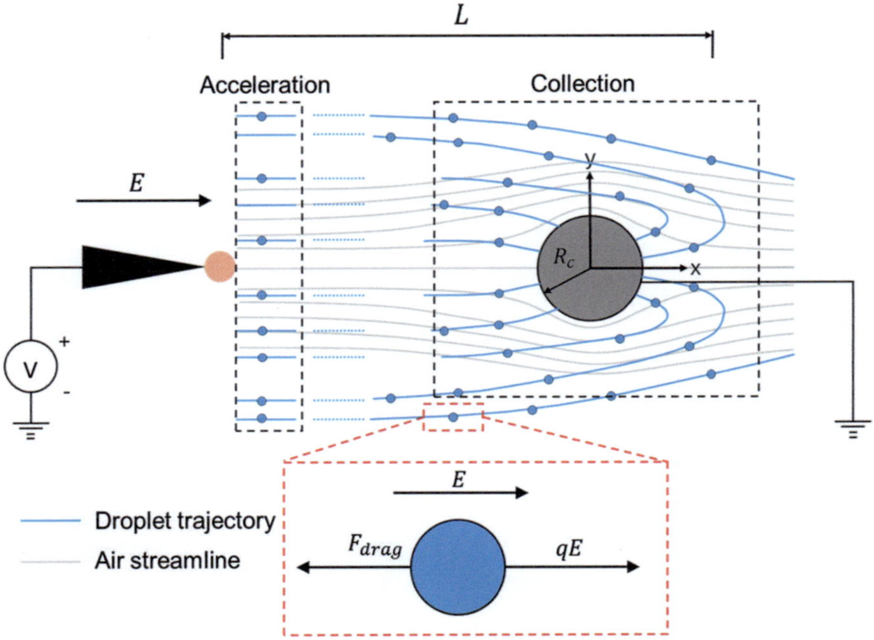

Fig. 12 Electrostatic fog harvesting, where fog droplets are charged via corona discharge, and collected using a charged collector. *Inset* the electric field overcomes the drag force and accelerates the droplets leading to enhanced deposition efficiency

rapid removal of droplets as opposed to purely gravity driven shedding of conventional fog collectors. However, recovery of water after being absorbed on the cotton is challenging. Thus, another approach explored preparation of single membrane having contrasting wettability on the upstream and downstream side. This involves silanization of the front surface of an aluminum sheet, followed by boehmitization[1] of the back, leading to SHB front and SHL back (Ren et al. 2017). In addition, the pores were made conical by laser machining, which promoted faster droplet removal due to Laplace pressure gradient. Again, the Janus membrane collected maximum amount of water, followed by purely SHL and purely SHB aluminum membranes respectively.

The coupling of Janus effect with aerodynamics of conventional fog collecting meshes is important from the perspective of system level design of Janus membrane incorporated fog collectors. To delineate such a coupled problem, systematic study by varying the hole size of the mesh and therefore the shade coefficient is important (Lee et al. 2022). Janus treatment of different combinations i.e., superhydrophobic-hydrophilic (SHB-HL), hydrophilic-superhydrophilic (HL-SHL),

[1] Boehmitization is a process where aluminum reacts with water heated to a temperature ~70° C leading to formation of $Al_2O_3.\gamma\ H_2O$ on its surface, rendering it superhydrophilic.

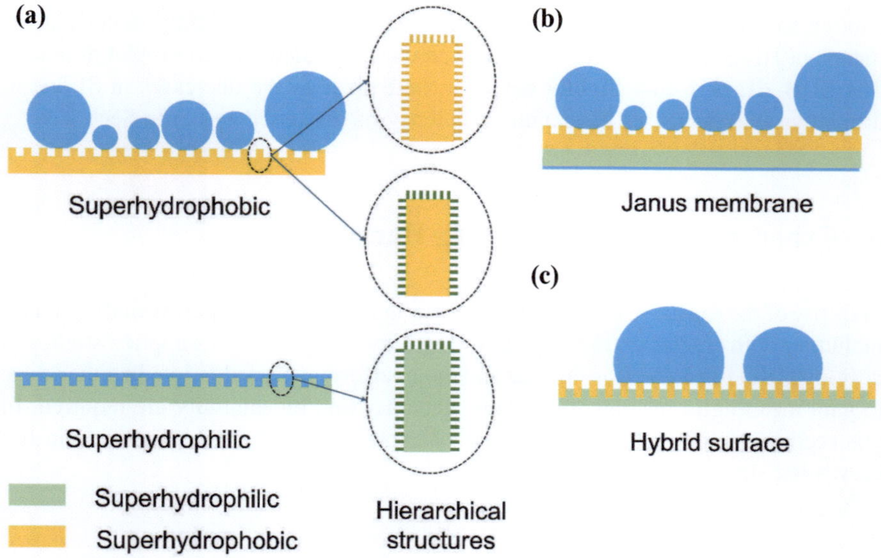

Fig. 13 Different approaches for tuning surface properties to prevent clogging and enhance droplet transport, **a** Superhydrophobic and superhydrophilic surfaces with or without hierarchical structures **b** Janus surface with a front-to-back wettability gradient **c** hybrid surface with wettability gradient in the same plane

hydrophilic-superhydrophobic (HL-SHB) and hydrophilic-superhydrophilic (HL-SHL) were applied. Meshes used were classified as clogged mesh ($p < \lambda_c$), unclogged mesh ($p > \lambda_c$) and intermediate ($p \cong \lambda_c$), depending on the mesh opening size p and capillary length scale λ_c. For each size regime, four kinds of Janus treatments were tested as described above. The fog collection results show that for very small pores ($p < \lambda_c$), SHB-HL performed better than other combinations because of the high front to back wettability gradient, and although it is unclear whether such gradients are strong enough to detach liquid bridges formed due to clogging, such gradients are expected to at least accelerate the transport of captured water. For intermediate pores sizes, i.e., $p \approx \lambda_c$, the performance of SHB-HL and uncoated meshes are found to be worst while HL-SHL and SHL-HL are found to perform well. This is due to absence of clogging on SHL-HL on which water forms a microfilm on the front surface with enhanced drainage enhances the aerodynamic efficiency considerably while uncoated and SHB-HL have droplets forming instead of films which may eventually lead to lateral coalescence and clogging. The HL-SHL surface combines the advantages of relatively less clogging and wettability gradient to give a better performance than the uncoated mesh. Thus, at the intermediate sizes, the lack of clogging due to superhydrophilicity is a more dominating factor for better performance as compared to the wettability gradient-based water transport. In case of large sizes ($p > \lambda_c$), HL-SHL performs slightly better than other combinations, although this increase is much less than at smaller sizes, since the pore diameters are now large enough to prevent clogging in uncoated state as well. The slightly better performance of HL-SHL can be attributed to wettability gradient driven droplet drainage. The SHB-HL performs slightly worse at these sizes as the decreases in St due to droplet growth decreases the capture efficiency, as discussed earlier in Sect. 2.3.

6 Experimental Methods in Fog Harvesting

Majority of the fog harvesting studies in existing literature are experimental in nature, including both studies on the physics of fog harvesting as well as applied studies to develop novel fog collection devices and their performance evaluation. In either case, to gain meaningful insights, reasonable accuracy and repeatability are required. In this section, we discuss the importance of experimental methods employed in fog harvesting studies.

6.1 Boundary Conditions in Fog Harvesting Experiments

The pioneering studies of fog harvesting, using the Standard Fog Collectors (SFC) were performed by Schemenauer et al. (1988), in mountainous coasts of Chile in the 1980s. Unlike modern investigations discussed in the previous sections, the objectives of these studies were to maximize water yield by placing standard two-ply raschel meshes in regions with high water content fog, rather than to enhance the efficiency of the meshes themselves. As such, these were done in the open environment, where the fog collectors were mounted sufficiently high to avoid boundary layer effects of the ground. The governing parameters important for fog harvesting, i.e., the fog flow velocity, and liquid water content (LWC), were obtained from local meteorological stations. This approach suffers from two major drawbacks, namely, the highly fluctuating nature of weather conditions, and lack of repeatability, which made controlled experiments difficult. Particularly, some geographic locations experience fog only for two to three winter months per year, and even such months typically have substantial number of no-fog days, i.e., when the water content is not sufficient for measurable water collection experiments (Shi et al. 2018). Thus, under such circumstances, it becomes very difficult to perform long term experiments to evaluate the performance of novel fog harvesting devices. The need for performing year-round, controlled, repeatable fog harvesting experiments thus led to development of several lab-scale experimental facilities.

Essentially, the problem of designing controlled fog harvesting experiments boils down to maintaining four boundary conditions: (i) controlled flow conditions (typically turbulent) with uniform velocity inlet boundary conditions, (ii) controlled, easily measurable LWC and (iii) infinite flow conditions. In fluid dynamics experiments, the wind tunnel is an apparatus often used to achieve controlled, infinite flow conditions (Bradshaw and Mehta, 1979). The wind tunnel consists of a contraction cone, a test section where the mesh may be mounted and a diffuser. An air flow is created by a fan, or a blower and artificial fog is mixed with it, and the design of the wind tunnel, in combination with honeycombs and screen meshes, filters out unwanted fluctuations generating a steady uniform inlet flow. To maintain infinite flow conditions, the test section should be designed to be much larger than the characteristic length of the sample, such that the wall effects do not interfere with the aerodynamic interactions of the fog with the sample. In fog harvesting, many researchers have developed such wind tunnel like systems (Park et al. 2013; Shahrokhian et al. 2020; Azeem et al. 2020) to obtain uniform fog flow. However, this method may introduce errors into the system, as dry air mixing with fog may lead to evaporation of some fog due to moisture gradient between the two mixing streams. This can be mitigated by placing the wind tunnel and fog generating system in an environmental chamber with controlled humidity and temperature (Jiang et al. 2019), where two humidifiers A and B are used to create fog flow and maintain humidity respectively. At first, humidifier B is turned on till the humidity reaches 100% at given temperature, then, humidifier A is started to add fog droplets into the humid air stream. This ensures that no fog is lost in maintaining humidity, leading to accurate measurement of efficiencies.

Next, we look at different ways of artificial fog generation. The most common method of artificial fog generation is by atomization. We know that fluid forced to flow through small orifices breaks up into very small droplets. Specially designed nozzles are used to create water droplets of micron size, which can then be suspended in the air flow (Shi et al. 2020a, b). However, such systems typically require a very high pumping power leading to bulky and costly design. A cheaper alternative is ultrasonic atomization, as shown in Fig. 14a, b. This method consists of placing a piezoelectric actuator submerged in a water bath. This actuator consists of a piezoelectric membrane which vibrates at ultrasonic frequencies in response to voltage input. These vibrations are transmitted into the water bath, leading to formation of surface waves in the water. These surface waves create an oscillating jet on the surface of water as shown in Fig. 14b. This oscillating jet experiences strong shear forces from the surrounding air, leading to droplets detaching from it. These detached droplets undergo further break up, to form very fine particles, represented by the light blue region around the jet. These particles are then delivered by a fan into the main air flow, leading to mixing and formation of fog. Due to low cost and easy operation, ultrasonic atomization is the most popular method of generating artificial fog in the fog harvesting literature. One drawback suffered by atomization-based fog generators is the droplet distribution of the fog produced is often unrealistic. As we have seen in Sect. 2.3, droplet size strongly affects the efficiency of fog capture. Secondly, the fog produced by atomization has very high-water content (~10 g/m^3, as opposed to natural fog having an LWC of ~ 0.1–1 g/m^3).

One method of creating artificial fog closely mimicking the natural one is by utilizing psychrometric principles of mixing and condensation. In this method, a hot, moist air stream is mixed with a cold dry air stream leading to heat and mass exchange between them. If the two streams are saturated at their respective temperatures, the

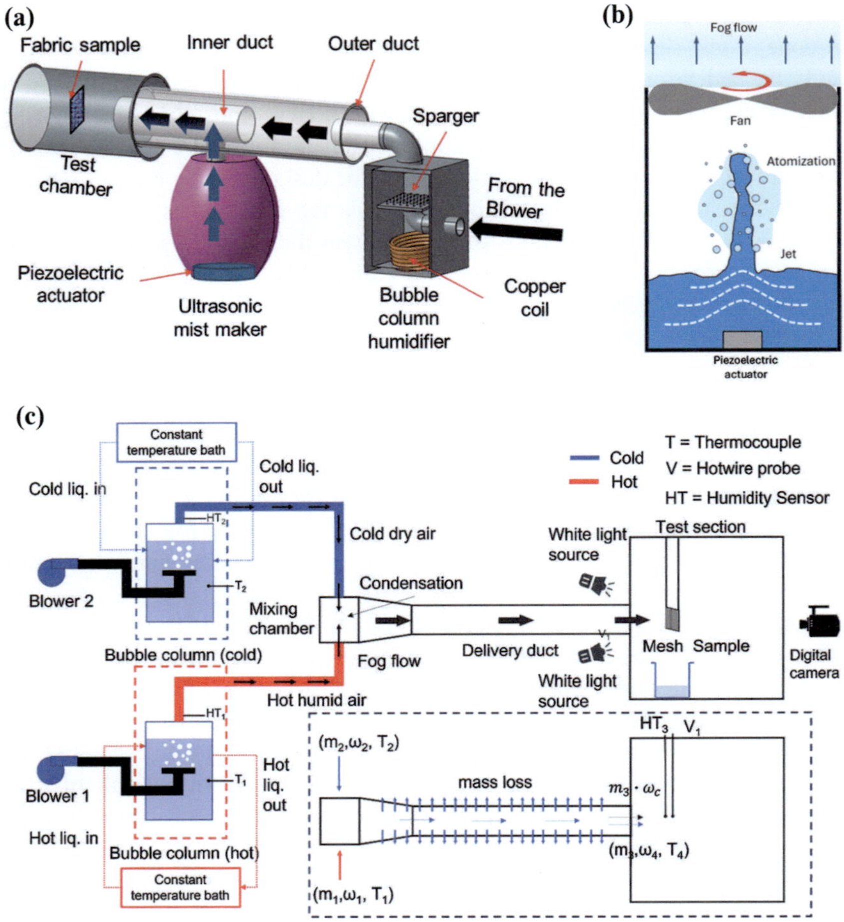

Fig. 14 **a** Ultrasonic atomizer-based fog harvesting setup. The bubble column at the inlet ensures only humid air is drawn such that no fog is lost in evaporation. **b** Details of working of an ultrasonic atomizer. **c** A condensation-based fog harvesting facility

resulting mixture becomes supersaturated, and condensation occurs until a saturated state is achieved at the temperature of the resulting mixture. This condensation occurs with the help of particulates present in the airstreams acting as condensation nuclei and form fog. This process, mimicking natural fog formation, is often used by climatologists for studying cloud formation and their dynamics. An experimental facility utilizing condensation-based fog generation and collection is shown in Fig. 14c, similar to that developed by Goswami et al. (2023). It consists of a bubble column humidification-dehumidification system, which creates saturated hot and cold air streams. The hot air stream, by virtue of its high temperature has a high specific humidity while the cold stream has a much lower specific humidity even though both are saturated. The two air streams mix, leading to fog formation, which is then delivered to the fog collection chamber. In the fog collection chamber, the fog comes in with uniform free stream velocity profile and is intercepted by the suspended sample. Fog is collected in a collector placed appropriately to prevent water loss by splashing, which is placed on a high precision weighing balance to estimate fog collection. The fog collection chamber has optical access for flow and drop dynamics visualization.

6.2 *Measurement of Fog Harvesting Efficiencies*

The multiscale coupled nature of fog harvesting phenomenon makes it challenging to deduce the role of design parameters in governing the fog harvesting efficiency. Thus, while the measurement of overall efficiency is quite straightforward, the three efficiencies must be decoupled, and separate experimental approaches are required to study each stage of efficiency. As we have seen in Sect. 2.1, the aerodynamic efficiency, η_a can be estimated with the knowledge of either the velocity just upstream of the mesh, u or the area of cross section of the flow which spreads out to cover the mesh surface area due to the aerodynamic resistance of the mesh. The measurement of u can be done by hotwire anemometry (Marchand et al. 2024) at different locations of the mesh in spanwise direction followed by averaging it over a line just upstream of the mesh. However, intrusive measurements like HWA may lead to errors due flow disturbance upstream of the sample itself. Secondly it becomes challenging to measure the freestream velocity, U_∞ and u simultaneously, as intrusive measurement of the former will inevitably disturb the latter. A suitable alternative is to perform particle image velocimetry (PIV), which is non-intrusive and allows simultaneous measurement of both u. and U_∞ (Azeem et al. 2020). PIV is particularly suitable for fog harvesting experiments because the particles of fog themselves act as passive tracer particles necessary for such measurements, and additional seeding arrangements are not required. Several authors have also used PIV to study salient flow structures formed by novel fog harvesting devices in addition to simply measuring the aerodynamic efficiency (Azeem et al. 2020; Li et al. 2021).

To measure capture efficiency, η_c, the system has to be simplified from a fibre assembly to a single fibre. This reduces the complex problem of flow past a porous bluff body into flow past a cylinder, which is typically laminar because of the smaller length scale of the fibre. To isolate the draining efficiency, superhydrophilic coating maybe added such that majority of the water captured is collected (Jiang et al. 2019). The method is similar to icing studies performed in ice wind tunnels, where a cylinder is placed in ice flow, and the thickness of ice layer is measured as a function of angular distance from the stagnation point (Mackonnen et al. 2018). Finally, draining efficiency can be measured by weighing the sample before and after the experiment, and adding the difference to the collected water. This then gives the total water captured. The ratio of collected water to residual captured water on mesh surface gives the draining efficiency (Shi et al. 2020a, b; Knapczyk-Korczak et al. 2020).

6.3 Visualisation of Drop Dynamics

As seen in the previous sections, droplet dynamics, i.e., droplet growth, coalescence and droplet motion play a significant role in determining fog harvesting efficiency. Thus, development of experimental techniques for quantification of drop dynamics on fog harvesting fibres becomes necessary. Experimental techniques in drop dynamics are primarily based on direct visualisation of droplets, followed by post processing of acquired image sequences to track droplet trajectories and evaluate fluid mechanical parameters like velocity and strain rate. Temporal resolution of such dynamics, especially for coalescence, requires high speed visualisation to resolve the millisecond scale events like bridge formation, coalescence induced oscillation etc. (Khandekar and Muralidhar 2020).

As we have discussed earlier, contact angle and its hysteresis is crucial to droplet shedding. The general methodology to evaluate static contact angles involves deposition of droplets on surfaces, followed by evaluation of slope of the droplet profile at the extremities of the triple contact line (TCL). This is followed by a sequence of image processing operations (see Fig. 15a): first the droplet is distinguished from its background by thresholding, followed by identification of the droplet profile using edge detection algorithms. Once the droplet profile is identified, a suitable curve fitting operation is performed to find the equation of the droplet profile in two dimensions. Typically, equations of circle, ellipse and Young–Laplace equations are used for such fitting operations. Finally, the value of slope of the curve at its point of intersection with the TCL gives the tangent of contact angle. Experimentally, static contact angles on either side of a droplets are not equal, and an average value of the two is considered as the static contact angle. Measurement of hysteresis can be done in two ways. The first is by continuously increasing and decreasing the volume of a deposited droplet and recording the contact angle just before the TCL starts moving using the above-mentioned image processing methodology. The image frame just before onset of TCL motion gives the advancing contact angle when the droplet volume increases, while in case of decreasing droplet volume, the receding contact angle maybe evaluated. The second method of evaluating the advancing and receding contact angle is given by sloping the surface in incremental steps and measuring the advancing and receding contact angle from the image acquired just before the droplet rolls of. When a droplet is in motion, the advancing and receding contact angles are different from those observed just before onset of droplet motion. These are called dynamic and static contact angles respectively. The measurement of dynamic contact angles becomes challenging when the droplet is continuously deforming. In such cases, it is difficult to predict the entire droplet profile. Thus, in such cases, post edge detection, sufficiently large number of pixel locations are taken near the TCL, and a polynomial fitting is performed (see Fig. 15a, *inset*), and the slope of polynomial at its point of intersection with the TCL gives the instantaneous advancing and receding contact angle (Jaiswal and Khandekar 2021). The velocity of droplet motion is given by locating the centre of mass of the droplet and tracking its location in subsequent frames (Gilet et al. 2010). Somwanshi et al. (2020), performed detailed study of the

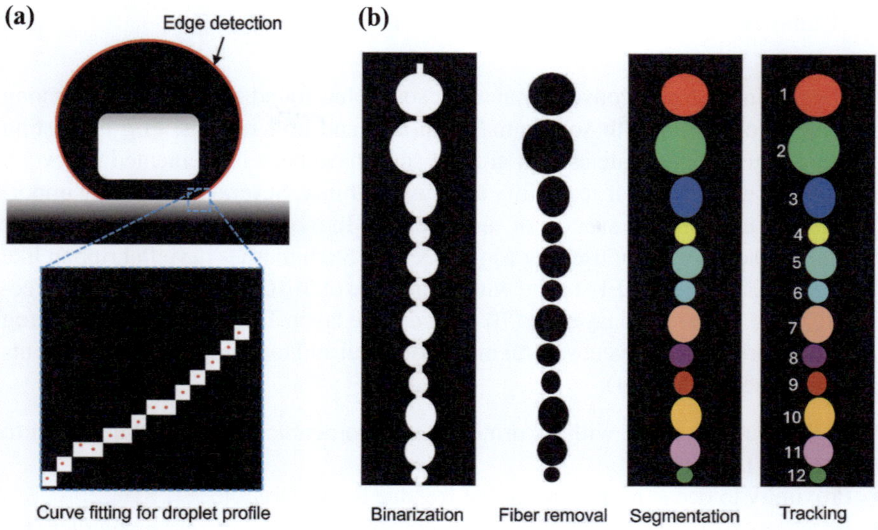

Fig. 15 **a** Schematic representation of image processing operations for estimating droplet volume and contact angle. Red dots indicate centroid of pixels. **b** Schematic representation of image processing scheme for tracking multiple droplets on fibres

coalescence dynamics of sessile and pendant droplets using high speed videography and image processing techniques. The time series data for location of the centre of mass of the merging liquid mass was obtained from image processing. This time series was then numerically differentiated to obtain velocity as a linear function of time. Once the velocity is obtained, shear rates can be obtained by approximating a linear velocity profile inside the liquid. Evaluation of shear stress is very important for analysing the durability of coating. If the shear stresses overcome the yield stress of the coating, then substantial degradation is also observed (Somwanshi et al. 2018).

In the context of fog harvesting, often a collection of large number of droplets are formed on fog harvesting fibres. Under such circumstances, it becomes important to distinguish between different droplets and track them individually. Also, on such fibres, multiple regimes of coalescence and droplet motion can be observed (Kowalski and Boreyko 2022). This can be handled by modifying the image processing algorithms as shown in Fig. 15b to have high sensitivity for distinguishing closely spaced droplets, while tracking each droplet separately. Schwarzwälder et al. (2023), utilized an approach which consisted of a segmentation called watershed transform, and introduced a sensitivity parameter to resolve the edges of closely spaced droplets. Further, a Kalman filter was applied, which assigned each droplet visible in a given frame with a unique tracking ID, which was discarded once the droplet left the frame, or coalesced with another droplet. This approach enabled the authors to track the individual dynamics of multiple droplets.

7 Feasibility and Sustainability of Fog Harvesting

Fog harvesting as a non-conventional water source has found great popularity among developing countries with substantial semi-arid and arid regions. Fog harvesting projects, both at pilot scale as well at large scale have been implemented in several countries to establish their feasibility and sustainability. Several factors play important roles in ensuring the success of such projects. In most cases, standard two-layer raschel meshes have been used for fog collection. Such meshes have flat ribbon like fibres with diameters of 0.1–1 mm, with a pore size of 5–10 mm, leading to an effective SC of 35% for single layer and 70% for double layer. The site selection for a fog harvesting project must meet several meteorological and topographical requirements (Schemeneuer et al. 2004):

- High mountain range with its principal axis perpendicular to wind direction to intercept fog flow.
- Proximity to ocean to prevent loss of fog due to evaporation.
- Broad basin on the leeward side of the mountain to enhance pressure gradient for large wind velocities.

Thus, most of the fog harvesting projects have been executed in dry tropical and subtropical regions with high elevation, where conventional sources of water are expensive. These include primarily, the western coast of South America, southwest and northwest coasts of Africa. In addition to topographical requirements, another crucial requirement is that of availability of fog for at least 4–6 months annually. Accordingly, several units of Standard Fog Collector (area: 1 m^2) and Large Fog collector (area: 40 m^2) can be installed according to the needs of beneficiary communities. Table 1 summarizes the data recorded for several large and pilot scale fog harvesting projects in different regions of the world, including their location, fog collection area, and water productivity in terms of volume collected per unit area per day. We can see that fog collection area as large as 4000 m^2 have been utilized, while water productivities as large as 30 $L/m^2/day$ have been achieved. Most of these large-scale projects have been successful in supplying 25–30 L per day per person to the targeted beneficiaries. Most of such beneficiaries belong to low-income groups, and these projects have greatly benefitted the societies, leading to even reverse migration from cities to villages, reducing the pressure on resources. In addition to the quantity, quality of fog water has to be taken into account for such feasibility studies. Being situated at rural settings, these projects have been able to achieve the potable water standards of the WHO. Besides household usage, fog harvesting has been largely successful in drip irrigation and reforestation applications as well. The sustainability of fog harvesting practices have been possible due to several techno-socio-economic factors (Fessehaye et al. 2014). From a technical perspective, the simplicity of installation, operation and maintenance, coupled with lack of requirement of external energy supply contribute greatly to the success of these systems in resource constrained ecosystems. However, being a decentralized water system, these projects require strong community participation from the beneficiaries, with at least 80 people per community including 20–30 adults (Klemm et al. 2012). Such participation includes training the beneficiary population in maintenance and supervision of the fog harvesting systems. Large fog collection meshes often get damaged due to large wind pressure and require prompt repair for continued productivity. Successful projects have seen the local communities' perception towards such projects transition into a joint entrepreneurial venture. For example, in Chile, local community developed the water collection systems to support local liquor production ventures, leading to economic development of the population. On the other hand, several projects have seen local politics and lack of community support hinder the progress, ultimately leading to stoppage of operation. From a financial aspect, the costs can be divided into capital expenditure to survey possible project sites, material and shipping, installation and technology transfer costs, and operating expenditures, including cost of distribution and transport. Table 2 shows a summary of capital expenditure of setting up of fog harvesting system in Eritrea, Africa (financial report, Vision Eritrea), where the total cost per beneficiary comes to around USD 21. According to Schemeneuer 1994, a typical project with 100 LFCs, the capital expenditure is around USD 40000. In most of the large-scale projects, initial capital expenditure, as well as technical assistance is provided by NGOs through foreign aids. Once established, the system has a lifespan of the order of ten years. Operation costs are very low, requiring USD 2.5–3 per m^3 in Chile (Cereceda et al. 1992). One must also appreciate the limitations suffered by fog harvesting as a water technology. These include seasonality, geographical restrictions and requirement of community participation in maintenance and supervision, limiting the widespread use of this technology. Thus, researchers are prompted towards innovation enhancing collection per unit area per unit mass of fog available, as well as reducing the dependence on fog characteristics like velocity and droplet size, as outlined over all previous sections.

8 Closure

In this chapter we discuss the potential development of fog harvesting into a sustainable water technology to supplement the conventional freshwater reserves to meet the ever-growing demand. The fog harvesting process is governed by the coupled effects of aerodynamics of fog flow and dynamics of collected droplets. The most fundamental trade-off in designing fog harvesting systems is the aerodynamic losses at the system scale governed by the geometric properties of shade coefficient, pore size and material properties like permeability. The aerodynamics at microscale is governed by the disturbance created by the presence of fibres, such that droplets with higher inertia strike the fibres and get captured, while the air with lower inertia gets filtered out. The surface wettability and roughness determine the contact angle and its hysteresis, which in turn governs the shedding and collection of captured droplets. Thus, strategies to enhance fog harvesting efficiencies revolve around enhancing higher surface

Table 1 Fog collection data from scientific and commercial projects (Klemm et al. 2012; Fessehaye et al. 2014)

Location	Collection period (months)	Total area (m^2)	Water productivity $L/m^2/day$	Source
El Tofo, Chile	12	4000	3.75	Schaumeneur et al. (1988)
Ecuador	7–8	1900	12	Henderson and Falk (2001)
Guatemala	4–6	1900	4.5	Garcia-Santos et al. (2011)
Oman	2	16.8	30	Abul-Wahab et al. (2010)
Yemen	3	–	4.5	Schemenauer (2004)
Canary Islands	12	2	8–10	Marzol et al. (2010)
Morocco	4	2	10.5	Marzol et al. (2010)
Croatia	6	11	4	Mileta and Likso (2010)
Iberian Peninsula, Spain	3	0.12	7	Estrella (2008)
Mount Machos, Spain	4	2500	3.3	Estrella (2009)

Table 2 Capital expenditure break of two independent fog harvesting systems in Eritrea, Africa (*Source* Vision Eritrea financial report)

Items	Unit	Number of units	Total cost (USD)
Establishment of 20 LFC units	m^2	900	28,193
PVC pipeline system	km	1.2	14,762
Water tankers and distribution system	m^3	30	14,278
Training of water committee (five members)	LS	1	1200
Implementation (personnel and transportation)			5485
Total	(USD)		**63,918**
Number of direct beneficiaries			**3045**
Cost per beneficiary			**21**

area for droplet-fibre collision without increasing the pressure-drop, followed by appropriate surface treatments to reduce the droplet size, accelerate droplet motion, reduce droplet losses due to clogging and re-entrainment. As such, engineered porous media with optimal geometric properties and surface properties need to be developed for the next generation of high performing fog harvesting devices. Future works may be focussed on theoretical understanding of aerodynamics of fog flow through general porous media other than screen meshes, reducing dependence of fog collection efficiency on fog characteristics and droplet size, theoretical understanding of contact angle hysteresis on textured porous surfaces along with the ensuing droplet dynamics, followed by establishing scalability of such surface engineering techniques.

Nomenclature

A	Area (m^2)
E	Attenuation function (−)
F	Force (N)
L	Length (m)
S	Pitch (m)
V	Voltage (V)
$\overline{V}$	Volume (m^3)
d	Diameter (m)
m	Mass (kg)
n	Number of layers (−)
p	Pressure (Pa)
r	Radius (m)
u	Velocity (m/s)

Greek Alphabets

α	Angle of fibre intersection
β	Angle of attack
δ	Vortex size
$\in$	Electrical permeability
γ	Surface tension
μ	Viscosity
ρ	Density
ψ	Stream function
χ	Filtered fraction
θ	Contact angle
ϕ	Incident fraction
Γ	Vortex size
λ	Length scale

Subscripts

∞	Properties of undisturbed flow
air	Properties of air

bump	Properties of millimetric bump
bulk	Properties of bulk fluid
cb	Cassie-Baxter state
c	Capture
col	Collecting wire
cr	Critical
d	Drainage
drop	Properties of collected water droplet
e	Entrainment
f	Fibre
g	Gravity
lv	Liquid vapor interface
p	Fog particle
r	Receding
w	Water
W	Wenzel state
1	Properties at region 1
2	Properties at region 2

Acronyms

CAH	Contact angle hysteresis
LWC	Liquid water content
LFC	Large fog collector
PA6	Polyamide-6
PLA	Polylactic acid
PS	Polystyrene
SHL	Superhydrophilic
SHB	Superhydrophobic
TCL	Triple Contact Line
USD	US Dollar
WHO	World health organisation

Non-dimensional Numbers

Bo	Bond number $\left(\frac{\rho_w \overline{V} g}{2\pi \gamma_{lv} d_f}\right)$
Ca	Capillary number $\left(\frac{\mu_w u_{drop}}{\gamma_{lv}}\right)$
Ke	Electrical number $\left(\frac{d_p \in V^2}{\mu_{air} L_{col} r_{col} u_\infty}\right)$
Oh	Ohnesorge number $\left(\frac{\mu}{\sqrt{\rho \gamma g}}\right)$

Re	Reynolds number $\left(\frac{\rho_{air}u_{\infty}L}{\mu}\right)$, where subscript denotes length scale
SC	Shade Coefficient
St	Stokes Number $\left(\frac{\rho_p d_p^2 u_{\infty}}{18\mu_{air}d_f}\right)$

References

Abdul-Wahab SA, Al-Damkhi AM, Al-Hinai H, Al-Najar KA, Al-Kalbani MS (2010) Total fog and rainwater collection in the Dhofar region of the Sultanate of Oman during the monsoon season. Water Int 35(1):100–109

Adam NK (1937) Detergent action and its relation to wetting and emulsification. J Soc Dyers Colour 53(4):121–129

Azeem M, Guérin A, Dumais T, Caminos L, Goldstein RE, Pesci AI, De Dios Rivera J, Torres MJ, Wiener J, Campos JL, Dumais J (2020) Optimal design of multilayer fog collectors. ACS Appl Mater Interfaces 12(6):7736–7743

Azizi F (2019) On the pressure drop of fluids through woven screen meshes. Chem Eng Sci 207:464–478

Beard KV, Pruppacher HR (1969) A determination of the terminal velocity and drag of small water drops by means of a wind tunnel. J Atmos Sci 26(5):1066–1072

Brundrett E (1993) Prediction of pressure drop for incompressible flow through screens. ASME J Fluids Eng 115(2):239–242

Brun RJ, Lewis W, Perkins PJ, Serafini JS (1955) Impingement of cloud droplets on a cylinder and procedure for measuring liquid-water content and droplet sizes in supercooled clouds by rotating multicylinder method. NASA Rep. No. 1215

Cao M, Xiao J, Yu C, Li K, Jiang L (2015) Hydrophobic/hydrophilic cooperative Janus system for enhancement of fog collection. Small 11(34):4379–4384

Carvajal D, Silva-Llanca L, Larraguibel D, González B (2020) On the aerodynamic fog collection efficiency of fog water collectors via three-dimensional numerical simulations. Atmos Res 245:105123

Cereceda P, Schemenauer RS, Suit M (1992) An alternative water supply for chilean coastal desert villages. Int J Water Resour Dev 8(1):53–59. https://doi.org/10.1080/07900629208722533

Chaudhury MK, Whiteside GM (1992) How to make water run uphill 256(5063):1539–1542

Chou TH, Hong SJ, Liang YE, Tsao HK, Sheng YJ (2011) Equilibrium phase diagram of drop-on-fiber: coexistent states and gravity effect. Langmuir 27(7):3685–3692

Comanns P, Buchberger G, Buchsbaum A, Baumgartner R, Kogler A, Bauer S, Baumgartner W (2015) Directional passive liquid transport: the Texas horned lizard as a model for a biomimetic 'liquid diode'. J R Soc Interface 12(109):20150415. https://doi.org/10.1098/rsif.2015.0415

Cumberbatch E (1982) Two-dimensional flow past a mesh. Q J Mech Appl Math 35(3):335–344

Damak M, Varanasi KK (2018) Electrostatically driven fog collection using space charge injection. Sci Adv 4(6):eaao5323

Davies RM, Taylor GI (1944) Aerodynamics of porous sheets. Rep Memoranda Aeronautic 2236:159–162

Dawson TE (1998) Fog in the California redwood forest: ecosystem inputs and use by plants. Oecologia 117:476–485

De Coninck J, Dunlop F, Huillet T (2017) Contact angles of a drop pinned on an incline. Phys Rev E 95(5):052805

ElSherbini AI, Jacobi AM (2004) Liquid drops on vertical and inclined surfaces: II. A method for approximating drop shapes. J Colloid Interface Sci 273(2):566–575

Eral HB, De Ruiter J, De Ruiter R, Oh JM, Semprebon C, Brinkmann M, Mugele F (2011) Drops on functional fibres: from barrels to clamshells and back. Soft Matter 7(11):5138–5143

Estrela MJ, Valiente JA, Corell D, Millán MM (2008) Fog collection in the western Mediterranean basin (Valencia region, Spain). Atmos Res 87(3–4):324–337
Estrela MJ, Valiente JA, Corell D, Fuentes D, Valdecantos A (2009) Prospective use of collected fog water in the restoration of degraded burned areas under dry mediterranean conditions. Agric for Meteorol 149(11):1896–1906
Extrand CW, Kumagai Y (1995) Liquid drops on an inclined plane: the relation between contact angles, drop shape, and retentive force. J Colloid Interface Sci 170(2):515–521
Fessehaye M, Abdul-Wahab SA, Savage MJ, Kohler T, Gherezghiher T, Hurni H (2014) Fog-water collection for community use. Renew Sustain Energ Rev 29:52–62
Friedlander SK (2000) Smoke, dust, and haze, vol 198. Oxford University Press, New York
García-Santos G, Bruijnzeel LA (2011) Rainfall, fog and throughfall dynamics in a subtropical ridge top cloud forest, National Park of Garajonay (La Gomera, Canary Islands, Spain). Hydrol Processes 25(3):411–417
Ghosh R, Patra C, Singh P, Ganguly R, Sahu RP, Zhitomirsky I, Puri IK (2020) Influence of metal mesh wettability on fog harvesting in industrial cooling towers. Appl Therm Eng 181:115963
Gilet T, Terwagne D, Vandewalle N (2010) Droplets sliding on fibres. Eur Phys J E 31(3):253–262
Goldstein S (1965) Modern developments in fluid dynamics vol. 1. Dover
Goswami S, Sidhpuria RM, Khandekar S (2023) Effect of droplet-laden fibers on aerodynamics of fog collection on vertical fiber arrays. Langmuir 39(50):18238–18251
Henderson B, Falk D (2001) Fog water collection in ecuador: An appropriate technology for the rural poor. In: Proceedings of the 2nd international conference on fog and fog collection (vol 1520)
Hinds WC, Zhu Y (2020) Aerosol technology: properties, behavior, and measurement of airborne particles. Wiley
Idelchik IE (1986) Handbook of hydraulic resistance, fourth ed. Washington
Jaiswal AK, Khandekar S (2021) Drop-on-drop impact dynamics on a superhydrophobic surface. Langmuir 37(43):12629–12642
Jiang Y, Savarirayan S, Yao Y, Park KC (2019) Fog collection on a superhydrophilic wire. Appl Phys Lett 114(8):083701
Ju J, Bai H, Zheng Y, Zhao T, Fang R, Jiang L (2012) A multi-structural and multi-functional integrated fog collection system in cactus. Nat Commun 3:1246–1247
Kalugin VT, Epikhin AS, Chernukha PA, Kalugina MD (2021) The effect of perforation on aerodynamic characteristics and the vortex flow field around a flat plate. IOP Conferen Series: Mater Sci Eng 1191(1):012007
Katz J, Plotkin A (2001) Low-speed aerodynamics, vol 13. Cambridge university press
Khandekar S, Muralidhar K (eds) (2020) Drop dynamics and dropwise condensation on textured surfaces. Springer, Cham, Switzerland
Klemm O, Schemenauer RS, Lummerich A, Cereceda P, Marzol V, Corell D, Fessehaye GM (2012) Fog as a fresh-water resource: overview and perspectives. Ambio 41:221–234
Knapczyk-Korczak J, Ura DP, Gajek M, Marzec MM, Berent K, Bernasik A, Chiverton JP, Stachewicz U (2020) Fiber-based composite meshes with controlled mechanical and wetting properties for water harvesting. ACS Appl Mater Interfaces 12(1):1665–1676
Koo JK, James DF (1973) Fluid flow around and through a screen. J Fluid Mech 60(3):513–538
Kowalski NG, Shi W, Kennedy BS, Boreyko JB (2021) Optimizing fog harps. ACS Appl Mater Interfaces 13(32):38826–38834
Kowalski NG, Boreyko JB (2022) Dynamics of fog droplets on a harp wire. Soft Matter 18(37):7148–7158
Labbé R, Duprat C (2019) Capturing aerosol droplets with fibers. Soft Matter 15(35):6946–6951
Langmuir I, Blodgett K (1946) A mathematical investigation of water droplet trajectories. Army Air Forces Headquarters, Air Technical Service Command, United States
Ledda PG, Boujo E, Camarri S, Gallaire F, Zampogna GA (2021) Homogenization-based design of microstructured membranes: wake flows past permeable shells. J Fluid Mech 927:A31

Lee JH, Lee YJ, Kim HY, Moon MW, Kim SJ (2022) Unclogged janus mesh for fog harvesting. ACS Appl Mater Interfaces 14(18):21713–21726

Lei J, Guo Z (2020) A fog-collecting surface mimicking the Namib beetle: its water collection efficiency and influencing factors. Nanoscale 12(13):6921–6936

Li J, Ran R, Wang H, Wang Y, Chen Y, Niu S, Arratia PE, Yang S (2021) Aerodynamics-assisted, efficient and scalable kirigami fog collectors. Nat Commun 12(1):5484

Lorenceau É, Quéré D (2004) Drops on a conical wire. J Fluid Mech 510:29–45

Makkonen L, Zhang J, Karlsson T, Tiihonen M (2018) Modelling the growth of large rime ice accretions. Cold Reg Sci Technol 151:133–137

Marchand OC, Ramananarivo S, Duprat C, Josserand C (2024) Three-dimensional flow around and through a porous screen. J Fluid Mech 987:A20

Marzol MV, Sánchez J, Yanes A, Derhem A, Bargach J (2010) Meteorological patterns and fog water in Morocco and the Canary Islands. In: Proceedings of the 5th international conference on fog, fog collection and dew, vol 2530

Marcu B, Meiburg E, Newton PK (1995) Dynamics of heavy particles in a Burgers vortex. Phys Fluids 7(2):400–410

Marzol-Jaen MV (2011) Historical background of fog water collection studies in the Canary Islands. In: Bruijnzeel LA, Scatena FN, Hamilton LS (eds) Tropical montane cloud forests: science for conservation and management. Cambridge University Press, pp 352–358

Mehta RD, Bradshaw P (1979) Design rules for small low speed wind tunnels. Aeronaut J 83(827):443–453. https://doi.org/10.1017/S0001924000031985

Mileta M, Likso Z (2010) Fog water collection with SFC on the mountain Velebit (Croatia) during the period 2000–2009. In: Proceedings of the 5th international conference on fog, fog collection and dew, vol 2530

Nioras D, Ellinas K, Constantoudis V, Gogolides E (2021) How different are fog collection and dew water harvesting on surfaces with different wetting behaviors? ACS Appl Mater Interfaces 13(40):48322–48332

Park KC, Chhatre SS, Srinivasan S, Cohen RE, McKinley GH (2013) Optimal design of permeable fiber network structures for fog harvesting. Langmuir 29(43):13269–13277

Parker AR, Lawrence CR (2001) Water capture by a desert beetle. Nature 414:33–34

Pezzulla M, Strong EF, Gallaire F, Reis PM (2020) Deformation of porous flexible strip in low and moderate Reynolds number flows. Phys Rev Fluids 5(8):084103

Protiere S, Duprat C, Stone HA (2013) Wetting on two parallel fibers: drop to column transitions. Soft Matter 9(1):271–276

Quéré D (1999) Fluid coating on a fiber. Annu Rev Fluid Mech 31(1):347–384

Ren F, Li G, Zhang Z, Zhang X, Fan H, Zhou C, Wang Y, Zhang Y, Wang C, Mu K, Su Y, Wu D (2017) A single-layer Janus membrane with dual gradient conical micropore arrays for self-driving fog collection. J Mater Chem A 5(35):18403–18408

Rivera JDD (2011) Aerodynamic collection efficiency of fog water collectors. Atmos Res 102(3):335–342

Roshko A (1954) A new hodograph for free-streamline theory (No. NACA-TN-3168)

Sauret A, Bick AD, Duprat C, Stone HA (2014) Wetting of crossed fibers: multiple steady states and symmetry breaking. EPL 105(5):56006

Schemenauer RS, Fuenzalida H, Cereceda P (1988) A neglected water resource: the Camanchaca of South America. Bull Am Meteorol Soc 69(2):138–147

Schemenauer RS, Osses P, Leibbrand M (2004) Fog collection evaluation and operational projects in the Hajja Governorate, Yemen. In: Proceedings of the third international conference on fog, fog collection and dew, vol 1115

Schwarzwälder A, Dittler A, Meyer J (2023) Experimental analysis of fundamental water droplet coalescence mechanisms on a single vertical fiber. Exp Fluids 64:103

Semprebon C, Brinkmann M (2014) On the onset of motion of sliding drops. Soft Matter 10(18):3325–3334

Shahrokhian A, Feng J, King H (2020) Surface morphology enhances deposition efficiency in biomimetic, wind-driven fog collection. J R Soc Interface 17(166):20200038
Shi W, Anderson MJ, Tulkoff JB, Kennedy BS, Boreyko JB (2018) Fog harvesting with Harps. ACS Appl Mater Interfaces 10(14):11979–11986
Shi W, van der Sloot TW, Hart BJ, Kennedy BS, Boreyko JB (2020a) Harps enable water harvesting under light fog conditions. Adv Sustain Syst 4(6):2000040
Shi W, De Koninck LH, Hart BJ, Kowalski NG, Fugaro AP, van der Sloot TW, Boreyko JB (2020b) Harps under heavy fog conditions: superior to meshes but prone to tangling. ACS Appl Mater Interfaces 12(42):48124–48132
Showket J, Majumder S, Kumar N, Sett S, Mahapatra PS (2024) Fog harvesting on micro-structured metal meshes: effect of surface ageing. Micro Nano Eng 22:100236
Somwanshi PM, Muralidhar K, Khandekar S (2018) Coalescence dynamics of sessile and pendant liquid drops placed on a hydrophobic surface. Phys Fluids 30(9):092103
Somwanshi PM, Muralidhar K, Khandekar S (2020) Coalescence of vertically aligned drops over a superhydrophobic surface. Phys Fluids 32(5). https://doi.org/10.1063/5.0007419
Steiros K, Hultmark M (2018) Drag on flat plates of arbitrary porosity. J Fluid Mech 853:R31–R311
Turner MR, Sear RP (2023) Critical scaling law for the deposition efficiency of inertia-driven particle collisions with a cylinder in high Reynolds number air flow. Phys Rev Fluids 8(1):014302
Wieghardt KEG (1953) On the resistance of screens. Aeronaut Q 4(2):186–192
Weyer F, Duchesne A, Vandewalle N (2017) Switching behavior of droplets crossing nodes on a fibre network. Sci Rep 7(1):13309
Xu M, Patruno L, Lo YL, de Miranda S (2020) On the use of the pressure jump approach for the simulation of separated external flows around porous structures: a forward facing step. J Wind Eng Ind Aerodyn 207104377. https://doi.org/10.1016/j.jweia.2020.104377
Yu DI, Doh SW, Kwak HJ, Kang HC, Ahn HS, Park HS, Kiyofumi M, Kim MH (2015) Wetting state on hydrophilic and hydrophobic micro-textured surfaces: thermodynamic analysis and X-ray visualization. Appl Phys Lett 106(17):171602
Yu Z, Zhang H, Huang J, Li S, Zhang S, Cheng Y, Mao J, Dong X, Gao S, Wang S, Chen Z, Jiang Y, Lai Y (2021) Namib desert beetle inspired special patterned fabric with programmable and gradient wettability for efficient fog harvesting. J Mater Sci Technol 61:85–92. https://doi.org/10.1016/j.jmst.2020.05.054
Zeng MJ, Qu ZG, Zhang JF (2023a) Experimental study on water collection performance of wire-to-plate electrostatic fog collector at various fog generation rates and fog flow velocities. Sep Purif Technol 305:122465
Zeng MJ, Qu JG, Zhang JF (2023b) Design and optimization of an electrostatic fog collection system for water harvesting: modelling and experimental investigation. Sep Purif Technol 323:124422
Zheng Y, Bai H, Huang Z, Tian X, Nie FQ, Zhao Y, Zhai J, Jiang L (2010) Directional water collection on wetted spider silk. Nature 463(7281):640–643
Zhu P, Chen R, Zhou C, Tian Y, Wang L (2021) Asymmetric fibers for efficient fog harvesting. Chem Eng J 415:128944

Solar Thermal Water Purification System

Tarun Kulshrestha and Malay K. Das

1 Introduction

This chapter discusses water purification systems driven by solar thermal energy called solar stills, briefly described in Chap. 2. The chapter outlines a detailed overview consisting of various solar still configurations proposed, factors dictating freshwater productivity, and mathematical models for evaluating thermal performance and comparing them with experimentally obtained results. Furthermore, the utility of components such as heat pipes and evacuated tube solar collectors, which enhance the water productivity of solar stills, has also been discussed. Finally, the chapter concludes with the challenges involved in conventional solar stills and the future course of action required for developing a low-cost, portable solar thermal desalination system with enhanced freshwater production for remote regions receiving abundant solar insolation.

1.1 Solar Stills

Solar stills are a passive, simple device that treats impure water (saline/brackish) and converts it into potable drinkable water, as shown in Fig. 1. The following components constitute the basic design of a solar still.

T. Kulshrestha
Department of Engineering, School of Computing, Engineering and Mathematical Sciences, La Trobe University, Bendigo, VIC, Australia

T. Kulshrestha · M. K. Das (✉)
Department of Mechanical Engineering, Indian Institute of Technology Kanpur, Kanpur, Uttar Pradesh, India
e-mail: mkdas@iitk.ac.in

R. Narayanaswamy et al. (eds.), *Sustainable Water Harvesting*, Springer Tracts in Mechanical Engineering, https://doi.org/10.1007/978-981-95-3373-2_8

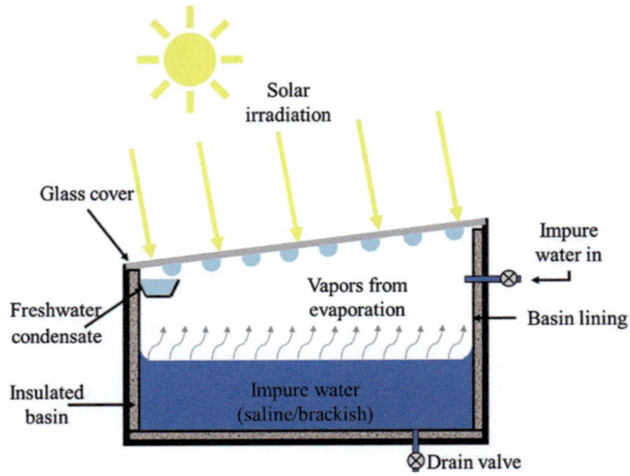

Fig. 1 Schematic of freshwater production in a conventional solar still

1.1.1 Basin

A reservoir for impure (saline/brackish) water, mainly trapezoidal, made of concrete, plexiglass, wood, or metals like aluminum, along with proper insulation (glass wool, polyurethane foam) on the inner surface for mitigation of parasitic heat losses. The inner surfaces are also painted black for maximum absorption of the incident solar energy.

1.1.2 Glass Cover

The top surface of the basin is kept inclined, and a transparent glass, see Fig. 1, also known as the glass cover, is placed over it. The angle of inclination is chosen based on the geographical latitude of the location, allowing solar insolation to fall perpendicularly and ensuring maximum absorption. The bottom surface of the glass cover should be relatively smooth to maintain an easy flow of the freshwater condensate into the collection chamber.

First, a filling valve is used to charge the basin with impure water, see Fig. 1, and the solar still is kept in the sun. The incident solar radiation is transmitted through the glass cover due to its transparent nature, and is absorbed by the impure water, resulting in bulk heating. Water molecules near the air–water interface are agitated by bulk heating, which evaporates the impure water near the air–water interface. Evaporation leads to increased temperatures in the vicinity of the air–water interface, which sets up convection currents between the glass cover and the air–water interface due to density gradients. The freshwater distillate is collected from the vapor condensation on the bottom surface of the glass cover; see Fig. 1. All contact surfaces of the solar still susceptible to any leakage should be adequately sealed to ensure the proper functioning of the solar still. Generally, a drain valve is also attached to the basin to drain the impure water, if required.

A broad classification of solar stills is generally done into passive solar stills and active solar stills, primarily based on their energy source. Thermal energy generated from direct sunlight incident on the glass cover, see Fig. 1, acts as the energy source for impure water evaporation in passive solar stills. Though solar stills are independent of impure feed water quality, their freshwater productivity is inherently poor due to the low evaporation rate of impure feed water. The low evaporation rate is due to significant interior and exterior heat losses from the solar still. These include heating of the impure (saline/brackish) bulk water, convection, and radiation losses from the air–water interface, glass cover, and the basin surface to the ambient. Moreover, the reflection of solar irradiation at the air–water interface, see Fig. 2a, also adds to the poor evaporation rate. Hence, conventional solar stills have been far from being commercially adapted due to the abovementioned bottlenecks and require performance enhancement. The performance enhancement of passive solar stills involves identifying the critical factors responsible for their operation and the scope for improvement. Mohsenzadeh et al. (2021) identified four categories: heat receiving, absorption, storage, and condenser design, which provide performance enhancement of passive solar stills, leading to high freshwater productivity. Some categories mentioned above require additional energy from external sources, and solar stills operating with external energy sources are termed active solar stills. External sources include flat plate solar collectors, photovoltaic (PV) panels, parabolic troughs, and solar towers for concentrated solar power. The parameters affecting the performance and the freshwater productivity of conventional solar stills are explained next.

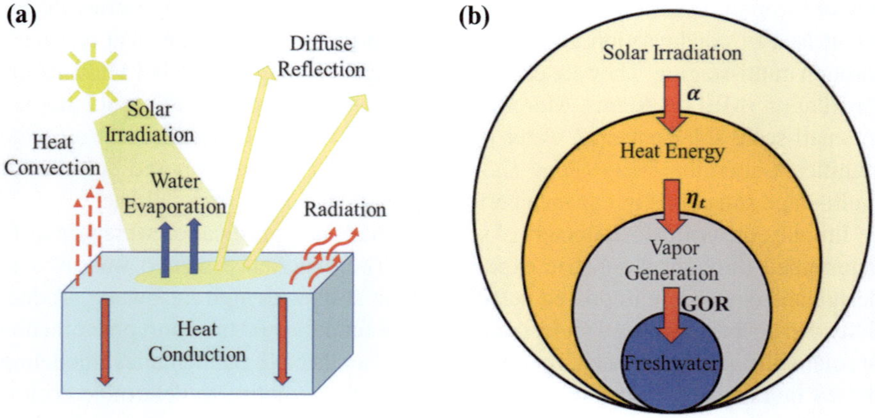

Fig. 2 **a** Energy balance diagram of bulk water evaporation in conventional solar thermal water purification systems. **b** Important parameters for an efficient solar thermal water purification system

2 Performance-Influencing Factors of Solar Stills

A parameter termed Specific Water Productivity (SWP) is used to evaluate the performance of solar thermal water purification systems; see Fig. 2b. SWP is defined as the volume of freshwater produced per unit area of solar irradiation, per unit time. It indicates the efficiency with which impure water (saline/brackish) is converted into potable or freshwater using the energy available from solar irradiation. For practical applications, solar irradiance is generally considered equivalent to one sun. SWP is expressed using Eq. (1) (Wang et al. 2019),

$$\mathrm{SWP} = \frac{I(t)}{h_{\mathrm{lv}}}\alpha\eta_t\mathrm{GOR} \tag{1}$$

where $I(t)$ is the solar irradiation (kW/m^2), h_{lv} is the total enthalpy change (kWh/l), α is the solar absorptivity of the solar thermal desalination system (fraction of solar insolation converted to heat), η_t is the heat-to-vapor conversion efficiency (fraction of heat used for vapor generation), GOR or the Gain Output Ratio (fraction of a kilogram of freshwater produced per kilogram of the vapor produced), i.e., the reutilization extent for the latent heat of condensation in freshwater production. This is because the thermal energy required to produce 1 kg of water vapor from impure water in the liquid state at its saturation temperature is constant. Moreover, the first two efficiencies mentioned above cannot exceed unity. However, one can achieve a GOR greater than unity by reusing the latent heat of condensation (usually dissipated into the ambient) to preheat the impure bulk water, leading to enhanced evaporation rates of impure water. The technology for generating freshwater by retrieving the latent heat of condensation already exists for large-scale water purification plants through multi-staging. They are called Multi-Stage Flashing (MSF) and Multi-Effect Distillation (MED) systems. Moreover, latent heat recovery through multi-staging in small-scale solar-powered water purification systems (solar stills) has received significant attention lately (Peng and Sharshir 2023). Some proposed designs for multi-stage solar stills to enhance performance are discussed in detail later.

Initially, the critical parameters of solar thermal water purification were primarily determined through parametric experiments. The experience gained was used to design and develop an improved version of solar stills with high freshwater productivity. However, a thorough understanding of the fundamental transport phenomenon in solar stills at a device scale (or component-wise) through mathematical modeling (based on energy balance) has been the preferred alternative to determine critical parameters and propose superior designs for enhanced performance. The availability of thermal models and their analytical and numerical solutions enables a quicker assessment of the key parameters of both simple and complex design configurations with high fabrication costs and long timescales. Moreover, meaningful results from thermal models are obtained only by adopting realistic assumptions, formulating the governing equations properly, and applying initial and boundary conditions appropriately. The governing equations, along with the assumptions, for thermal models of solar stills that encompass the fundamental transport phenomena are discussed next.

2.1 Mathematical Modeling

This section discusses the formulation of thermal models for a conventional single-basin single-slope (SBSS) solar still, as shown in Fig. 3. These solar stills have a simple design, and the heat and mass transport governing equations will be easier for the reader to grasp. The governing equations in the mathematical model of solar stills are derived by applying energy conservation to each of its key components: the glass cover, the basin, and the impure water bulk. The energy conservation for each component is written on a per unit area basis at the average temperature between the two successive components. First, we describe the assumptions required for the governing equations of the thermal model:

1. Negligible vapor and impure water leak from the SBSS solar still.
2. The quantity of impure water evaporated is insignificant compared to the quantity of impure water in the basin.
3. Negligible temperature gradient along the depth of impure water.
4. The absorptance, heat capacity, and inclination of the glass cover with respect to the horizontal are considered negligible.
5. The area of the SBSS solar still basin, surface area of impure water, and glass cover area are equivalent.

First, we describe the heat transfer mechanisms in solar stills, followed by their thermal analysis. Heat transfer processes are classified as steady or unsteady based on

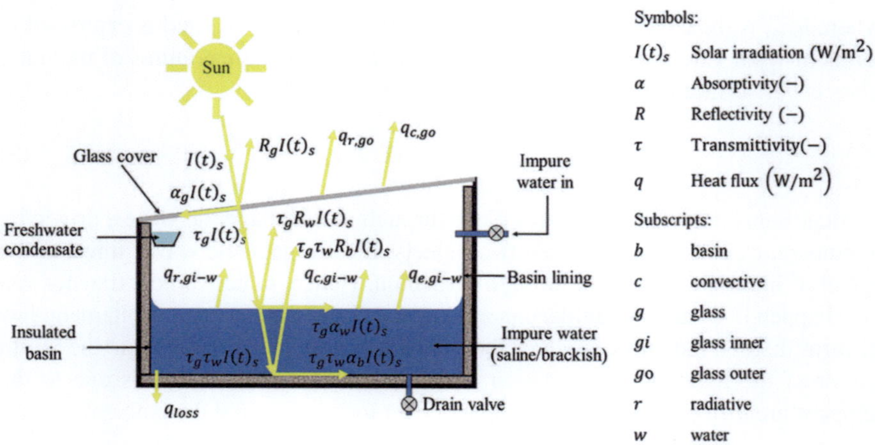

Fig. 3 Schematic showing the flow of energy in a conventional solar still

the time dependence of quantities, such as the rate of heat transfer and temperature, over time. Processes with a constant rate of heat transfer and temperature with respect to time are termed steady-state heat transfer. However, solar stills undergo unsteady heat transfer because the quantities mentioned above vary over time. The analysis of unsteady heat transfer is complex, but it can be analyzed using some presumed steady conditions. The heat transfer of a solar still is further classified into two categories, viz., external heat transfer and internal heat transfer respectively (Elango et al. 2015). The classification is based on the thermal energy flow both inside and outside of the SBSS solar still. These processes are discussed below:

2.1.1 External Heat Transfer

The heat lost from an SBSS solar still to the ambient is termed external heat transfer. The solar still loses heat to the ambient independently through the top, side, and bottom surfaces, respectively. The top heat loss occurs between the top surface of the glass cover and the ambient, whereas side and bottom heat losses occur between the solar still basin and the ambient. These heat losses comprise fundamental heat transfer modes like conduction, convection, and radiation. Top heat loss from the top of the glass cover to the ambient is governed by convection and radiation. Convective heat transfer is conduction heat transfer associated with bulk fluid motion. Heat conduction in bodies transfers kinetic energy within molecules through lattice vibrations. Thermophysical properties of the bulk fluid, the bulk motion, and the geometry and roughness of the solid body associated with bulk fluid motion are key factors for convective heat transfer. The convective heat loss due to the temperature difference between the top surface of the glass cover and the ambient is calculated from Eq. (2),

$$q_{c,\mathrm{go}} = h_{c,\mathrm{go}}\left(T_{\mathrm{go}} - T_a\right) \tag{2}$$

where $h_{c,\mathrm{go}}$ is the convective heat transfer coefficient (W/m^2 K) and is expressed in terms of wind velocity (v) by Eq. (3), T_{go} and T_a are the temperatures of the outer glass cover surface and ambient (°C) (Elango et al. 2015).

$$h_{c,o} = 2.8 + (3.0 \times v) \tag{3}$$

Heat transfer by radiation takes place through electromagnetic waves, driven by a temperature difference between two objects. Electromagnetic waves travel at the speed of light and can travel through a vacuum. Hence, radiative heat transfer can also happen in a vacuum, unlike conduction and convection. Stefan–Boltzmann law governs thermal radiation, in which the heat transfer rate is proportional to the fourth power of the temperature. Equation (4) shows the radiative heat loss due to the temperature difference between the glass cover top surface and the ambient,

$$q_{r,\mathrm{go}} = h_{r,\mathrm{co}}\left(T_{\mathrm{go}} - T_a\right) \tag{4}$$

where $h_{r,co}$ is the radiative heat transfer coefficient between the glass cover surface and ambient (W/m^{2}K), expressed by Eq. (5) (Elango et al. 2015).

$$h_{r,go} = \varepsilon_g \sigma \left[\frac{(T_{go} + 273)^4 - (T_{sky} + 273)^4}{(T_{go} - T_a)} \right] \tag{5}$$

where ε_g is the emissivity of the glass cover, σ is the Stefan–Boltzmann constant (5.67×10^{-8} W/m^2 K^4), and T_{sky} is the temperature of the sky (K) and is calculated from Eq. (6).

$$T_{sky} = T_a - 6 \tag{6}$$

The total top heat loss from the glass cover is the summation of Eqs. (2) and (4),

$$q_{t,go} = q_{c,go} + q_{r,go} = h_{t,go}(T_{go} - T_a) \tag{7}$$

$$h_{t,go} = h_{c,go} + h_{r,go} \tag{8}$$

where $h_{t,go}$ is the total heat transfer coefficient between the top glass cover and the ambient, see Eq. (8), and can be calculated from the summation of Eqs. (3) and (5). The total heat transfer coefficient between the top glass cover and ambient can also be obtained from Eq. (9), having a combined convective and radiative effect through wind velocity. Elango et al. (2015) suggest that the difference in $h_{t,go}$ obtained from Eqs. (8) and (9) is negligible.

$$h_{t,go} = 5.7 + (3.8 \times v) \tag{9}$$

The overall top heat transfer coefficient from impure bulk water surface through the glass cover to the ambient can be calculated from Eq. (10),

$$\frac{1}{U_{top}} = \frac{1}{h_{t,\text{w-gi}}} + \frac{L_g}{K_g} + \frac{1}{h_{t,go}} \tag{10}$$

where $h_{t,\text{w-gi}}$ is the total internal heat transfer coefficient between impure bulk water and the interior side of the glass cover (W/m^2 K), and is defined in the next section, L_g is the length of the glass cover (m), K_g is the thermal conductivity of the glass cover (W/m K).

The heat lost from the impure bulk water to the ambient through the basin wall liner and insulation of an SBSS solar still constitutes the side and bottom external heat transfer. The following heat transfer modes are responsible: convection between the impure bulk water and the basin wall liner, conduction through the basin wall liner and the subsequent insulation, and a combination of convection and radiation to the ambient air. There is zero bottom convective and radiative heat loss to the ambient for ground-mounted solar stills.

The rate of convective heat loss between the impure bulk water and the basin wall liner is given by Eq. (11)

$$q_w = h_w(T_w - T_b) \tag{11}$$

where h_w is the heat transfer coefficient between impure bulk water and basin wall liner (W/m^2 K), T_w and T_b are temperatures of the impure bulk water and the basin wall liner (°C). Equation (12) calculates the conductive heat transfer between the basin wall liner and the ambient through the insulation,

$$q_b = h_b(T_b - T_a) \tag{12}$$

where h_b is the heat transfer coefficient between the basin wall liner and ambient through the insulation (W/m^2 K), calculated from Eq. (13),

$$\frac{1}{h_b} = \frac{L_{\text{ins}}}{K_{\text{ins}}} + \frac{1}{h_{t,b\text{-}a}} \tag{13}$$

where L_{ins} is the length of insulation (m), K_{ins} is the thermal conductivity of insulation (W/m K), and $h_{t,b\text{-}a}$ is obtained from Eq. (8). The value of $h_{t,b\text{-}a}$ is zero if the solar still is grounded. Equation (14) calculates the total bottom heat loss coefficient between impure bulk water and ambient.

$$\frac{1}{U_b} = \frac{1}{h_w} + \frac{1}{h_b} \tag{14}$$

Considering A_{sw} and A_b as the side wall area and area of the basin, respectively, the total side heat loss coefficient between impure bulk water and ambient is given by Eq. (15). The value of U_{sw} can be neglected for small depths of impure bulk water due to the small side wall area available for heat loss compared to the basin area.

$$U_{\text{sw}} = \left(\frac{A_{\text{sw}}}{A_b}\right) U_b \tag{15}$$

The overall heat loss from the bottom and sides of the solar still is calculated using Eq. (16).

$$U_{\text{bs}} = U_b + U_{\text{sw}} \tag{16}$$

Finally, the total external heat transfer coefficient from the solar still to the ambient can be calculated from Eq. (17).

$$U_t = U_{\text{top}} + U_{\text{bs}} \tag{17}$$

2.1.2 Internal Heat Transfer

The heat loss between the impure bulk water and the interior side of the glass cover is termed internal heat transfer in SBSS solar stills. The following heat transfer modes, viz., convective, radiative, and evaporative, occur within the solar still. Convective heat transfer exchanges heat between the impure bulk water and the interior side of the glass cover through the moist air in between. Equation (18) calculates the internal convective heat transfer rate,

$$q_{c,\text{w-gi}} = h_{c,\text{w-gi}}\left(T_w - T_{\text{gi}}\right) \tag{18}$$

where $h_{c,\text{w-gi}}$ is the convective heat transfer coefficient between impure bulk water and the interior side of the glass cover (W/m^2 K) and calculated by Eq. (19) (Elango et al. 2015). T_w and T_{gi} are the temperatures of the impure bulk water and the interior side of glass cover (°C).

$$h_{c,\text{w-gi}} = 0.884 \times \left[\left(T_w - T_{\text{gi}}\right)\frac{\left(P_w - P_{\text{gi}}\right)(T_w + 273.15)}{268.9 \times 10^3 - P_w}\right]^{1/3} \tag{19}$$

Here P_w and P_{gi} are the saturation vapor pressures at the air–water interface of impure bulk water and the interior face of the glass cover, respectively. The following expression can be used to determine the vapor pressures,

$$P = \exp\left[25.317 - \left(\frac{5144}{T + 273}\right)\right] \tag{20}$$

The radiative heat transfer inside an SBSS solar still occurs between the impure bulk water and the interior face of the glass cover. However, the view factor plays a crucial role in radiative heat transfer. Since we have assumed a negligible inclination of the glass cover with respect to the horizontal, the view factor of impure bulk water with respect to the interior face of the glass cover will be unity. The radiative heat transfer between impure bulk water and the interior face of the glass cover is obtained from Eq. (21),

$$q_{r,\text{w-gi}} = h_{r,\text{w-gi}}\left(T_w - T_{\text{gi}}\right) \tag{21}$$

where $h_{r,\text{w-gi}}$ is the radiative heat transfer coefficient between impure bulk water and the interior face of the glass cover (W/m^2 K), and calculated by Eq. (22),

$$h_{r,\text{w-gi}} = \varepsilon_{\text{eff}}\sigma\left[(T_w + 273)^2 + \left(T_{\text{gi}} + 273\right)^2\right]\left(T_w + T_{\text{gi}} + 546\right) \tag{22}$$

where ε_{eff} is the effective emittance between the impure bulk water and the interior face of the glass cover, expressed as Eq. (23), and σ is the Stefan–Boltzmann constant

(5.67×10^{-8} W/m^2 K^4).

$$\frac{1}{\varepsilon_{\text{eff}}} = \frac{1}{\varepsilon_w} + \frac{1}{\varepsilon_g} - 1 \tag{23}$$

Here, ε_{eff} and ε_g are the emissivities of impure bulk water and glass cover, respectively.

At a specific liquid temperature, if the vapor pressure of the liquid is lower than the corresponding saturation pressure at that particular temperature, the liquid molecules near the liquid–vapor interface tend to cross the interface. This phenomenon, in which liquid molecules pass through the liquid–vapor interface, leading to the generation of water vapor, is termed evaporation. The evaporative heat loss between the air–water interface and the interior face of the glass cover occurs in addition to convective and radiative heat losses. Equation (24) calculates the evaporative heat transfer inside a solar still,

$$q_{e,\text{w-gi}} = h_{e,\text{w-gi}}\left(T_w - T_{\text{gi}}\right) \tag{24}$$

where $h_{e,\text{w-gi}}$ is the evaporative heat transfer coefficient between impure bulk water mass and the interior face of the glass cover (W/m^2 K) and is obtained by using Eq. (25), also known as Dunkle's model (Elango et al. 2015).

$$h_{e,\text{w-gi}} = 16.273 \times 10^{-3} \times h_{c,\text{w-gi}}\left[\frac{P_w - P_{\text{gi}}}{T_w - T_{\text{gi}}}\right] \tag{25}$$

The total internal heat transfer within the SBSS solar still is obtained by the summation of Eqs. (18), (21), and (24). Equation (26) calculates the total internal heat transfer.

$$q_{t,\text{w-gi}} = h_{t,\text{w-gi}}\left(T_w - T_{\text{gi}}\right) \tag{26}$$

where

$$h_{t,\text{w-gi}} = h_{c,\text{w-gi}} + h_{r,\text{w-gi}} + h_{e,\text{w-gi}} \tag{27}$$

Thermal conduction occurs across the glass cover and is obtained from Eq. (28).

$$q_{\text{cd,gi-go}} = \frac{K_g}{L_g}\left(T_{\text{gi}} - T_{\text{go}}\right) \tag{28}$$

Finally, the freshwater distillate obtained per hour from the solar still can be calculated from Eq. (29). One should note that the following expression is written based on the assumption that the mass of the evaporated impure water is equal to the mass of the freshwater distillate collected. Equation (30) calculates the overall thermal efficiency of the solar still.

$$m_{\text{fw}} = \frac{q_{e,\text{w-gi}}}{h_{\text{fg}}} \times 3600 = \frac{h_{e,\text{w-gi}}\left(T_w - T_{\text{gi}}\right)}{h_{\text{fg}}} \times 3600 \tag{29}$$

$$\eta_{\text{SS}} = \frac{m_{\text{fw}} \times h_{\text{fg}}}{I(t)_s \times A_s \times 3600} \tag{30}$$

2.1.3 Equations for Energy Balance

This section discusses the equations for energy balance at critical locations in solar stills are discussed in this section. Key locations include the glass cover exterior and interior face, basin wall liner, and impure bulk water. Energy balance for the glass cover exterior face is written as follows:

Energy flow rate incident at the interior side of the glass cover face through conduction = Heat loss rate from the exterior glass cover face to the ambient through convection and radiation

$$q_{\text{cd,gi-go}} = q_{c,\text{go}} + q_{r,\text{go}} = q_{t,\text{go}} \tag{31}$$

Substituting the values from Eqs. (7) and (28), we get,

$$\frac{K_g}{L_g}\left(T_{\text{gi}} - T_{\text{go}}\right) = h_{t,\text{go}}\left(T_{\text{go}} - T_a\right) \tag{32}$$

The energy balance for the glass cover interior face is written as follows:

Energy flow rate absorbed from solar insolation + Energy flow rate received at the interior glass cover face from impure bulk water through convection, radiation, and evaporation = Heat loss rate to the exterior glass cover face through conduction

$$\alpha_g I(t)_s + q_{c,\text{w-gi}} + q_{r,\text{w-gi}} + q_{e,\text{w-gi}} = q_{\text{cd,gi-go}} \tag{33}$$

$$\alpha_g I(t)_s + q_{t,\text{w-gi}} = q_{\text{cd,gi-go}} \tag{34}$$

Substituting the values from Eqs. (26) and (28), we get,

$$\alpha_g I(t)_s + h_{t,\text{go}}\left(T_{\text{go}} - T_a\right) = \frac{K_g}{L_g}\left(T_{\text{gi}} - T_{\text{go}}\right) \tag{35}$$

The energy balance for the basin wall liner is written as:

Energy flow rate absorbed from solar insolation = Rate of heat lost to the impure bulk water through convection + Rate of bottom heat rejected to the ambient

$$\alpha_b \tau_g \tau_w I(t)_s = q_w + q_b \tag{36}$$

Substituting the values from Eqs. (11) and (12), we get

$$\alpha_b \tau_g \tau_w I(t)_s = h_w(T_w - T_b) + h_b(T_b - T_a) \tag{37}$$

where

$$\alpha_b \tau_g \tau_w = \alpha_b\big(1 - \alpha_g - R_g\big)(1 - \alpha_w - R_w) \tag{38}$$

The energy balance for impure bulk water is written as:

Energy flow rate absorbed from solar insolation + Energy flow rate received from basin wall liner through convection + Energy flow rate supplied by external devices = Energy flow rate stored + Heat loss rate to the interior face of the glass cover through convection

$$\alpha_w \tau_g I(t)_s + q_w + Q_u = m_w C_w \frac{\mathrm{d}T_w}{\mathrm{d}t} + q_{c,\text{w-gi}} + q_{r,\text{w-gi}} + q_{e,\text{w-gi}} \tag{39}$$

Or,

$$\alpha_w \tau_g I(t)_s + q_w + Q_u = m_w C_w \frac{\mathrm{d}T_w}{\mathrm{d}t} + q_{t,\text{w-gi}} \tag{40}$$

Substituting the values from Eqs. (11) and (26), we get,

$$\alpha_w \tau_g I(t)_s + q_w + Q_u = m_w C_w \frac{\mathrm{d}T_w}{\mathrm{d}t} + h_{t,\text{w-gi}}\big(T_w - T_{\text{gi}}\big) \tag{41}$$

where

$$\alpha_w \tau_g = \alpha_w\big(1 - \alpha_g - R_g\big) \tag{42}$$

For passive solar still, $Q_u = 0$. The values of T_{gi} and T_b can be substituted from Eqs. (32), (35), and (37) into Eq. (41), and it will reduce to a non-homogeneous first-order differential equation in T_w, of the form shown in Eq. (43).

$$\frac{\mathrm{d}T_w}{\mathrm{d}t} + aT_w = f(t) \tag{43}$$

In this section, we discuss the thermal energy balance for a single-basin single-slope (SBSS) solar still, which is also regarded as a simple passive solar thermal water purification system. The mathematical model can be used to streamline the design of a solar thermal water purification system with maximum productivity that suits a particular set of parameters, such as geographical location and availability of solar energy, etc. Next, we discuss lab-scale experiments in the literature to determine critical parameters affecting freshwater production in a solar still.

2.2 *Experimental Studies on Passive Solar Stills*

In the previous section, we discussed that the freshwater productivity of solar thermal systems is influenced by atmospheric conditions such as solar insolation, ambient temperature, relative humidity, wind velocity, and sky temperature. These factors are beyond human control. However, various design parameters significantly impact the evaporation and condensation rates in solar thermal purification systems. These include the tilt angle of the glass cover, the depth of the impure bulk water, the orientation of the solar still, the area of the absorber plate, the initial temperature of the impure bulk water, the temperature gradient between the air–water interface and the glass cover, the materials used for each component of the solar still, and the insulation material (Muthu Manokar et al. 2014).

2.2.1 Transmission and Absorbance of Solar Energy

The solar energy spectrum incident on the surface of Earth comprises a wide range of wavelengths ranging from 300 nm to 2.5 μm, which is further classified into the ultraviolet region (300–400 nm), the visible light region (400–700 nm) and the infrared region (700 nm–2.5 μm), as shown in Fig. 4. The infrared region accounts for the majority of the solar spectrum (52%) incident on the surface of Earth, followed by the visible light region (45%). Freshwater distillate generated from solar stills is determined by the transmittance of the glass cover for the incident solar energy, followed by the absorption of the transmitted energy in the basin, and impure bulk water, leading to its evaporation and condensation. Hence, the transmittance and absorbance of solar energy are critical factors that determine the efficiency of a solar still.

Generally, the transmittance value of the glass cover should tend to unity for maximum transmission of incident solar energy to the impure bulk water in the

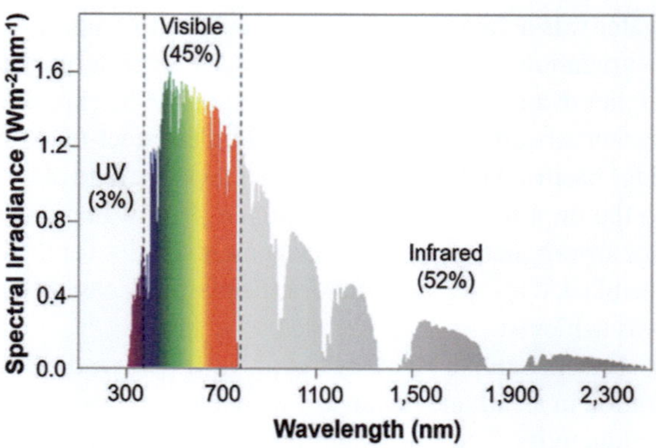

Fig. 4 Solar spectral irradiance on the surface of Earth across the wavelength range from 300 to 2500 nm (reproduced with permission from Chen et al. 2019, Elsevier)

basin. Solar transparent coatings like anti-reflective coatings, solar-selective coatings, low-iron glass coatings, and nanostructured coatings are designed to increase the transmission of sunlight through the glass cover. The use of such coatings reduces the radiative heat loss from the exterior glass cover face to the ambient, see Eq. (4), leading to enhanced evaporation rates. Subsequently, the freshwater output is also enhanced. High reflection of light is observed due to a significant difference in the refractive index (n) of the two media, say 1 for air and 1.5 for glass. Silica-based porous single-layer anti-reflection coatings have been commercialized worldwide to mitigate the heat loss from high reflection. Equation (44) is used to determine the intermediate refractive index (n_i) of the anti-reflection coating between the two media having a respective refractive index of n_1 and n_2. However, challenges like durability and pore clogging in hot and humid climates have shifted attention toward multi-layer anti-reflection coatings. The effective intermediate refractive index of such coatings can be easily tuned using various combinations of high and low refractive index materials, leading to better performance and increased durability. Although multi-layer coatings lead to higher production costs, their superior performance and longevity make them promising candidates for future commercial adoption, surpassing single-layer coatings in solar thermal applications (Law et al. 2023).

$$n_i = \sqrt{n_1 n_2} \tag{44}$$

The absorption rate of basin water refers to how effectively the impure bulk water heats up by absorbing the incident solar energy, leading to the evaporation of impure bulk water and condensation of the generated vapors. About 11% of the total solar irradiation falling on a typical solar still remains unutilized in the basin and is reflected (Muthu Manokar et al. 2014). Reflection loss in bulk water is generally mitigated by increasing the absorption rate by adding solar energy-absorbing materials like chemical dyes, black rubber mats, ethanol, etc., in the impure water. Rajvanshi (1981), added various dyes, viz. black, green, and red, in the impure bulk water to investigate the effect of dyes on the absorption rate of solar energy along the depth of the impure bulk water. The maximum absorption of solar energy throughout the depth of bulk water was achieved with black dyes, due to a significant reduction in photochemical degradation, which led to increased temperatures near the air–water interface, and higher evaporation rates. Akash et al. (1998) experimentally evaluated the effect of various absorbing materials on a stainless steel-made single-basin solar still, having a 3 m^2 basin area. Black rubber mat, black dye, and black ink were added independently to the impure bulk water to compare freshwater production with solar stills that do not absorb materials. An enhancement of 38% for the black rubber mat case, 60% for the black dye case, and 45% for the black ink case in the daily freshwater productivity was achieved.

Panomwan Na Ayuthaya et al. (2013) investigated the effect of ethanol concentration in an ethanol solar still with a finned basin surface to achieve increased water productivity. The freshwater productivity rises with an increase in solution concentration within the 10–50% v/v range, as the mass transfer is influenced by both convection and diffusion. However, freshwater output decreased as concentration increased. This is because diffusion becomes the dominant process. Additionally, a further shift in the boiling point of the solution was reported with the decrease in concentration. Abed et al. (2022) experimentally investigated the influence of chemical dyes on improving the freshwater productivity of passive solar stills. The aim was to identify an appropriate chemical or dye to make the process of impure water purification in solar stills technically and economically viable. The use of thymol blue solution resulted in the highest freshwater production at a depth of 2.5 cm, and 40% increase the freshwater productivity was reported with the use of chemicals. The results also suggest that the impure bulk water depth influences water productivity, and its effect is discussed next.

2.2.2 Impure Water Depth

Tiwari and Tiwari (2006) investigated the parametric effect of impure bulk water depth through experiments carried out in the summer season. They concluded that the freshwater productivity and overall efficiency of the solar still have an inverse proportionality relation with the impure bulk water depth. This is because the heat capacity of the impure water mass is inversely proportional to its depth. Hence, higher temperatures of impure bulk water are obtained at lower water depths due to enhanced solar absorbance. Various studies have obtained similar results on active solar stills (Tripathi and Tiwari 2005) and passive solar stills with plexiglass as the basin material (Phadatare and Verma 2007). Plexiglass has been used to tackle and overcome the challenge of corrosion on metallic surfaces like aluminum and galvanized iron from saline water. Wooden basins have low durability and longevity when used with saline water.

Shirsath et al. (2018) experimentally investigated the effect of impure bulk water depth on the freshwater production of a doubly inclined solar still, as shown in Fig. 5a, with water depths of 10 mm, 30 mm, and 50 mm, respectively. Three feed water samples, viz. pure water, salt water, and saline water, respectively, were also used for the parametric investigations. A doubly inclined solar still is superior in water collection to a conventional single inclination solar still, as the freshwater condensate flows to a single point in the former, compared to a line in the latter. They also concluded that freshwater production is highest for the smallest feedwater depth. However, the duration of freshwater production is longer at the highest feed water depth due to higher temperatures. However, the cumulative water produced was still high at the smallest feed water depth, as shown in Fig. 5b. This is because very high bulk water evaporation occurs at small water depths. Results suggest that both high inlet water temperatures and high air–water interface temperature promote evaporation, and their effects are discussed next.

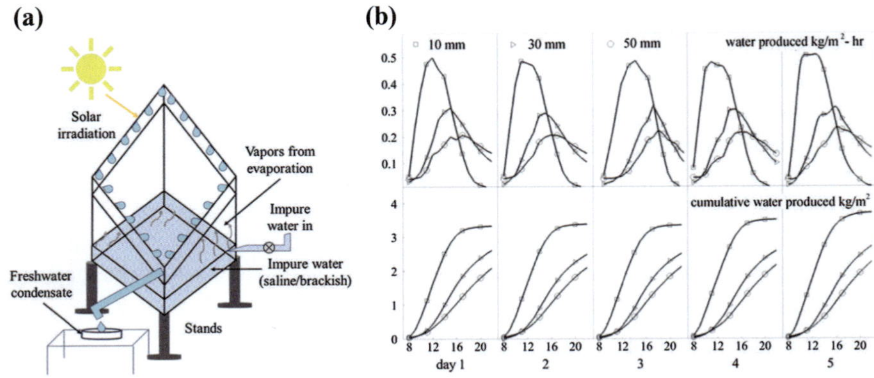

Fig. 5 Schematic of the designed **a** doubly inclined solar still and **b** plot showing the effect of impure bulk water depth on the individual (top row) and cumulative (bottom row) daily freshwater productivity (reproduced with permission from Shirsath et al. 2018, Elsevier)

2.2.3 Temperatures of Feed Water at Inlet and Air–Water Interface

It is evident from the previous sections that the inlet temperature of impure feed water and its temperature at the air–water interface inside the basin are crucial factors in the operation of solar stills. The solar stills were charged with wastewater from the thermal and chemical power plants. The parametric effect of feed water temperature has been studied in the literature by maintaining various patterns of feed water flow inside the solar still basin. These were divided into solar stills that were charged once a day (Sodha et al. 1981), having intermittent flow (Tiwari and Garg 1985) and constant flow (Sodha et al. 1981; Yadav and Kumar 1991) of waste hot water, as shown in Fig. 6a, respectively. Solar stills with intermittent and continuous feedwater flows reported higher freshwater production than those filled once a day. An increase in the mass flow rate of waste feed water reduces the daily distillate output and the solar-to-water conversion efficiency. This is due to a reduction in the residence time for waste hot water inside the still basin, resulting in a lower maximum water temperature.

The effect of initial bulk water temperature on the daily freshwater output and solar-to-water conversion efficiency inside a solar still. The daily freshwater output and solar-to-water conversion efficiency increase monotonically with the increasing bulk water temperature, due to enhanced evaporation rates. Higher evaporation rates result in higher collection of freshwater distillate. Thus, it can be concluded that higher feedwater temperature within the basin results in higher freshwater production. In addition to feeding the solar still basin with hot waste/impure water, various basin design modifications have been proposed to enhance the performance of solar stills.

Hussain and Rahim (2001) placed an aluminum plate 20 mm below the air–water interface to improve the performance of the solar still. The water temperature at the air–water interface increases as the impure bulk water reservoir present at the top of

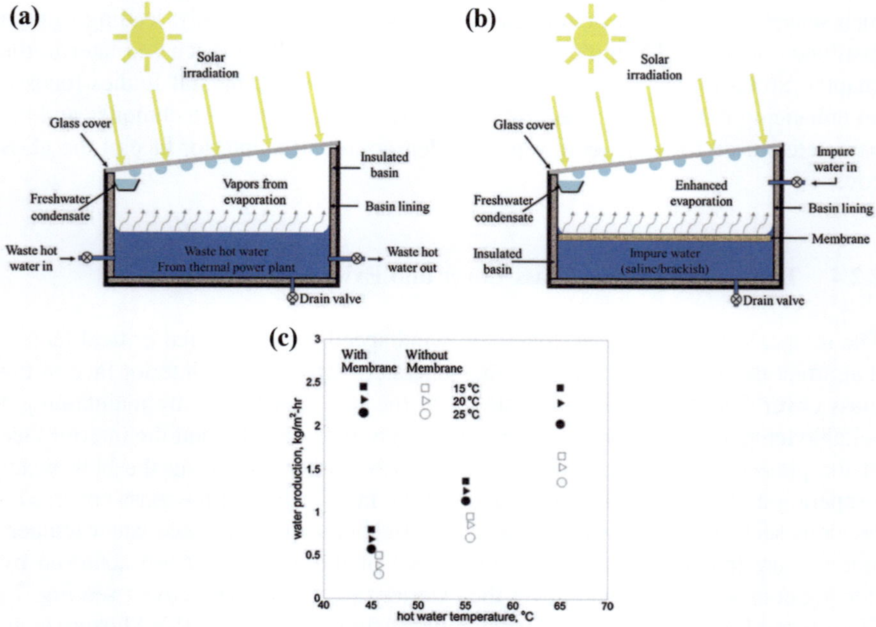

Fig. 6 Schematic showing **a** the constant supply of waste hot water from the thermal power plant to the solar still and **b** hydrophobic membrane floating over the impure bulk water surface of a passive solar still. **c** Effect of feed water temperature on freshwater production with and without a membrane (reproduced with permission from Shirsath et al. 2020, Elsevier)

the aluminum plate mimics a solar still with shallow water depth. Hence, the freshwater production improves due to enhanced evaporation rates. Moreover, the impure bulk water reservoir under the aluminum plate serves as a heat storage reservoir. Thus, the temperature of the impure bulk water remains higher even after sunset, which aids in evaporation. Similar techniques have been adopted in the literature for increasing freshwater productivity by raising the temperature of bulk water at the air–water interface, viz., floating a black-coated perforated aluminum sheet (Nafey et al. 2002), finned black-coated perforated aluminum (Valsaraj 2002), mica plate (El-Sebaii et al. 2000), etc., over the bulk water surface.

Shirsath et al. (2020) evaluated the thermal performance of a lab-scale variable air gap solar distillation system, both with and without a hydrophobic membrane floating on the surface of the impure bulk water, as shown in Fig. 6b. The distillate collection obtained, see Fig. 6c, was higher for the former case for all hot water temperature cases studied. Another technique for improving the performance of solar stills is by floating hydrophilic porous media structures over the impure bulk water surface. The impure water wicks to the top surface of the porous structure due to the action of capillary forces. A large number of tiny air–water interfaces are formed between the pores of the structure. Thus, the surface area available for the evaporation of impure water increases significantly due to the formation of such meniscuses. Hence, the evaporation rate is enhanced significantly, leading to high distillate collection. Examples of such techniques have been discussed later in the chapter. So far in this chapter, we have discussed the experimental studies focused on enhancing evaporation rates. Next, we discuss experimental techniques and the parameters used for enhancing vapor condensation on the interior face of the glass cover.

2.2.4 Temperature of the Glass Cover and External Wind Speed

The temperature of the glass cover and wind speed are considered critical factors that affect the condensation rate of the generated vapors on the interior face of the glass cover. The freshwater productivity of the still is enhanced by maintaining a suitable temperature difference between the impure bulk water and the interior face of the glass cover. This is typically achieved by either increasing the bulk water temperature or lowering the temperature of the interior face of the glass cover. The previous section has already discussed investigations with high feed water temperatures. Low temperatures at the interior face of the glass cover are achieved by flowing cold water in a jacket over the exterior face of the glass cover; see Fig. 7a (Tiwari and Rao 1984; Lawrence et al. 1990; Arunkumar et al. 2012; Shirsath et al. 2020). Such modifications lead to an increase in the freshwater productivity of solar stills compared to passive solar stills. However, the cold-water flow through the water jacket reduces the impure bulk water temperature in the basin; see Fig. 7b. One should note that a lower glass cover temperature enhances condensation efficiency but must be balanced to avoid excessive heat loss. Proper material choice and insulation are crucial to managing this temperature. Also, using a chiller for freshwater production in a solar still affects its thermal efficiency. Figure 7b shows how the chiller affects the variation in thermal efficiency of a solar still over time, for both clear (top row) and cloudy (bottom row) days. The dashed line represents data without the chiller, while the solid line indicates the cold surface temperature maintained at 10 °C. The energy used by the chiller consistently reduces the thermal efficiency, though the impact is relatively small. The efficiency peak observed during the early morning is diminished due to the ongoing energy consumption by the chiller (Fig. 7b).

Rajvanshi (1981), Zurigat and Abu-Arabi (2004), and El-Sebaii (2011) reported an increase in freshwater productivity at higher wind speeds. Convective heat loss from the glass cover increases at high wind speeds. This also leads to a lower temperature of the glass cover. While this temperature drop enhances condensation by increasing the temperature difference between the impure water and the glass cover, excessive wind may cause overcooling and reduce the efficiency of the system. This section discusses critical parameters that influence the productivity of passive solar water purification systems, termed solar stills. The following section outlines multiple performance enhancement techniques implemented to improve the poor productivity of conventional solar stills.

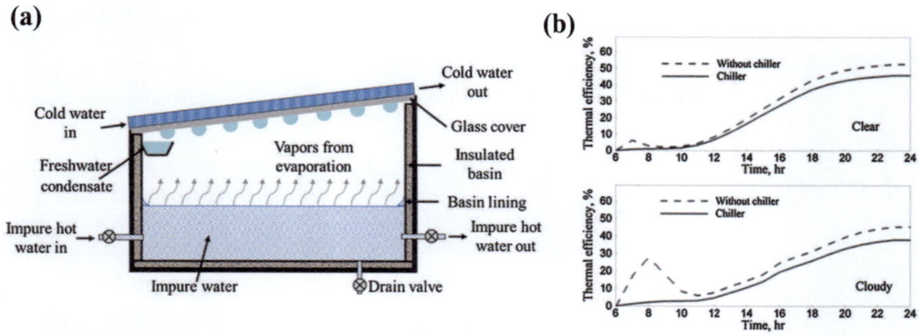

Fig. 7 Schematic of **a** an active solar still using a cold-water jacket on the glass cover surface. **b** Effect of chiller on the variation of thermal efficiency of a solar still with respect to time for clear (top row) and cloudy (bottom row) days (reproduced with permission from Shirsath et al. 2020, Elsevier)

3 Performance Enhancement of Solar Stills

In the previous section, we discussed experimental studies on passive solar stills. A few analytical, numerical, and active solar still studies were also reviewed to enhance the freshwater and thermal performance of passive solar stills. Freshwater distillate production can be improved by redesigning the glass cover for enhanced condensation. Generally, water jackets are employed on the glass cover of solar stills that maintain a suitable temperature differential between the vapor-generating and condensing surfaces. The air gap between the vapor-generating and condensing surfaces is also a critical parameter affecting the efficiency of the solar still. Apart from enhancing the source, i.e., the solar energy receiver, and the sink, i.e., the condenser of the solar stills, heat absorption and storage also play a crucial role in freshwater production. Adequate heat absorption is achieved by modifying the basin geometry or employing accessories, such as porous media, in the bulk water. Further, heat storage is achieved through proper insulation or by incorporating multiple heat storage materials like pebbles, sand beds, water tanks, and phase change materials. Figure 8 summarizes various design advancements and modifications implemented to improve the performance of passive solar stills. They have been classified into four major categories, viz., isolation collection (modifying the design of the receiver), heat absorption (modifying the design of the basin), heat storage (modifying the design of the basin/walls), and vapor condensation (modifying the design of the condenser) of a passive solar still. Each of them has been discussed in detail next.

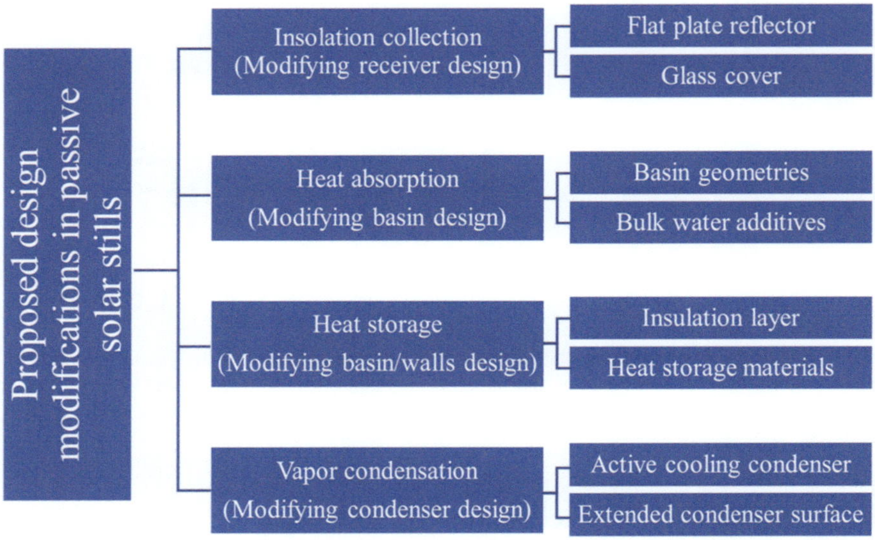

Fig. 8 A summary of different design advancements and modifications implemented to improve the performance of passive solar stills

3.1 Insolation Collection

The primary process in freshwater production using a passive solar still is the process of vapor generation by evaporating the impure bulk water. The generation of water vapor requires considerable solar energy due to the parasitic losses to the bulk and ambient; see Fig. 2a. Hence, the first step is to collect all the solar radiation that falls on Earth's surface. Investigations from various researchers propose the integration of flat plate reflectors to capture the maximum solar energy available for vapor generation; see Fig. 9. The flat plate reflectors are either installed inside the solar still on the side walls, or outside over the edge of the glass cover, or a combination of both. Bataineh and Abbas (2020) installed flat plate reflectors on the side walls of the basin along with a finned base for its performance enhancement, as shown in Fig. 9a. An increment of about 22 to 34% in freshwater production was achieved with the above modification to the passive solar still. Karimi Estahbanati et al. (2016) installed internal reflectors in the basin and reported an increase of 65% in freshwater during the winter season. The vapor-generating rate during the winter season is low due to a large amount of solar irradiation falling on the side walls. This is because the sun follows a lower trajectory across the sky due to the tilt of the Earth's axis. Higher-altitude regions are the most affected by this phenomenon in winter. Hence, we experience a reduction in the rate of freshwater generation. Integrating reflectors on the side walls increases the amount of solar irradiation falling on the impure bulk water, thus improving the vapor generation rate.

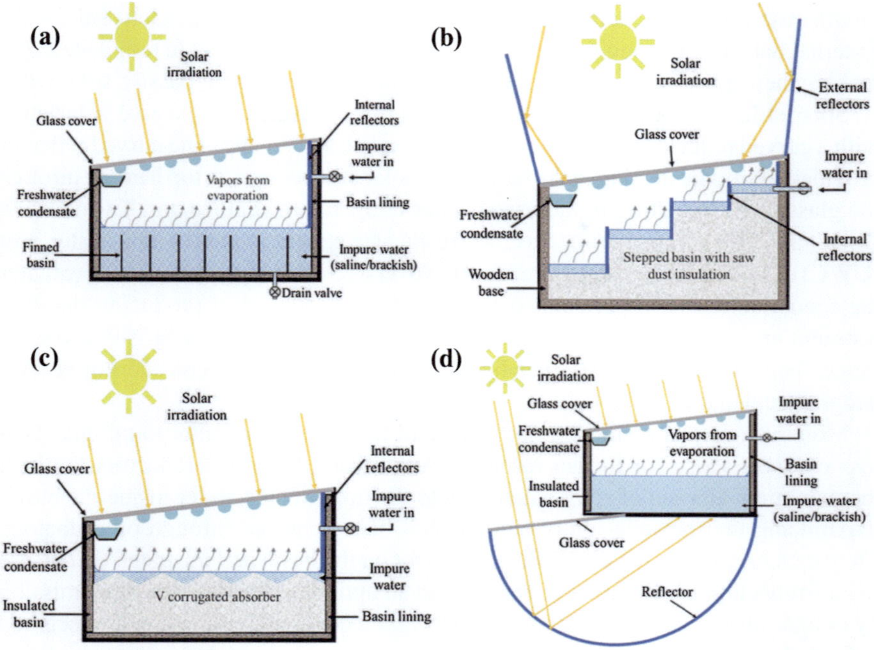

Fig. 9 Schematic of a conventional solar still **a** integrated with flat plate internal reflectors on the inner walls of the basin with a finned base, **b** stepped basin, **c** V corrugated wick basin absorber, and **d** inverted absorber solar still, respectively

Omara et al. (2014, 2016) reported an enhancement in water productivity by conducting experiments on a stepped basin solar still by integrating both internal and external flat plate reflectors, see Fig. 9b, and on a V corrugated wick basin absorber solar still, see Fig. 9c, respectively. However, one should be cautious about the aspect ratio of external flat reflectors and their inclination relative to the glass cover, as they can obstruct sunlight during the early morning and evening hours. Tanaka (2010) and Huang and Chang (2017) used computer modeling to determine the optimal monthly angle for an external top reflector. They concluded that using a reflector at this optimal angle, with a fifty percent basin size, could increase annual water yield by 67% and 26% compared to a reflector fixed vertically. However, the installation of reflectors outside the solar still increases the temperature of the glass cover. The increased glass cover temperature provided by the reflectors compensates for the reduced ambient and water temperatures. This maintains or even boosts the evaporation and condensation rates, leading to improved water productivity of solar stills. In extremely hot climates, excessive heat an reduce system efficiency due to overcooling or thermal saturation, where the temperature difference becomes less pronounced. The temperature rise is aggravated during summer due to the increased accumulation of distillate droplets. Moreover, external flat reflectors have been recommended for remote regions as they can function smoothly with a manual adjustment every month instead of using motors for optimum daily settings (Mohsenzadeh et al. 2021). External flat collectors require additional space, which is abundant in remote regions, making them a perfect solution for enhancing the performance of passive solar stills.

Solar stills with an inverted absorber combine conventional passive solar stills with a curved reflector attached to the basin bottom; see Fig. 9d. The curved reflector provides additional heating at the base of the solar still, alongside top heating through the glass cover, resulting in increased water temperature (Suneja and Tiwari 1999; Dev et al. 2011). A 192% increase in freshwater generation and a unit water cost (UWC) of US$0.015 per liter were reported over a 15-year life cycle. UWC evaluates the economic feasibility of solar stills and refers to the cost of producing one liter (or cubic meter) of distilled water. The UWC of a conventional solar still typically ranges between ~ US$0.0163 per liter to US$0.034 per liter, considering various geographical and climatic factors.

Moreover, insulating the bottom surface of the inverted reflector to mitigate heat loss provides scope for further research. Apart from the use of reflectors to direct the maximum amount of solar irradiation to the impure bulk water inside the basin, maximizing the transmission of sunlight has been achieved through other factors. They include varying the tilt of the glass cover on the basin and its material. The use of anti-reflection or selective coatings is also an option to maximize the transmission by mitigating the radiative heat loss from the glass cover to the ambient, as discussed in Sect. 2.2.1.

3.2 Heat Absorption

We know that the freshwater productivity of passive solar stills is low due to inherent parasitic losses, as shown in Fig. 2a. The efficient conversion of solar energy falling on the glass cover into heat is one of the key factors for judging the performance of a passive solar still; see Fig. 2b. This is accomplished by absorbing ~ 95% of the solar insolation incident on the impure bulk water, resulting in high water temperatures. High bulk water temperatures result in higher freshwater productivity as discussed in Sect. 2.2.3. Several investigations have explored various geometrical designs for constructing the basin with an expanded surface area for enhanced evaporation rates, which include configurations like stepped, see Fig. 9b, inclined stepped basin, see Fig. 10c, V-corrugated, see Fig. 9c, and concave, see Fig. 10d, respectively (Omara et al. 2014, 2016; Elango and Kalidasa Murugavel 2015; Kabeel 2009). These configurations increase the basin area available for vapor generation and reduce the moist air gap between the air–water interface and the glass cover. Hence, such configurations have reported higher freshwater productivity (~ 20%) than passive solar stills by enhancing the vapor generation rates (Omara et al. 2014).

Moreover, the heat absorption has been enhanced by increasing the area available for vapor generation by adding fins on the inner surface of the basin (Bataineh and Abbas 2020; Jani and Modi 2019). See Figs. 9a and 10a. The use of fins in the basin not only enhances the evaporation area but also improves the heat capacity of the

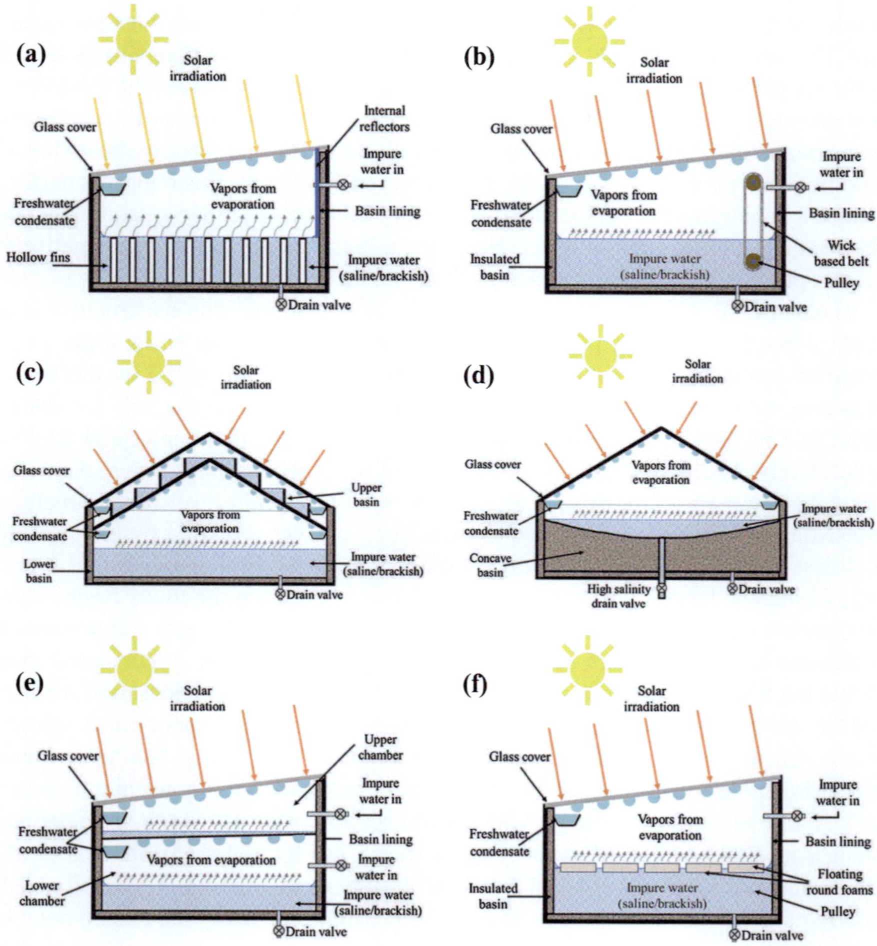

Fig. 10 Schematic of solar still with **a** hollow fins with square cross-section, **b** rotating wick fabric belt, **c** double-stage double-slope stepped basin, **d** concave basin, **e** double-stage inclined basin, and **f** floating porous foams, respectively

basin. Thus, vapor generation continues inside the basin even after sunset, as the impure bulk water temperature remains high due to the enhanced heat capacity of the bulk. However, metals generally corrode when in contact with saline/brackish water. Investigations on solar still basins integrated with fins made of anti-corrosion materials like mica or glass have reported similar enhancement in freshwater productivity in comparison to the metallic fins (El-Sebaii and El-Naggar 2017; Velmurugan et al. 2008). Jani and Modi (2019), the thermal performance of a single-basin dual-slope solar still equipped with hollow fins of circular and square cross-sections was analyzed, as illustrated in Fig. 10a. Their study revealed that circular hollow fins produced 54% more water yield than square fins of the exact dimensions and quantity. This result is somewhat unexpected, as similar-sized fins in both shapes would typically perform comparably. Optimizing fin parameters such as height, thickness, and density (the number of fins per unit basin area) is crucial for maximizing freshwater output in solar stills with integrated fins. Taller and thicker fins have been shown to significantly enhance water yield. The increased mass of the fins improves performance by enlarging the contact area, increasing heat storage capacity, and reducing the heat capacity of impure water. However, a high density of fins can negatively impact productivity due to shadowing effects on the basin surface.

Haddad et al. (2017) incorporated a motorized rotational porous structure in a vertical position inside a passive solar still; see Fig. 10b. The basin walls were made of a transparent material, allowing maximum sunlight to fall on the wick. However, transparent basin walls incurred parasitic losses from the still. Enhancement of 15% and 51% in freshwater yield were observed in summer and winter, respectively. Additionally, the unit water cost of operation was high due to the extra input energy required to drive the prototype motor. A multi-level basin structure consisting of two basin levels, as shown in Fig. 10c, was proposed by Elango and Kalidasa Murugavel (2015). The upper basin, constructed from a transparent material, facilitates the efficient transmission of solar radiation to the lower basin. This configuration effectively doubles the exposed surface area of the bulk water without increasing the overall basin area for solar radiation absorption, as compared to a single flat basin. Additionally, the upper basin functions as a condensation surface for the lower level, transferring the heat released during water vapor condensation to the bulk water above. This enhances performance by capturing heat that would otherwise be rejected to the ambient from the glass cover of a conventional solar still. Another technique for improving the contact area is the use of porous materials along with the modified basin designs, as shown in Fig. 10d–f (Arunkumar et al. 2018; Jani and Modi 2019; Modi and Modi 2019). Arunkumar et al. (2018) added multiple porous floating sponges directly into basin water rather than using a wick material layer on the basin surface. The impure water wicks to the top surfaces of these sponges through the action of capillary forces. Thus, a thin layer of dirty water forms on the top surface of these porous materials and is directly exposed to the incident solar irradiation.

Wiener et al. (2024) proposed a cost-effective method to enhance the vapor generation rate and freshwater production of a solar still by floating polyurethane rollers wrapped with textile fabrics on the impure water surface, as shown in Fig. 11. Polyurethane rollers coated with black textile fabric showed the best performance among all other fabrics used for vapor generation. Floating porous structures at the air–water interface offer the advantage of focusing the solar energy on the thin layer of impure water formed at the top surface of the porous structures. Vapor generation through such a phenomenon is known as "interfacial solar heating", which the next chapter discusses in detail. However, the authors employed active methods to produce freshwater condensate from the generated vapors. A maximum thermal efficiency of ~ 62.16% and a UWC of US$0.023 per liter of freshwater production were obtained. The use of readily available materials like cotton fabric and polyurethane, combined

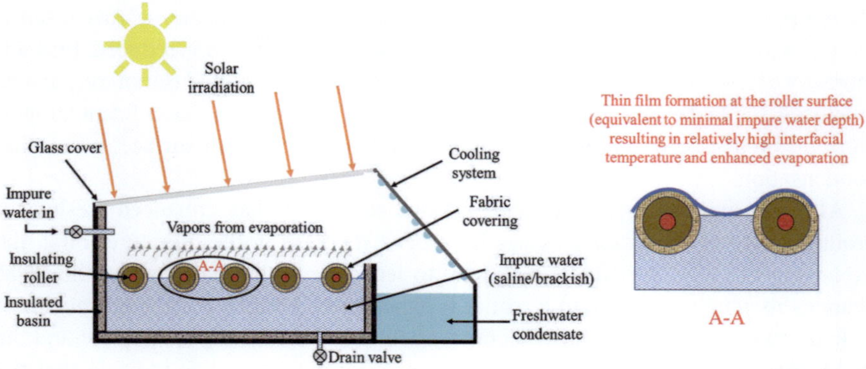

Fig. 11 Schematic of solar still using polyurethane rollers wrapped with textile fabrics showing the formation of a thin film at the roller surface (equivalent to minimal impure water depth), resulting in relatively high interfacial temperature and enhanced evaporation (adapted from Wiener et al. 2024, Springer Nature under Creative Commons CC BY license)

with the low operational cost of the system, suggests potential for sustainable and scalable deployment of the proposed solar still design in water-scarce regions.

3.3 Heat Storage

Conventional solar stills usually produce freshwater during the daytime, with the solar irradiation availability being the most critical factor. With sunlight intermittent on cloudy days, freshwater generation is significantly reduced, or may be zero at night. The peak temperature of impure bulk water is generally attained between 11 AM and 1 PM during the daytime, as the maximum solar irradiance (~ 1000 W/m^2) falls on the glass cover of a passive solar still. When the incoming solar radiation decreases, the temperature of the bulk water also drops, leading to a lower evaporation rate. This occurs because the system receives less energy while continuing to lose heat through the cover and sidewalls. Since passive solar stills lack any energy storage medium to offset this loss, the evaporation process is affected. A common temperature trend in passive solar stills begins with the system at ambient temperature in the early morning, then rises rapidly to a peak shortly after midday. In the afternoon, the temperature gradually declines and continues to drop at a moderate rate, eventually returning to ambient temperature in the evening.

Phase change materials (PCMs) like paraffin wax, hydrated salts, stearic acid, etc., have attracted significant attention from the research community to overcome the bottleneck of low water production in passive solar stills. PCMs store excess heat absorbed during the daytime as latent heat, which can then be utilized for water production after sunset, thereby continuing the efficient period of freshwater generation. This process lowers the peak temperature observed around midday and stretches the temperature profile. As a result, heat loss during midday decreases because the temperature gradient between the system and the surrounding air is reduced. Furthermore, by extending water production into the nighttime, the condensation rate can be enhanced due to the cooler nighttime temperatures. This creates a significant temperature difference between the water vapor and the condensation surface, promoting condensation.

Although integrating PCMs with passive solar stills has enhanced freshwater productivity, few operational issues have been reported in the literature. First, the expansion of PCM as it melts may lead to leakage. Additionally, when the PCM changes its phase from liquid to solid, it experiences volume reduction, making it difficult to ensure an ideal surface contact between the solar absorber plate and the PCM (Al-Kayiem and Lin 2014). Moreover, Simulation studies indicate that the phase change process of phase change materials (PCM) is often assumed to occur without a temperature gradient (Ansari et al. 2013). In this regard, Cheng et al. (2019) combined a pyramid-shaped solar still with a shape-stabilized PCM made from paraffin wax, high-density polyethylene (HDPE), and an expanded graphite composite material, as shown in Fig. 12a. This PCM maintained its shape, resulting in a 43% increase in freshwater yield compared to conventional solar stills. However, there was a potential contamination risk from impure bulk water due to the direct contact between PCM and the bulk water, which could lead to melting. Moreover, Yousef and Hassan (2019) enhanced the thermal conductivity of the composite PCM by integrating it with pin fins and steel wool fibers to boost the productivity of passive solar stills, as shown in Fig. 12b. An instantaneous energy efficiency of approximately 80% and an average daily energy efficiency of around 18% were recorded for the configuration with PCM and pin fins, indicating strong short-term thermal performance during peak solar hours. However, a comprehensive exergy and economic analysis demonstrated that the configuration utilizing PCM in conjunction with steel wool fiber was more effective overall, owing to its superior thermodynamic efficiency and cost-effectiveness, making it the more viable option for practical solar water purification applications.

Sarath et al. (2023) gave a thorough overview of the progression in understanding the physics involved, mathematical modeling, and computational techniques used

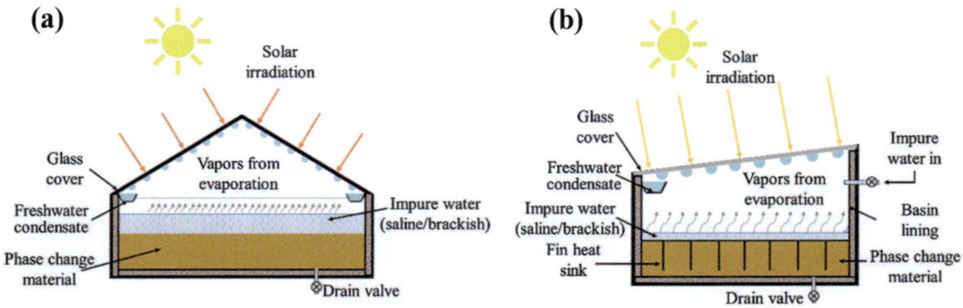

Fig. 12 Schematic of **a** a pyramid solar still integrated with a new shape-stabilized phase change material and **b** a single-slope solar distillation system incorporating a PCM-based pin fin heat sink

to represent physical phenomena in their review article. Traditional experimental methods provide limited data on transient temperature measurements and basic visualizations at specific locations. However, with the emergence of advanced flow visualization techniques such as particle image velocimetry (PIV), it is now possible to clearly explain Rayleigh–Bénard convection during phase change processes. The review also covered modern experimental approaches for investigating thermal convection in PCM. Moreover, accurately predicting the evolution of flow variables can help researchers develop more precise mathematical models for estimating convection currents in complex latent heat storage systems.

Sensible heat storage involves storing energy by raising the temperature of a material, typically a fluid or a solid, without changing its phase. Sensible heat storage is a cheap yet viable option to address the issue of bulk water contamination in the case of direct contact between impure bulk water and latent heat storage materials like PCM (Deshmukh and Thombre 2017; Kabeel et al. 2018). Typical examples of sensible heat storage materials include black-painted stones or pebbles, sand, concrete blocks, bricks, metal pieces (such as mild steel or aluminum chips), and gravel. (Murugavel et al. 2010). A 5% increase in water yield was reported when a 19 mm layer of quartzite rock was placed over the basin, due to sensible heat storage. A similar study (Kabeel et al. 2018) suggested the use of graphite as a sensible heat storage material due to its superior thermal conductivity in comparison to the abovementioned alternatives. The use of sensible heat storage material also offers better thermal stability by reducing inconsistent temperature fluctuations. Hence, incorporating sensible heat storage materials into solar stills provides a more consistent and prolonged water production throughout the day by better utilization of the available solar energy.

The insulation layer of the basin and walls is also vital for storing heat and increasing the performance of a solar still. Proper insulation ensures that more energy is available for heating the water, leading to higher evaporation rates by reducing heat loss, increasing freshwater production, and improving the overall efficiency of solar stills. Polystyrene foam (EPS), bubble wrap, and sawdust are some of the insulating materials used in various investigations to mitigate parasitic losses from the solar still (Bait and Si-Ameur 2017; Arunkumar et al. 2018). All the sections discussed for the performance enhancement of solar stills focus on improving the evaporation rate of impure bulk water and storing excess heat. However, vapor condensation is also critical in freshwater production in solar stills. The following section will discuss investigations carried out to improve vapor condensation. Next, we discuss improving distillate collection through multi-staging by recovering the latent heat of condensation lost to the surroundings of the passive solar still.

3.4 Vapor Condensation and Multi-staging

Vapor condensation occurs when the ambient or the surface temperature is below the saturation temperature corresponding to the equilibrium pressure. Thus, vapors undergo clustering near the condensing surface, leading to nucleation and growth of water droplets. The condensation phenomenon is classified into two categories: filmwise condensation and dropwise condensation. Filmwise condensation occurs when the condensing vapors lead to the formation of a continuous liquid film on the condensing surface (usually hydrophilic). In dropwise condensation, the vapors condense into droplets of varying sizes instead of forming a film on the condensing surface (usually hydrophobic). The liquid film formed in filmwise condensation provides extra resistance to the heat transfer between vapors and the condensing surface. However, there is no film formation in dropwise condensation because droplets slide after reaching a critical size. Hence, the heat transfer coefficient is superior for the case of dropwise condensation than for filmwise condensation.

In solar thermal water purification systems, efficient condensation of vapors generated from the evaporation of impure bulk water plays a critical role in their thermal performance and freshwater productivity. Shirsath et al. (2019) studied the condensation of water vapors on various inclined patterned surfaces mimicking the interior face of a glass cover. Figure 13a shows the schematic of the setup in which both the vapor-generating and condensing surfaces are temperature-controlled. In this regard, three surfaces were prepared: bare copper, a pillared surface on a copper substrate produced by wire EDM, and a laser ablated surface. Figure 13b shows the droplet spreading patterns obtained on these surfaces when a droplet is placed on a dry substrate (first column) and the condensation patterns observed during the experiment (second column).

Filmwise condensation occured on the bare copper substrate because the droplets reached a Wenzel state. On the other hand, dropwise condensation was observed on both textured surfaces due to the Cassie–Baxter state of the droplets on the substrates. However, it was observed that the contact angles on both textured surfaces during condensation were slightly lower than those obtained when a droplet was placed on a dry substrate. This is because the first droplet resting on the condensing surface transitions from the Cassie–Baxter state to the Wenzel state upon nucleation of the second droplet. However, surface hydrophobicity is temporarily restored after the droplet slides down the condensing surface. The effect of tilt angle on droplet slide-off time and freshwater productivity at various condensing surface temperatures is shown in Fig. 13c and d. The droplet slide-off time reduces, increasing freshwater productivity as the condensing temperature is lowered. Both these parameters are enhanced when the slope of the condensing surface is increased from 15° to 35° for each case (wire EDM being the best). However, the freshwater productivity for the plain substrate case was best when the slope of the condensing surface was increased to 55° at all condensing surface temperatures. Hence, the tilt angle of the condensing surface plays a vital role in the performance of solar thermal water purification systems.

Apart from altering the surface morphology to enhance condensation rates on flat condensing surfaces, various glass cover designs have been proposed to increase the condensation area. Zhao et al. (2024) summarized some typical condenser designs available in the literature for solar thermal water purification systems in their review, followed by the scope and challenges in their commercialization. The limitation of

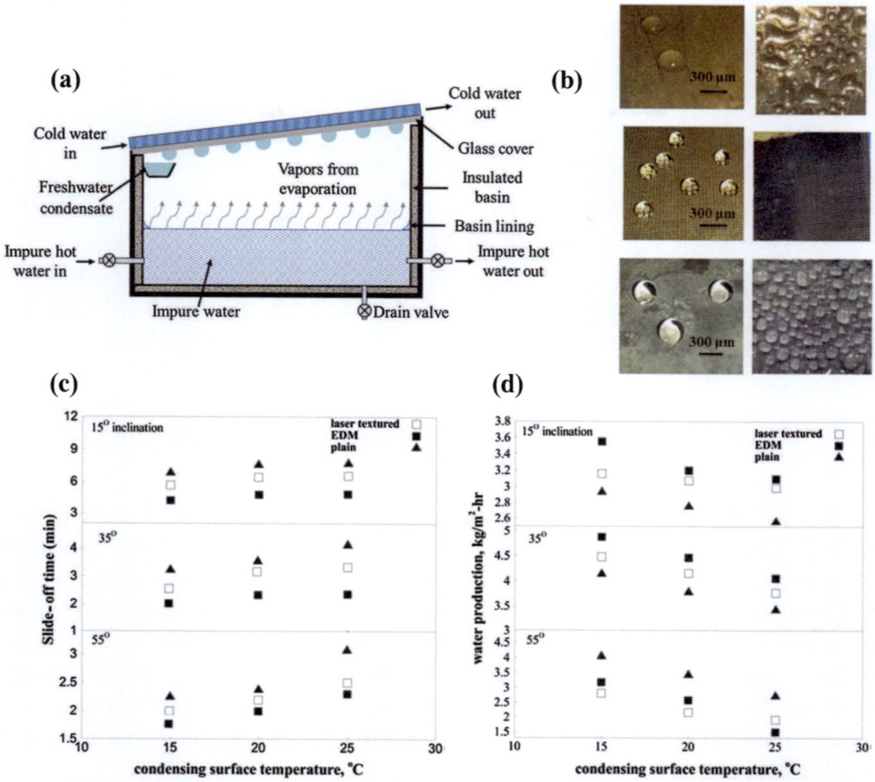

Fig. 13 **a** Schematic of experimental setup with temperature-controlled evaporation and condensation. **b** Photograph showing various dropwise condensation patterns on plain (first row) and textured surfaces using wire EDM (second row) and laser ablation (third row), respectively. Effect of surface texturing on **c** slide-off time and **d** water production with respect to condensing surface temperature at different inclinations of the condensing surface, respectively (reproduced with permission from Shirsath et al. 2019, Elsevier)

the high dependence of single-slope solar stills on the angle of incidence of sunlight is overcome by incorporating symmetry in glass cover design, as shown in Figs. 10c, d and 12a. Various investigations have been carried out on solar stills having pyramid-shaped glass covers to overcome this limitation, as shown in Fig. 14a (Kabeel and Abdelgaied 2020; Dhaoui et al. 2024; Pandey and Naresh 2024). A pyramid-shaped glass cover reduces the dependency on solar tracking mechanisms and overcomes the limitations of light refraction in double-slope solar stills, leading to consistent evaporation rates. Moreover, the high slope of the pyramid allows easy flow of distillate to the distillate collection chamber compared to the conventional solar stills. Dhaoui et al. (2024) reported an enhancement of ~ 17–18% in the overall energy of the pyramid solar still than the double-slope solar still with and without angled fins.

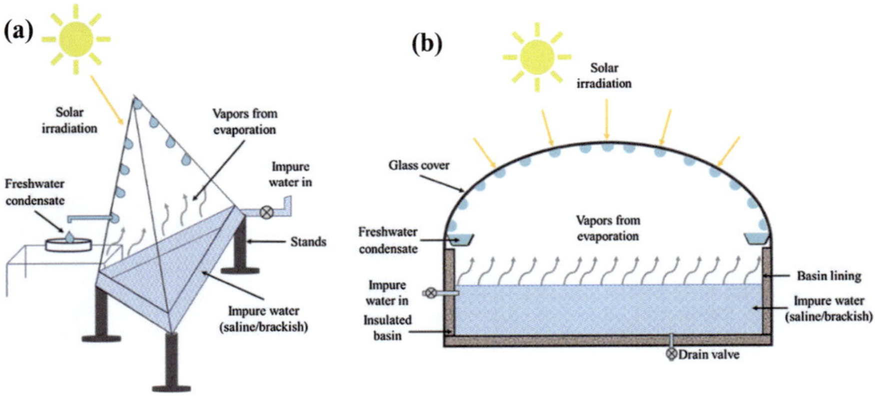

Fig. 14 Schematic of solar still with **a** pyramid-shaped and **b** glass cover, respectively

Furthermore, spherical and hemispherical glass cover designs have been proposed, see Fig. 14b, to eliminate any refraction losses that occurred in the previous glass cover designs (Arunkumar et al. 2012; Hadi Attia et al. 2022). The curved surface of a hemispherical cover allows it to receive and distribute solar radiation more uniformly throughout the day. This is because the shape can capture sunlight from multiple angles, even at a lower incident angle of the incoming sunlight during early morning and evening, maximizing solar exposure. Also, the continuous curved surface minimizes the creation of shadows on the surface, ensuring that the covered area receives consistent sunlight.

3.5 *Latent Heat Recovery*

Conventional or modified design single-stage solar stills lose their latent heat of condensation to the surroundings because the condensation of water vapor occurs only once. Hence, there is a thermodynamic limit on the specific water productivity (refer to Eq. 1 for definition) of 1.32 l/m^2-h in the case of single-stage solar stills (Chiavazzo et al. 2018). The thermodynamic limit of single-stage stills can be breached by reusing the latent heat of condensation through multi-staging. Moreover, compared to large-scale commercial systems like Multi-Effect Distillation (MED) and Multi-Stage Flash (MSF), which operate solely in active mode, multi-stage solar stills can operate in both active and passive modes, offering greater design versatility. Additionally, multi-stage solar stills streamline the desalination process, enabling the creation of small-scale systems suitable for individual families or small communities in remote regions. Peng and Sharshir (2023) have reviewed several designs of multi-stage solar stills proposed in the literature, as shown in Fig. 15a–f, to enhance freshwater productivity and the scope for future research.

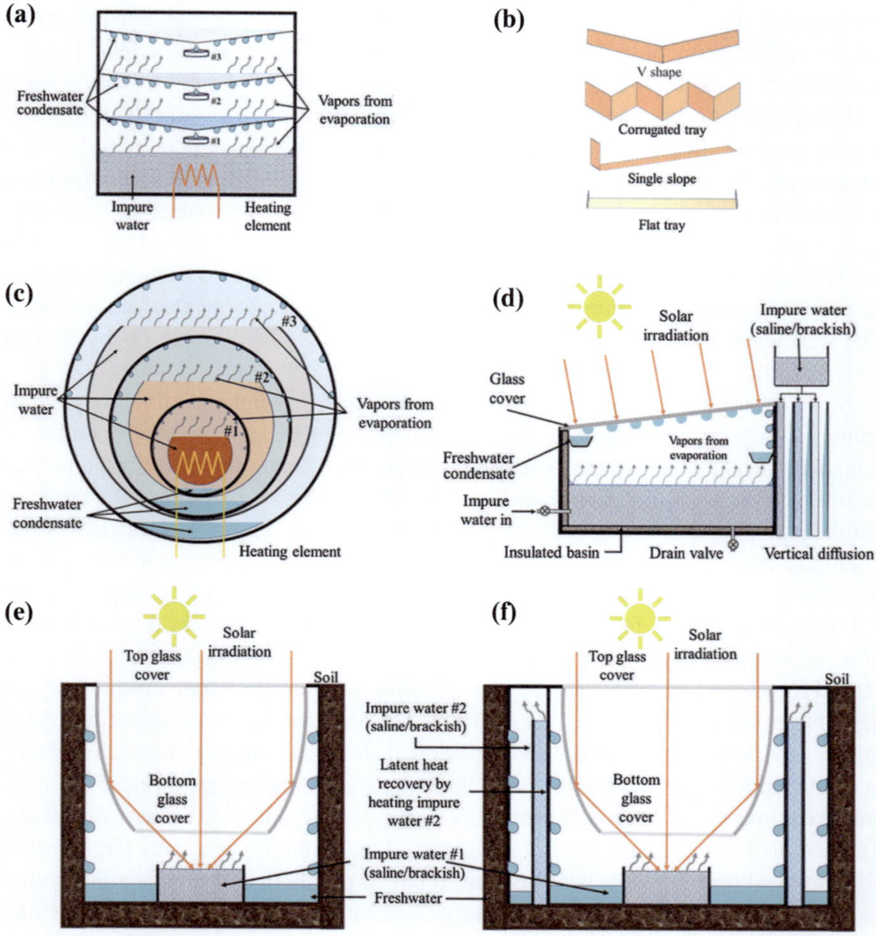

Fig. 15 Schematic of **a** stacked tray with **b** examples of various tray shapes, **c** tubular, and **d** vertical diffusion multi-stage solar still, respectively. Schematic of solar-concentrated stills embedded in the soil, **e** without and **f** with a latent heat recovery mechanism (reproduced with permission from Zhu et al. 2022, Elsevier)

Figure 15a shows a three-stage stacked active solar still. The evaporation of saline water during the initial stage is driven through a heat exchanger placed in the basin. The vapors generated rise above and condense below the inclined tray, denoted as #1, and the distillate is collected in the collection chamber shown below the tray. The latent heat of condensation released is used to heat and evaporate the saline water above tray #1. A similar evaporation and condensation process occurs in the two subsequent stages, #2 and #3. However, the vapor and saline water temperatures of each stage are lower than those of the preceding stage. The V-shaped tray shown in Fig. 15a can be replaced with the options shown in Fig. 15b. Suneja and Tiwari (1999) investigated the effect of impure bulk water depth in a two-stage inverted absorber solar still. They concluded that the freshwater output is considerably affected by the water depth in the first stage of the solar still. A large water depth in the first stage will lead to low evaporation in the first stage itself, resulting in less condensation and low evaporation in the subsequent stages. Hence, small water depths in multi-stage solar stills will increase freshwater productivity. Modi and Modi (2019) incorporated a small stack of wicks (made of jute and cotton) in the first stage of a double-stage inclined solar still to boost the evaporation rate. They observed increases of 18.03% and 21.46% in the freshwater yield at water depths of 0.01 m and 0.02 m, respectively, when using jute cloth wicks, compared to those using black cotton cloth. The freshwater productivity of a two-stage solar still typically improves by 30% to 50% over that of a conventional solar still.

Moreover, tubular-shaped solar stills were proposed to enhance the low yield of contemporary single-slope solar stills by providing more area for vapor condensation. Similarly, tubular-shaped solar sill designs have been proposed for latent heat recovery through multi-staging; see Fig. 15c. The tubular-shaped solar stills function similarly to the stacked tray solar stills described above. Zheng et al. (2013) experimentally investigated a tubular multi-stage solar still (two-effect and three-effect) by mimicking solar flux through electrical heating. Gain output ratios of 1.4 and 1.7 were obtained for the two cases. Yan et al. (2019) investigated the performance of a two-stage tubular active solar still. Three operating pressures, viz., 20, 60, and 101 kPa, were used to investigate the effect of vacuum on water productivity. A gain output ratio of 1.27, 0.7, and 0.57 was obtained for the abovementioned three ranges. The condensing surfaces are positioned above the vapor-generating surfaces in both stepped-type and tubular solar stills. Inclined condensing surfaces are typically required to collect the freshwater condensate, except in flat plate solar stills. This inclined design and water storage at each stage result in a relatively large system volume, leading to greater heat loss and reduced productivity gains as the number of stages increases. As a result, both stepped and tubular solar stills with more than five stages are seldom studied. Vertical diffusion solar stills shown in Fig. 15d have been proposed to overcome the limitations of stepped and tubular-type multi-stage solar stills. Generally, they incorporate the concept of interfacial vapor generation by utilizing wicks and porous metals to localize heat, thereby enhancing freshwater productivity. The following chapter discusses multi-staging integrated interfacial vapor generation in detail.

Zhu et al. (2022) stressed that operating solar stills through active methods by using secondary equipment only enhances the evaporation rates but does not resolve the inherent problems of conventional solar stills. Further, using a hydrophilic wick structure poses a new challenge of salt collection on the vapor-generating surface, hindering vapor generation. Hence, the authors proposed an integrated passive solar-concentrated still with heat recovery, as shown in Fig. 15e and f. The proposed design overcomes the hindrance to sunlight transmission through the glass cover caused by condensing droplets on its opposite end. A parabolic concentrator is also used to enhance the evaporation rate. The side wall losses occurring in conventional solar stills have been mitigated by allowing the vapors to condense on the side walls of the still; see Fig. 15e. The provision for latent heat recovery has also been proposed in the current design, see Fig. 15f, and the entire setup was buried in the soil to save floor space. Solar-to-water production efficiency of 38% and an increment of ~ 18% in fresh-water productivity was achieved through latent heat recovery.

4 Solar Collectors and Heat Pipes

A solar collector is a typical heat exchanger that converts incident solar energy into thermal energy. The thermal energy is transferred to a suitable working fluid, which is then used for a storage or direct use. Solar collectors are primarily categorized into two types: tracking and non-tracking, with each group having its own specific classification. Non-tracking, or stationary, collectors remain fixed, oriented at a set angle that remains constant regardless of the sun's position. Examples include flat plate collectors, Photo-Voltaic Therma (PV/T) systems, compound parabolic collectors, and evacuated tube collectors.

On the other hand, tracking collectors adjust their position throughout the day to keep the solar irradiance striking at a right angle for maximum absorption of solar energy. Unlike fixed-position collectors, these systems rely on sun-tracking mechanisms and are categorized into single-axis and dual-axis tracking systems. Examples of collectors utilizing single-axis tracking include parabolic trough collectors (PTCs) and linear Fresnel reflectors. Conversely, parabolic dish reflectors and circular Fresnel lenses belong to the dual-axis tracking group. Different types of solar collectors are optimized for specific temperature ranges, with evacuated tubes and flat plate collectors typically used for low-temperature applications below 100 °C, making them ideal for low heat flux processes such as desalination.

Solar water heaters can be subdivided into two main categories: active and passive systems. Active systems circulate the heat transfer fluid through a pump, while passive systems rely on gravity for fluid movement. Passive systems are classified into two types: (1) batch collectors, which combine the solar collector and storage tank into a single unit that functions both as the collector and the hot water storage system, and (2) thermosiphon systems, where the solar collector and storage tank are separate, with fluid circulation driven by natural convection. Solar collectors used in these systems may include evacuated tube collectors or flat plate collectors (Tang et al. 2010).

4.1 Flat Plate and Evacuated Solar Collectors

Figure 16a depicts the schematic of a typical solar flat plate collector, which transforms solar energy into heat. First, solar insolation incident on the glazing plates transmits sunlight to the absorber plate. These plates are typically made of high thermal conductivity metals. Typical examples include copper, aluminum, or steel.

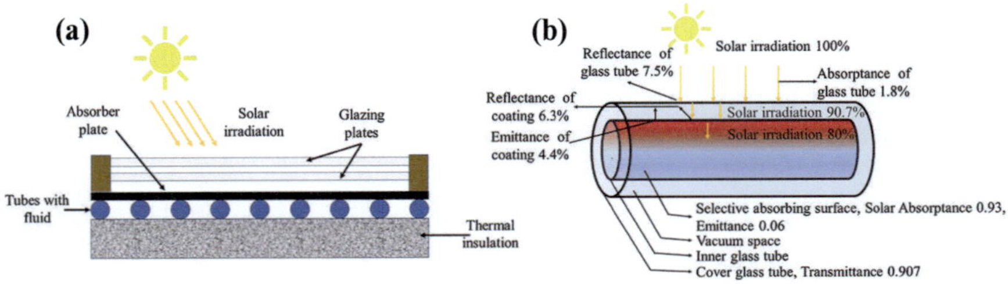

Fig. 16 Schematic of **a** a flat plate solar collector and **b** an evacuated tube solar collector

The top surface of the absorber plate is coated with selective materials to enhance the absorption of solar insolation. A cover sheet, or glazing, is placed over the absorber plate, allowing high-frequency solar radiation to pass while preventing the escape of low-frequency radiation emitted by the absorber. This cover also acts as a barrier, preventing the cooler surrounding air from contacting the absorber plate, and mitigating convective heat loss to the ambient. Glass is commonly used for glazing due to its effectiveness. However, certain plastic materials are a cheap alternative with less durability but higher transmittance. Insulating materials, such as fiberglass, are placed around the back and sides of the absorber plate to seal the system and further minimize heat loss. Generally, oil and water are the working fluids in the flat plate collectors. However, water is prone to freezing at low temperatures, damaging the collector plate. Anti-freeze mixtures are used to overcome this issue of freezing. Several types of flat plate solar collectors have been proposed in the literature (Cerón et al. 2015; Jiandong et al. 2015; Daghigh and Shafieian 2016; Zhu et al. 2017; He et al. 2024).

Evacuated tube solar collectors (ETSC) are solar thermal devices that capture and convert sunlight into heat, as shown in Fig. 16b. They consist of a series of glass tubes containing an absorber plate or a heat pipe enclosed in a vacuum-sealed environment. The vacuum between the annular regions of glass tubes acts as an insulator, significantly reducing heat loss through convection and conduction, allowing the system to maintain high efficiency even in cold or overcast conditions. The absorbed solar energy is transported to a working fluid, namely, water or a water–glycol mixture, which circulates through the system to deliver heat for applications like water heating, space heating, or industrial processes. Due to their high efficiency and ability to operate effectively under diverse weather conditions, evacuated tube solar collectors are extensively used in residential and commercial solar thermal systems.

Evacuated tube solar collectors offer distinct advantages over flat plate collectors, particularly in terms of efficiency and performance under varying conditions. Their vacuum-sealed design significantly reduces heat loss, making them more effective in cold climates and capable of maintaining high operating temperatures even in freezing conditions. Additionally, evacuated tube collectors excel on cloudy or overcast days, because they can efficiently absorb both direct and diffuse solar radiation. Their modularity enables for easy scalability and maintenance, with individual tubes replaceable without affecting the entire system. Furthermore, these collectors perform well at low solar angles, extending their effective operating hours throughout the day. The durability of the glass tubes also enhances their resistance to environmental factors, potentially leading to a longer lifespan compared to flat plate collectors. These characteristics make evacuated tube collectors a superior choice for highly efficient and reliable applications.

Evacuated tube solar collectors are typically divided into three categories—heat pipe ETSC, water-in-glass ETSC, and U-pipe ETSC—based on the method of heat extraction. (Tabarhoseini et al. 2022). Figure 17a depicts the schematic of a heat pipe ETSC, which consists of a metallic absorber plate inside a heat pipe. The solar energy falls on the evaporator section of the heat pipe, leading to the phase change of the working fluid. Due to pressure difference, the vapors formed travel to the condenser section of the heat pipe, which is connected to the manifold (tube with a large diameter in Fig. 17a). The vapors condense upon reaching the condenser section, heating the fluid in the manifold. The condensed working fluid returns to the evaporator section of the heat pipe, completing the thermodynamic cycle. Shafieian et al. (2018) and Alshukri et al. (2022) have extensively reviewed the current strategies and developments in the heat pipe-integrated ETSCs and their commercialization. The current challenges and the scope for future research have also been extensively discussed in these review articles.

In a water-in-glass ETSC, as illustrated in Fig. 17b, a single glass tube with an inner selective coating material serves as the absorber. The working fluid flows into the tube from the open-ended bottom, where it absorbs direct solar radiation, increasing its temperature and decreasing its density. As the fluid heats up and density drops,

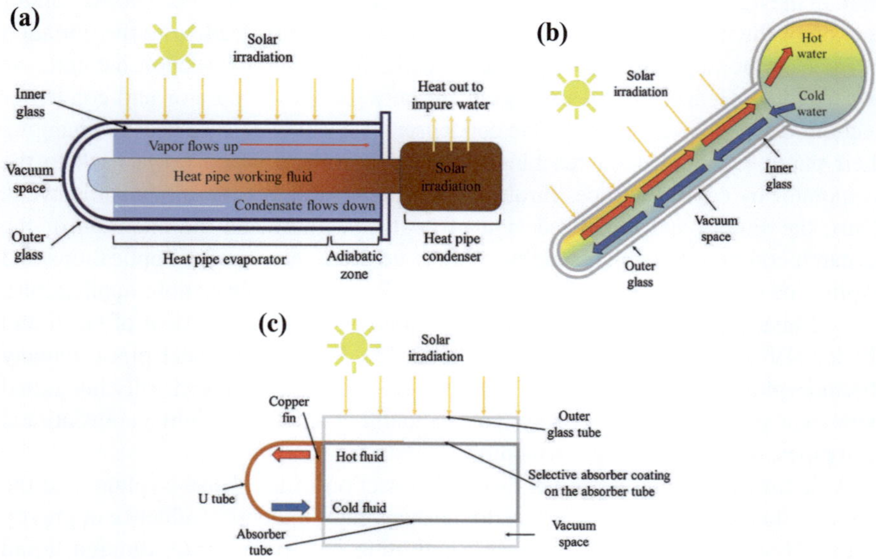

Fig. 17 Schematic of **a** heat pipe ETSC, **b** water-in-glass tube, and **c** U-type evacuated tube solar collector, respectively

it naturally flows toward the storage tank. Moreover, a U-type ESTC operates by harnessing solar energy to heat a fluid circulating through a U-shaped tube housed within each evacuated glass tube; see Fig. 17c. The U-shaped tube, usually made of a highly conductive metal like copper, is inserted into the evacuated tube, which is then sealed to create a vacuum. This vacuum minimizes heat loss by preventing heat conduction and convection. Solar insolation goes through the outer layer and is absorbed by the selective coating on the inner tube, converting it into heat. The working fluid is heated by the generated heat and begins to circulate within the U-tube. As the fluid heats up, it flows through the collector, transferring the absorbed heat to a storage system or directly to an application, such as water heating. The U-type design ensures efficient heat transfer and minimizes thermal losses, making the collector highly effective, even in less favorable weather conditions.

4.2 *Heat Pipes for Solar Thermal Desalination*

Heat pipes are regarded as devices that can transfer considerable heat passively between a source and a sink with negligible thermal resistance. A heat pipe is a hollow tube and comprises three main components: the evaporator section, the adiabatic section, and the condenser section. A wick structure is positioned along the length of the heat pipe, making contact with its inner surface. The heat pipe is then evacuated to remove all non-condensable gases in its casing and wick material. It is then charged with a suitable degassed working fluid and finally sealed. The choice of working fluid in the heat pipe depends on the intended operating temperature. The evaporator section of the heat pipe is exposed to a particular heat flux, leading to the initiation of nucleate boiling. The vapor pressure inside the evaporator section increases as vapors form. Hence, the pressure gradient between the evaporator and condenser sections drives the vapor toward the condenser. The vapors release heat by changing their phase to liquid upon reaching the condenser. The liquid flows back to the evaporator by capillary forces through the wick, closing the thermodynamic cycle. Thus, the simple and flexible design of heat pipes attracted the attention of the research community, which sought to integrate them into various applications and explore new designs. More details regarding the design, fabrication, applications, etc., of heat pipes and their types can be found in the review article of Groll and Rosier (1992) and the book by Mantelli (2021). Two special heat pipes, namely thermosyphon and pulsating heat pipe, see Fig. 18a, have been generally integrated with solar thermal desalination systems (Shoeibi et al. 2022). Unlike conventional heat pipes, neither has a wick structure.

A liquid pool is present in the evaporator section of the thermosyphon, and the working fluid condensate flows back to the evaporator under the influence of gravity forces. However, the heat transfer in a pulsating heat pipe occurs through liquid slugs and vapor plugs oscillating inside the meandering tube arrangement shown in Fig. 18b. Gravity forces no longer significantly affect its operation due to its small tube diameter. Hence, the diameter of the heat pipe tube casing is identified as a critical design parameter of heat pipes, thermosyphons, and pulsating heat pipes (Khandekar et al. 2003; Khandekar and Groll 2004). Regarding non-dimensionless numbers, the Bond number (Bo), defined as the ratio of gravity forces to surface tension forces, acts as the primary design parameter. The critical diameter of a heat pipe is calculated using Eq. (45),

$$\mathrm{Bo} = D_{\mathrm{crit}}\sqrt{\frac{g(\rho_{\mathrm{l}} - \rho_v)}{\sigma}} \leq 2 \tag{45}$$

where D_{crit} is the critical tube diameter (m), g is the acceleration due to gravity (m/s^2), ρ_{l} and ρ_v are the densities of the liquid and vapor phase (kg/m^3), and σ is the surface tension of the working fluid (N/m), respectively. It has been concluded that heat pipes with internal diameters less than the critical diameter operate as pulsating heat pipes. Heat pipes operate as thermosyphons when the internal tube diameter exceeds the critical diameter.

Heat pipes integrated with solar thermal water purification systems in the literature have been generally used to enhance freshwater productivity by increasing the bulk water temperature (Fallahzadeh et al. 2020; Patel et al. 2020; Shafieian et al. 2020). Pulsating heat pipes (PHPs) have been integrated with heat storage devices, such as phase change materials (PCMs), to enhance daily freshwater yield. Figure 19a and b shows a novel solar still integrated with a PHP and PCM to generate freshwater even after sunset (Khalilmoghadam et al. 2021, 2024). The evaporator of the PHP is combined with the PCM, and the condenser section of the PHP is attached to the impure bulk water. During the daytime, the vapors formed from impure water evaporation condense on the aluminum side walls, see Fig. 19a, releasing their latent heat of condensation, which is simultaneously stored in the PCM. After sunset, the

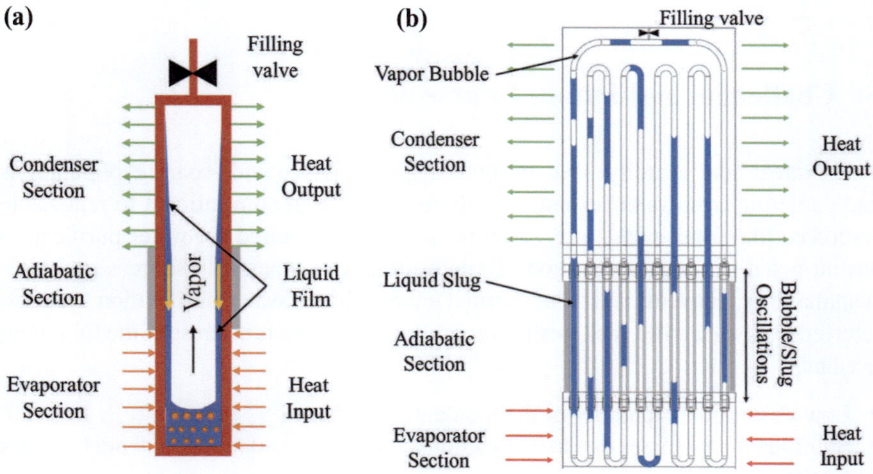

Fig. 18 Schematic showing the operation of a closed **a** thermosyphon and **b** pulsating heat pipe, respectively

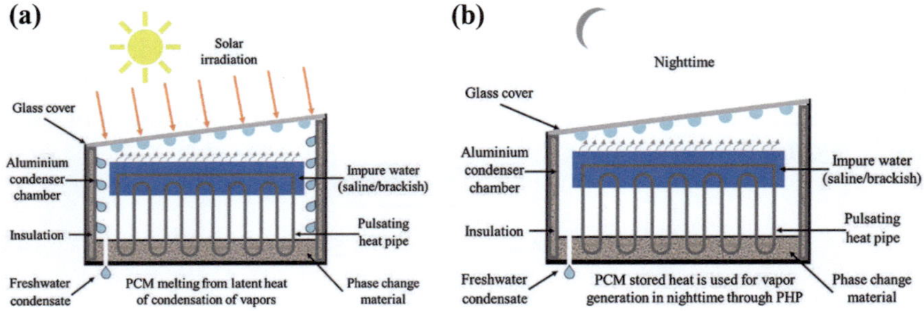

Fig. 19 Schematic of a pulsating heat pipe-integrated solar still operating during **a** daytime and **b** nighttime, respectively (reproduced with permission from Khalilmoghadam et al. 2024, Elsevier)

PCM releases its stored heat to the PHP evaporator, which is then transferred to the condenser. Hence, the impure water evaporates from this dissipated heat to generate freshwater during the nighttime, as shown in Fig. 19b. The experimental results demonstrated a significant enhancement in system efficiency, with thermal efficiency increasing from 23.7 to 48.5%. Additionally, the study presented a comprehensive thermal model that accounts for heat transfer and condensation on sidewalls, as well as the dynamic behavior of PHPs within a passive solar still system, offering a more accurate prediction of performance compared to conventional designs. Integrating various heat pipes into solar thermal water purification systems provides a promising approach to enhancing freshwater yield compared to traditional solar stills. However, such modifications also increase the unit water cost per liter of the generated freshwater. Hence, a system-level analysis of the unit water cost, efficiency, and exergy is essential for developing and commercializing such modified decentralized solar water purification systems.

5 Challenges and Future Course of Action

Water stress and scarcity have become a global issue, with soaring populations and depleting freshwater resources. The transition from conventional to renewable resources like solar, wind, etc., to fulfill the energy required for water purification has attracted significant attention. Though the use of renewable sources promotes sustainability, conventional solar thermal energy-driven water purification systems, referred to as solar stills, lack freshwater productivity. This is because of the following reasons:

- Low absorption capability of the incident sunlight
- Significant heat losses due to conduction, convection, and radiation from the glass cover
- Parasitic heat losses to the bulk water and side walls of the basin

- Reduced transmission of solar insolation along the glass covers due to condensed water droplets on the interior face of the glass cover during solar still operation.

Numerous solutions have been proposed in the literature through several parametric investigations to overcome the bottlenecks of freshwater production in solar stills. These investigations point out that solar intensity, wind speed, and ambient temperature significantly affect the production rate; however, these are meteorological parameters that cannot be controlled. Other important parameters include inclination angle, insulation thickness, and water depth. The inclination angle can be optimized according to the latitude angle of the location. The minimum water depth required to avoid drying out is preferred, and increasing insulation increases the output. Several unconventional designs like triangular stills, tubular stills, hemispherical, and multiple slopes have been tried without achieving a clear advantage. Additional usage of phase change material, condensers, nanoparticles like copper and aluminum oxide, reflectors, and vibratory harmonic effect has shown an increase in output, but with an increase in complexity and cost.

Moreover, reusing the latent heat of condensation has also been proposed in solar stills through multi-staging. However, most of them are still far from being commercialized. The incorporation of heat pipes with evacuated or vacuum-type solar collectors has been investigated for solar still configurations on a few occasions to heat the bulk water. Heat pipes have been integrated with phase change materials or thermoelectric devices; however, the performance improvement has not been satisfactory in terms of the additional unit water cost. Significant efforts have been made to enhance photo-thermal efficiency by increasing light absorption and reducing heat loss.

Interfacial Solar Vapor Generation technology has emerged as an innovative solution strategy in the last decade. In interfacial solar evaporation systems, a light-absorbing material is floated at the water–air interface, allowing only the air–liquid interface to be heated instead of the entire bulk water. This design significantly enhances photo-thermal efficiency, as depicted in Fig. 20. The principal objective of this technology involves the development of an advanced three-layer structure floating at the interface of seawater and air that serves the following purposes:

- Enhanced wide-spectrum light absorption
- Efficient light-to-heat transformation
- Localize heat on the evaporative surface
- Efficient water transport
- Good thermal insulation.

The next chapter discusses solar-based interfacial vapor generation in detail.

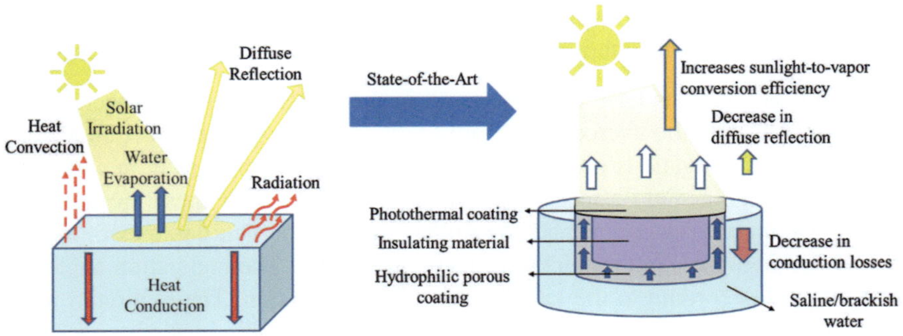

Fig. 20 Schematic showing the energy balance diagram of water evaporation (left) and an advanced strategy for interfacial solar vapor generation (right)

6 Closure

Production of small/medium-scale potable drinking water systems in a decentralized environment is vital for countries like India, which has many remote regions with high population density. Solar stills have been extensively investigated for desalination purposes for a long time. However, the inherently low efficiency of this technology limits its widespread commercialization. In this background, we reviewed vital parameters that effect the thermal performance and freshwater productivity of a conventional solar still. Various enhancement techniques were discussed in detail to enhance the performance and freshwater productivity. It is concluded that most enhancement techniques improve freshwater yield and increase the daily unit water cost produced, thereby limiting the widespread commercialization of such systems. Hence, a thorough understanding of the fundamental heat and mass transport of such systems is required to develop low-cost, robust, and potable solar water purification systems. Solar-based interfacial vapor generation systems appear to be an exciting solution to overcome the significant bottlenecks encountered in conventional solar stills.

Nomenclature

a	Differential equation constant (−)
A	Area (m^2)
C	Specific heat (kJ/kg K)
D	Diameter (m)
g	Acceleration due to gravity (m/s^2)
h	Heat transfer coefficient (W/m^2 K)
h_{fg}	Latent heat of vaporization (J/kg)
h_L	Sensible enthalpy change (J/kg)

h_{lv}	Total enthalpy change (J/kg)
$I(t)$	Solar irradiation (W/m^2)
K	Thermal conductivity (W/m K)
L	Length (m)
m	Mass (kg)
P	Pressure (Pa)
q	Heat flux (W/m^2)
Q_u	Internal energy (W)
R	Reflectivity (−)
R_g	Gas constant (J/kg K)
T	Temperature (K)
t	Time (s)
U	Overall heat transfer coefficient (W/m^2 K)
v	Wind velocity (m/s)

Greek Alphabets

α	Solar absorptivity (−)
ε	Emissivity (−)
η	Thermal efficiency (−)
ρ	Density (kg/m^3)
σ	Surface tension (N/m)
τ	Transmittivity (−)

Subscripts

a	Ambient
b	Basin
bs	Basin and side
c	Convective
cd	Conductive
e	Evaporative
eff	Effective
fw	Freshwater
g	Glass
gi	Glass inner
go	Glass outer
ins	Insulation
l	Liquid
r	Radiative
s	Solar

ss	Side
t	Total
top	Top
v	Vapor
w	Water

Acronyms

GOR	Gain output ratio
MED	Multi-effect desalination
MSF	Multi-stage flashing
PCM	Phase change material
PHP	Pulsating heat pipe
PV	Photovoltaic
SWP	Specific water productivity

Non-dimensional Numbers

Bo	Bond number $(D\sqrt{g(\rho_l - \rho_v)/\sigma})$

References

Abed FM, Ahmed AH, Hasanuzzaman M, Kumar L, Hamaad NM (2022) Experimental investigation on the effect of using chemical dyes on the performance of single-slope passive solar still. Sol Energy 233:71–83

Akash BA, Mohsen MS, Osta O, Elayan Y (1998) Experimental evaluation of a single-basin solar still using different absorbing materials. Renew Energy 14(1–4):307–310

Al-Kayiem HH, Lin SC (2014) Performance evaluation of a solar water heater integrated with a PCM nanocomposite TES at various inclinations. Sol Energy 109(1):82–92

Alshukri MJ, Kadhim Hussein A, Eidan AA, Alsabery AI (2022) A review on applications and techniques of improving the performance of heat pipe-solar collector systems. Sol Energy 236:417–433

Ansari O, Asbik M, Bah A, Arbaoui A, Khmou A (2013) Desalination of the brackish water using a passive solar still with a heat energy storage system. Desalination 324:10–20

Arunkumar T, Jayaprakash R, Denkenberger D, Ahsan A, Okundamiya MS, Kumar S, Tanaka H, Aybar HŞ (2012) An experimental study on a hemispherical solar still. Desalination 286:342–348

Arunkumar T, Kabeel AE, Raj K, Denkenberger D, Sathyamurthy R, Ragupathy P, Velraj R (2018) Productivity enhancement of solar still by using porous absorber with bubble-wrap insulation. J Clean Prod 195:1149–1161

Bait O, Si-Ameur M (2017) Tubular solar-energy collector integration: performance enhancement of classical distillation unit. Energy 141:818–838

Bataineh KM, Abbas MA (2020) Performance analysis of solar still integrated with internal reflectors and fins. Sol Energy 205:22–36

Cerón JF, Pérez-García J, Solano JP, García A, Herrero-Martín R (2015) A coupled numerical model for tube-on-sheet flat-plate solar liquid collectors. Analysis and validation of the heat transfer mechanisms. Appl Energy 140:275–287

Chen C, Kuang Y, Hu L (2019) Challenges and opportunities for solar evaporation. Joule 3(3):683–718

Cheng WL, Huo YK, Nian YL (2019) Performance of solar still using shape-stabilized PCM: experimental and theoretical investigation. Desalination 455:89–99

Chiavazzo E, Morciano M, Viglino F, Fasano M, Asinari P (2018) Passive solar high-yield seawater desalination by modular and low-cost distillation. Nat Sustain 1(12):763–772

Daghigh R, Shafieian A (2016) Thermal performance of a double-pass solar air heater. J Renew Energy Environ 3(2):35–46

Deshmukh HS, Thombre SB (2017) Solar distillation with single basin solar still using sensible heat storage materials. Desalination 410:91–98

Dev R, Abdul-Wahab SA, Tiwari GN (2011) Performance study of the inverted absorber solar still with water depth and total dissolved solid. Appl Energy 88(1):252–264

Dhaoui S, Choubani K, Hamdi F, Ghriss O, Bouabidi A, Alasiri A, El Hadi Attia M (2024) Experimental investigation of the effect of angled fins on the energy and exergy efficiencies of solar stills: a comparison between double slope and pyramid solar stills under Tunisia weather conditions. Process Saf Environ Prot 188:1069–1080

El Hadi Attia M, Kabeel AE, Zayed ME, Abdelgaied M, Sharshir SW, Abdulla AS (2022) Experimental study of a hemispherical solar distillation system with and without rock salt balls as low-cost sensible storage: performance optimization and comparative analysis. Sol Energy 247:373–384

Elango C, Gunasekaran N, Sampathkumar K (2015) Thermal models of solar still—a comprehensive review. Renew Sustain Energy Rev 47:856–911

Elango T, Kalidasa Murugavel K (2015) The effect of the water depth on the productivity for single and double basin double slope glass solar stills. Desalination 359:82–91

El-Sebaii AA (2011) On effect of wind speed on passive solar still performance based on inner/outer surface temperatures of the glass cover. Energy 36(8):4943–4949

El-Sebaii AA, Aboul-Enein S, Ramadan MRI, El-Bialy E (2000) Year-round performance of a modified single-basin solar still with mica plate as a suspended absorber. Energy 25(1):35–49

El-Sebaii AA, El-Naggar M (2017) Year round performance and cost analysis of a finned single basin solar still. Appl Therm Eng 110:787–794

Fallahzadeh R, Aref L, Gholamiarjenaki N, Nonejad Z, Saghi M (2020) Experimental investigation of the effect of using water and ethanol as working fluid on the performance of pyramid-shaped solar still integrated with heat pipe solar collector. Sol Energy 207:10–21

Groll M, Rösler S (1992) Operation principles and performance of heat pipes and closed two-phase thermosyphons. J Non-Equilib Thermodyn 17(2):91–151

Haddad Z, Chaker A, Rahmani A (2017) Improving the basin type solar still performances using a vertical rotating wick. Desalination 418:71–78

He Y, Yu H, Duan G, Wang Y, Yang Q, Feng L, Zhang J (2024) Thermal performance and experimental analysis of stainless steel flat plate solar collector with full-flow channels. Heliyon 10(7):e28255

Huang C-H, Chang T-R (2017) Determination of optimal inclination function for external reflector of basin type still for maximum distillate productivity. Energy 141:1728–1736

Rahim NHA (2001) Utilisation of new technique to improve the efficiency of horizontal solar desalination still. Desalination 138(1–3):121–128

Jani HK, Modi KV (2019) Experimental performance evaluation of single basin dual slope solar still with circular and square cross-sectional hollow fins. Sol Energy 179:186–194

Jiandong Z, Hanzhong T, Susu C (2015) Numerical simulation for structural parameters of flat-plate solar collector. Sol Energy 117:192–202

Kabeel AE (2009) Performance of solar still with a concave wick evaporation surface. Energy 34(10):1504–1509
Kabeel AE, Abdelgaied M (2020) Enhancement of pyramid-shaped solar stills performance using a high thermal conductivity absorber plate and cooling the glass cover. Renew Energy 146:769–775
Kabeel AE, Abdelgaied M, Eisa A (2018) Enhancing the performance of single basin solar still using high thermal conductivity sensible storage materials. J Clean Prod 183:20–25
Karimi Estahbanati MR, Ahsan A, Feilizadeh M, Jafarpur K, Ashrafmansouri S-S, Feilizadeh M (2016) Theoretical and experimental investigation on internal reflectors in a single-slope solar still. Appl Energy 165:537–547
Khalilmoghadam P, Kiyaee S, Rajabi-Ghahnavieh A, Warsinger DM, Behshad Shafii M (2024) An improved passive solar still integrated with pulsating heat pipes and phase change materials. Sol Energy 275:112612
Khalilmoghadam P, Rajabi-Ghahnavieh A, Shafii MB (2021) A novel energy storage system for latent heat recovery in solar still using phase change material and pulsating heat pipe. Renew Energy 163:2115–2127
Khandekar S, Dollinger N, Groll M (2003) Understanding operational regimes of closed loop pulsating heat pipes: an experimental study. Appl Therm Eng 23(6):707–719
Khandekar S, Groll M (2004) An insight into thermo-hydrodynamic coupling in closed loop pulsating heat pipes. Int J Therm Sci 43(1):13–20
Law AM, Jones LO, Walls JM (2023) The performance and durability of anti-reflection coatings for solar module cover glass—a review. Sol Energy 261:85–95
Lawrence SA, Gupta SP, Tiwari GN (1990) Effect of heat capacity on the performance of solar still with water flow over the glass cover. Energy Convers Manage 30(3):277–285
Mantelli MBH (2021) Thermosyphons and heat pipes: theory and applications, vol 1, no 1. Springer International Publishing, Switzerland, pp 1–413
Modi KV, Modi JG (2019) Performance of single-slope double-basin solar stills with small pile of wick materials. Appl Therm Eng 149:723–730
Mohsenzadeh M, Aye L, Christopher P (2021) A review on various designs for performance improvement of passive solar stills for remote areas. Sol Energy 228:594–611
Murugavel KK, Sivakumar S, Ahamed JR, Chockalingam KKSK, Srithar K (2010) Single basin double slope solar still with minimum basin depth and energy storing materials. Appl Energy 87(2):514–523
Muthu Manokar A, Kalidasa Murugavel K, Esakkimuthu G (2014) Different parameters affecting the rate of evaporation and condensation on passive solar still—a review. Renew Sustain Energy Rev 38:309–322
Nafey AS, Abdelkader M, Abdelmotalip A, Mabrouk AA (2002) Enhancement of solar still productivity using floating perforated black plate. Energy Convers Manage 43(7):937–946
Omara ZM, Kabeel AE, Abdullah AS, Essa FA (2016) Experimental investigation of corrugated absorber solar still with wick and reflectors. Desalination 381:111–116
Omara ZM, Kabeel AE, Younes MM (2014) Enhancing the stepped solar still performance using internal and external reflectors. Energy Convers Manage 78:876–881
Pandey N, Naresh Y (2024) Experimental study on the synergistic effects of phase change material, fins, and ultrasonic fogger in a pyramidal solar still. J Energy Storage 98:112997
Panomwan Na Ayuthaya R, Namprakai P, Ampun W (2013) The thermal performance of an ethanol solar still with fin plate to increase productivity. Renew Energy 54:227–234
Patel M, Patel C, Panchal H (2020) Performance analysis of conventional triple basin solar still with evacuated heat pipes, corrugated sheets and storage materials. Groundw Sustain Dev 11:100387
Peng G, Sharshir SW (2023) Progress and performance of multi-stage solar still—a review. Desalination 565:116829
Phadatare MK, Verma SK (2007) Influence of water depth on internal heat and mass transfer in a plastic solar still. Desalination 217(1–3):267–275
Rajvanshi AK (1981) Effect of various dyes on solar distillation. Sol Energy 27(1):51–65

Sarath KP, Feroz Osman M, Mukhesh R, Manu KV, Deepu M (2023) A review of the recent advances in the heat transfer physics in latent heat storage systems. Therm Sci Eng Prog 42:101886
Shafieian A, Khiadani M, Nosrati A (2018) A review of latest developments, progress, and applications of heat pipe solar collectors. Renew Sustain Energy Rev 95:273–304
Shafieian A, Rizwan Azhar M, Khiadani M, Kanti Sen T (2020) Performance improvement of thermal-driven membrane-based solar desalination systems using nanofluid in the feed stream. Sustain Energy Technol Assess 39:100715
Shirsath GB, Muralidhar K, Pala RGS (2020) Variable air gap membrane distillation for hybrid solar desalination. J Environ Chem Eng 8(3):103751
Shirsath GB, Muralidhar K, Pala RGS, Ramkumar J (2019) Condensation of water vapor underneath an inclined hydrophobic textured surface machined by laser and electric discharge. Appl Surf Sci 484:999–1009
Shirsath GB, Pala RGS, Muralidhar K, Khandekar S (2018) Effect of salinity and water depth on the performance of doubly inclined solar still. Desalin Water Treat 124:72–87
Shoeibi S, Mirjalily SAA, Kargarsharifabad H, Khiadani M, Panchal H (2022) A comprehensive review on performance improvement of solar desalination with applications of heat pipes. Desalination 540:115983
Sodha MS, Kumar A, Tiwari GN (1981) Utilization of waste hot water for distillation. Desalination 37(3):325–342
Suneja S, Tiwari GN (1999) Effect of water depth on the performance of an inverted absorber double basin solar still. Energy Convers Manage 40(17):1885–1897
Tabarhoseini SM, Sheikholeslami M, Said Z (2022) Recent advances on the evacuated tube solar collector scrutinizing latest innovations in thermal performance improvement involving economic and environmental analysis. Sol Energy Mater Sol Cells 241:111733
Tanaka H (2010) Monthly optimum inclination of glass cover and external reflector of a basin type solar still with internal and external reflector. Sol Energy 84(11):1959–1966
Tang R, Cheng Y, Wu M, Li Z, Yu Y (2010) Experimental and modeling studies on thermosiphon domestic solar water heaters with flat-plate collectors at clear nights. Energy Convers Manage 51(12):2548–2556
Tiwari AK, Tiwari GN (2006) Effect of water depths on heat and mass transfer in a passive solar still: in summer climatic condition. Desalination 195(1–3):78–94
Tiwari GN, Garg HP (1985) Effect of water flow over the glass cover of a single basin solar still with an intermittent flow of waste hot water in the basin. Energy Convers Manage 25(3):315–322
Tiwari GN, Rao VB (1984) Transient performance of a single basin solar still with water flowing over the glass cover. Desalination 49(3):231–241
Tripathi R, Tiwari GN (2005) Effect of water depth on intemal heat and mass transfer for active solar distillation. Desalination 173:187–200
Valsaraj P (2002) An experimental study on solar distillation in a single slope basin still by surface heating the water mass. Renew Energy 25(4):607–612
Velmurugan V, Gopalakrishnan M, Raghu R, Srithar K (2008) Single basin solar still with fin for enhancing productivity. Energy Convers Manage 49(10):2602–2608
Wang Z, Horseman T, Straub AP, Yip NY, Li D, Elimelech M, Lin S (2019) Pathways and challenges for efficient solar-thermal desalination. Sci Adv 5(7):eaax0763
Wiener J, Khan MZ, Shah K (2024) Performance enhancement of the solar still using textiles and polyurethane rollers. Sci Rep 14(1):5202
Yadav YP, Kumar A (1991) Transient analytical investigations on a single basin solar still with water flow in the basin. Energy Convers Manage 31(I)
Yan T, Xie G, Sun L, Du M, Liu H (2019) Experimental investigation on a two-effect tubular solar still operating under vacuum conditions. Desalination 468:114057
Yousef MS, Hassan H (2019) Energetic and exergetic performance assessment of the inclusion of phase change materials (PCM) in a solar distillation system. Energy Convers Manage 179:349–361

Zhao Z, Wang C, Wei D, Wang F (2024) Condensation device design represents a critical step for solar-driven water evaporation toward practical applications. Cell Rep Phys Sci 5(2):101794

Zheng H, Chang Z, Chen Z, Xie G, Wang H (2013) Experimental investigation and performance analysis on a group of multi-effect tubular solar desalination devices. Desalination 311:62–68

Zhu T, Diao Y, Zhao Y, Ma C (2017) Performance evaluation of a novel flat-plate solar air collector with micro-heat pipe arrays (MHPA). Appl Therm Eng 118:1–16

Zhu Z, Wang Q, Li Z, Chen M, Wang L, Zheng H (2022) Performance research and comparison of integrated passive solar-concentrated stills buried in soil: with/without heat recovery. Energy Convers Manage 256:115400

Zurigat YH, Abu-Arabi MK (2004) Modelling and performance analysis of a regenerative solar desalination unit. Appl Therm Eng 24(7):1061–1072

Interfacial Vapor Generation for Solar Desalination

Debartha Chatterjee and K. Muralidhar

1 Interfacial Vapor Generation-A New Paradigm

Solar stills are desalination devices driven by solar-thermal energy. They are ideal for setting up a small-scale decentralized potable water supply in arid regions. Although they are one of the most mature technologies available for desalination, they are yet to be commercially viable because of their low water productivity. Chapter 8 presents a detailed overview of state-of-the-art solar still technology. Active and passive methodologies adopted by researchers to enhance water productivity are discussed. However, these enhancement methodologies are primarily based on heating the bulk of saline or brackish water. The traditional solar still, as depicted in Fig. 1, has two inherent limitations that limit its water productivity. One is heating the bulk reservoir for vapor generation, and two is a transparent condensing surface for water collection. Traditional techniques that aim to enhance water productivity do not address these two critical limitations. Figure 1 shows how the heating of the bulk reservoir for vapor generation is associated with several parasitic losses, leading to low solar-to-vapor conversion efficiency. The incident solar spectrum comprises the infrared and visible portions, where the first can penetrate only a few mm of water depth, whereas the latter can be transmitted over several meters. In application, the visible part of the radiation is collected by a blackened absorber that heats up the rest of the water body. Hence, the visible portion of the radiation is used to heat the bulk of the reservoir via conductive and convective modes of heat transfer. This is clearly unnecessary since evaporation and vapor generation are surface phenomena. In addition, a significant portion of the incident solar irradiation is lost by diffuse reflection at

D. Chatterjee
Indian Institute of Technology Kanpur, Kanpur, Uttar Pradesh, India

K. Muralidhar (✉)
Department of Mechanical Engineering, Indian Institute of Technology Kanpur, Kanpur, Uttar Pradesh, India
e-mail: kmurli@iitk.ac.in

R. Narayanaswamy et al. (eds.), *Sustainable Water Harvesting*, Springer Tracts in Mechanical Engineering, https://doi.org/10.1007/978-981-95-3373-2_9

material interfaces. The transparent condensing surface, on the one hand, interferes with the incoming solar irradiation due to the formation of condensing droplets/films and provides no scope for latent heat recovery. Equation 1 has been developed to show the role of critical parameters contributing to the Specific Water Productivity (SWP) of a solar still (Ghasemi et al. 2014).

$$\mathrm{SWP} = \frac{Q_{\mathrm{solar}}}{L}\alpha\eta_t\mathrm{GOR} \tag{1}$$

Here, Q_{solar} is the incident solar irradiation, L is the latent heat of vaporization, α is the solar absorptivity, η_t is the solar-to-vapor conversion efficiency, and *GOR* is the Gain Output Ratio.

Interfacial vapor generation addresses the first limitation by drastically improving solar absorptivity and solar-to-vapor conversion efficiency using a three-layer strategy, thus augmenting specific water productivity. The three-layer strategy of Chen et al. (2019) has been adopted in the literature for interfacial vapor generation and is depicted in Fig. 2. Layer 1 is primarily a porous structure with interconnected pore networks that generate moderate-to-high capillary pumping power. This porous structure forms a saturated layer of saline or brackish water at the interface and is also in contact with the bulk reservoir, Fig. 2. Layer 2 is an insulating medium strategically placed between this interface and the top of the bulk reservoir (Wang 2018). This step decouples the vapor-generating surface and the bulk reservoir in order to minimize parasitic losses due to conduction. Layer 3 is a photo-thermal coating that sits in direct contact above the interface with the porous layer. The role of this layer is to efficiently across the complete solar spectrum and convert it into thermal energy (Kim et al. 2018). This heating is used to generate vapor from the liquid contained in the saturated layer of the porous structure. Thus, the three-layer strategy can localize the complete incoming solar irradiation along the interface and generate vapor with improved effectiveness (Zhu et al. 2019).

1.1 Photo-Thermal Materials

As the name suggests, photo-thermal materials convert light into the form of thermal energy. Solar radiation is mainly light and is a bundle of photons that propagate in the form of electromagnetic waves over a range of frequencies. When falling on a surface, it is absorbed, reflected, transmitted, and scattered. Absorption of solar energy in the visible spectrum manifests as an increase in the temperature of the object. This is commonly known as the photo-thermal effect. With the advent of microfabrication techniques in the last decade, several materials have shown to demonstrate the photo-thermal effect with high conversion efficiencies. Photo-thermal materials (layer 3 in the previous discussion) play a critical role in converting the incident solar irradiation to thermal energy, which is then utilized to generate water vapor. Photo-thermal materials proposed for interfacial vapor generation can be classified

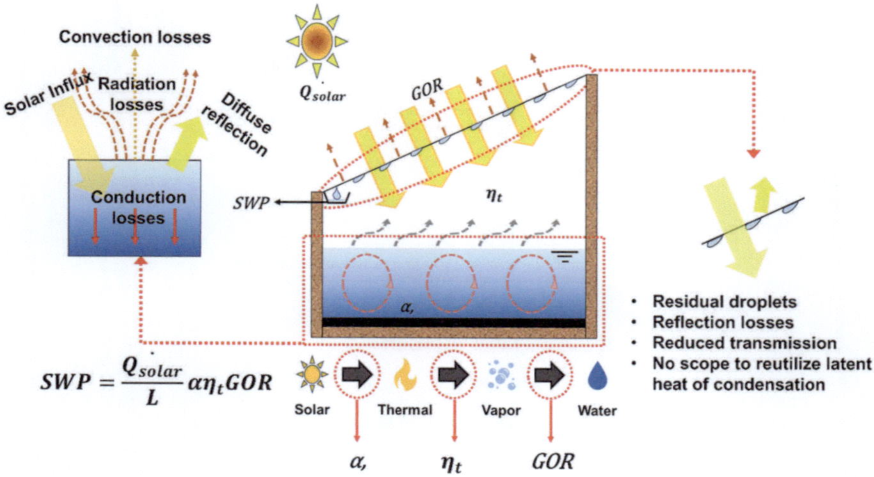

Fig. 1 Schematic depicting the losses inherent to a conventional solar still system, leading to low water productivity

into three broad categories, namely: (a) plasmonic materials, (b) semiconductors, and (c) carbonaceous or polymeric materials and also in a combination, Fig. 3.

Plasmonic materials are primarily based on metallic nanoparticles that play a crucial role in the photo-thermal effect using the principle of surface plasmon resonance (Adams and Zhang 2016). In this strategy, the photo-thermal effect is realized by selecting nanoparticles of gold, silver, or aluminum, which have electronic oscillation frequencies that match that of the incident light. The match in frequencies results in exciting the electrons that result in Joule heating, raising the temperature of the plasmonic material itself (Gao et al. 2016; Zhou et al. 2016) (Fig. 4).

Metal oxide semiconductors with a narrow bandgap are an excellent choice as photo-thermal materials. In semiconductors with broad bandgaps, the absorbed light energy is re-emitted as photons in the visible range, resulting in poor light-to-heat conversion efficiency. The narrow bandgap ensures that the absorbed light energy generates electron-hole pairs, which relax at the band edges and convert the extra energy by thermalization into heat transfer (Wang et al. 2017; Liu et al. 2018).

Carbonaceous and polymeric materials have received tremendous attention as photo-thermal materials mainly because of their low cost and high stability, in addition to their capability to absorb light and their subsequent conversion to thermal energy across a wide range of incident wavelengths (Ghasemi et al. 2014; Cui et al. 2018). These materials have many loosely held electrons that get excited from lower energy orbitals to higher energy levels when the photon energy of the incident light matches the required electronic transition. Heat is released when these excited electrons relax back to their original state. The most commonly used carbonaceous materials reported in the literature are graphene oxides, carbon black, carbon nanotubes, and 3D graphite structures (Yang et al. 2017; Wilson et al. 2019; Awad et al. 2018).

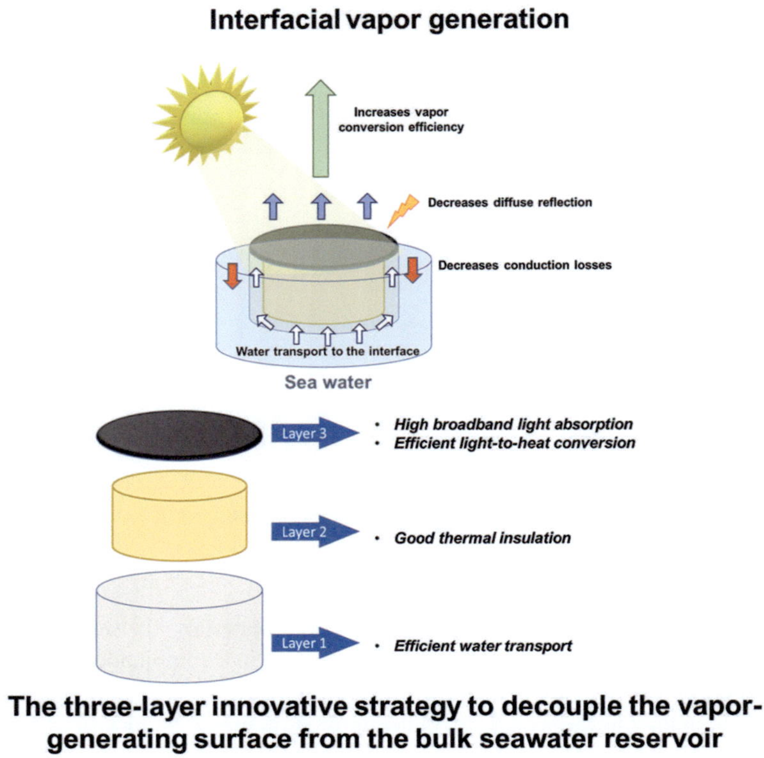

Fig. 2 Schematic showing the basic working principle of an interfacial vapor generation system and the strategy adopted to achieve heat localization along the interface for efficient solar-to-vapor conversion (Chatterjee et al. 2023, adapted with permission from Elsevier)

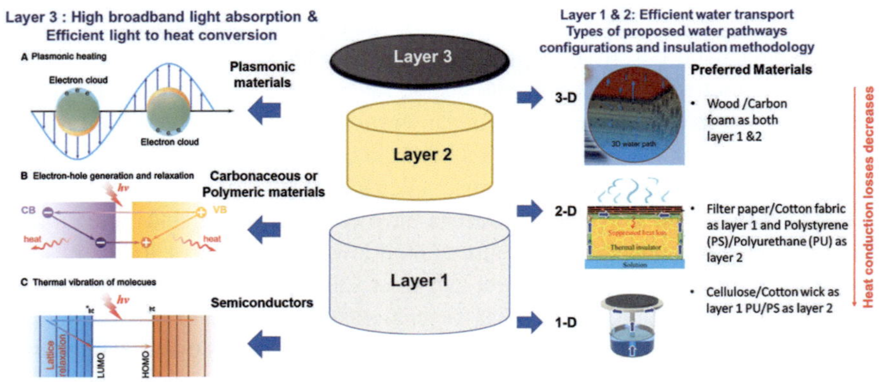

Fig. 3 Schematic showing the role of the three distinct layers in an interfacial vapor generation system and multiple strategies proposed in the literature to achieve them

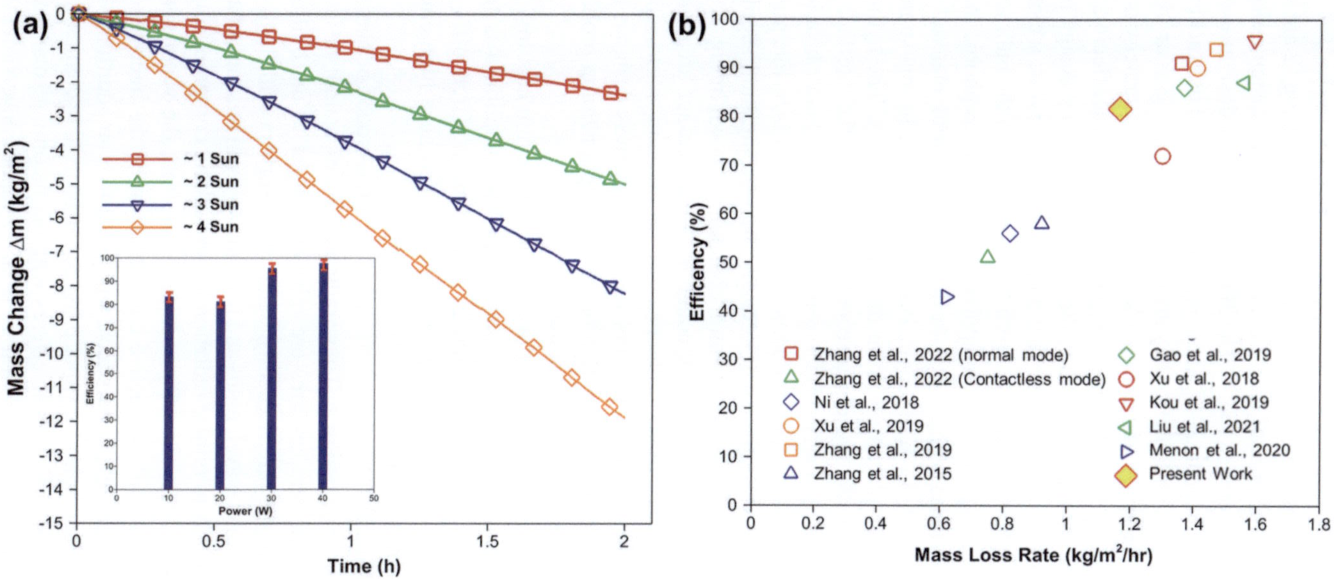

Fig. 4 **a** Effect of input heat flux on the vapor generation rate and vapor conversion efficiency for an interfacial vapor generation system. **b** Comparison of the vapor generation rate and vapor conversion efficiency achieved by various strategies proposed in the literature to avoid salt accumulation (Chatterjee et al. 2023, adapted with permission from Elsevier)

1.2 Water Pathway Design and Heat Localization Strategies

Water pathway design is concerned with how layers 2 and 3 of Fig. 2 are selected and combined to achieve the goal of heat localization. Pathway designs adopted by various researchers are depicted in Fig. 3. The 3D water pathway designs use porous materials such as wood (Liu et al. 2017) that help pump the saline or brackish water from the bulk reservoir to the interface. They also play a critical role as an insulating layer between the bulk reservoir and the interface. However, this strategy is limited for two reasons. First, the hydrophilic insulating layer has much less effective thermal conductivity than in its dry state. Secondly, the 3D water pathway design provides higher surface contact between the interface and the bulk reservoir, which limits the minimization of conduction heat loss from the interface into the bulk reservoir. The 2D water pathway design involves wrapping layer 2 (a hydrophobic insulating material such as polystyrene) with layer 1 (a hydrophilic porous material such as cellulose) (Peng et al. 2021; Li et al. 2016; Peng et al. 2020). In this strategy, the working fluid supply takes place only from the outer periphery. The hydrophobic insulating material in the center ensures a decoupling between the interface and the bulk reservoir by minimizing direct surface contact, leading to better heat localization. The surface contact between the interface and bulk reservoir is further minimized using a 1D water pathway design. In a 1D design, the working fluid supply using capillary pumping is allowed only via a central channel, as shown in Fig. 3. The 1D pathway is superior in heat localization; although the small passage area minimizes chances of heat loss via conduction, it may be futile for efficient salt transport (Ma et al. 2022a, b; Li et al. 2018). Hence, the choice of water pathway design is not only limited to ensuring proper heat localization but also an optimization problem that balances thermal and mass transport phenomena.

Heat localization is not only about minimizing the conduction heat loss into the bulk reservoir but also about minimizing convection and radiation loss from the vapor-generating surface, as shown in Fig. 2. According to Kirchhoff's Law in radiative heat transfer, materials with high absorptivity also have high emissivity, contributing to radiation losses (Figs. 5 and 6).

In systems such as interfacial vapor generation, all three modes of heat transfer are of the same order of magnitude. Thus, radiation losses cannot be neglected even though the operating temperatures are relatively closer to the boiling point of water at atmospheric pressure. Recently, selective absorbers with high absorptivity in the range of incident wavelengths and low emissivity in this range have been studied. Selective absorbers minimize radiation losses without compromising on the solar irradiation absorption process. Convection losses from the vapor-generating surface are inherent because of convective currents between the interface and the surrounding ambient. Many researchers have suggested that an additional layer of bubble wrap with embedded holes for vapor passage can be used to counter this limitation. However, any additional layer interferes with the incoming solar irradiation and a reasonable compromise that maximizes solar-to-vapor conversion efficiency is required (Wang et al. 2019).

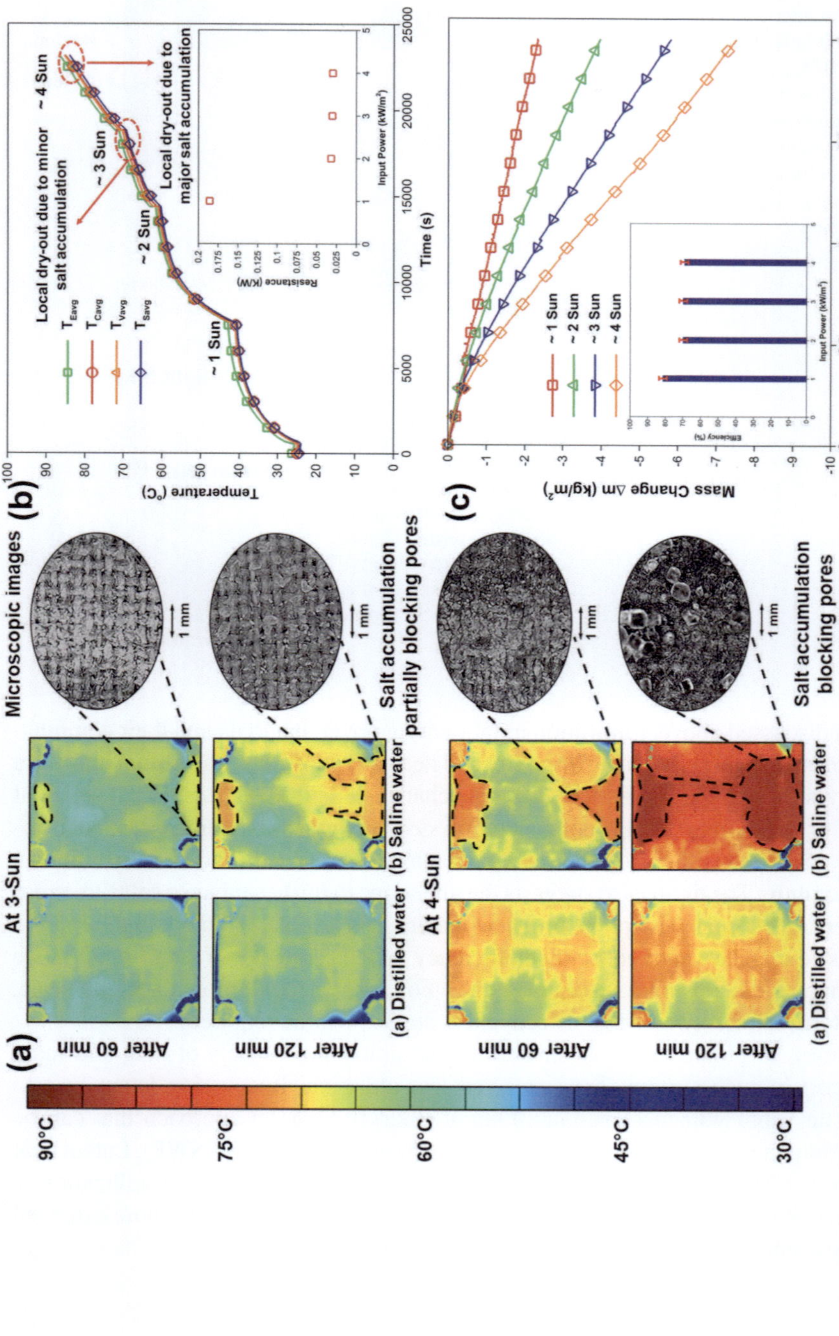

Fig. 5 **a** Infrared thermography showing the spatial temperature distribution and local emerging hot spots, **b** average temperature along the interface depicting the unsteady nature, **c** reduction in vapor generation rate and vapor conversion efficiency due to the accumulation of salt on the vapor-generating surface at higher input heat fluxes (Chatterjee et al. 2023, adapted with permission from Elsevier)

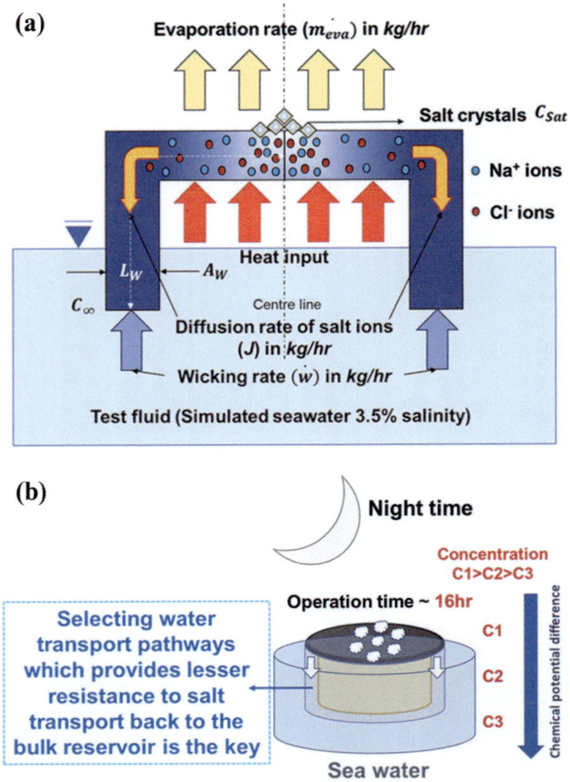

Fig. 6 Schematic of **a** the mechanism of salt accumulation on the vapor-generating surface and the salt transport mechanism from the interface back to the reservoir bulk (Chatterjee et al. 2023, adapted with permission from Elsevier); **b** the simplest salt mitigation strategy relying purely on concentration gradient-based diffusion process during the dark hours

As discussed above, innovation in fabricating novel materials and their combination ensures that high solar absorptivity and heat-to-vapor conversion efficiencies are jointly achieved. Figure 4a shows the mass change with time as a function of the input heat flux for an interfacial vapor generation system. Figure 4b compares the solar-to-vapor conversion efficiency and the rate of mass loss rate achieved by researchers in the literature. The figure also suggests that the scope to further enhance solar-to-vapor conversion efficiency by proposing new materials is limited, although high reported values of solar-to-vapor conversion efficiency may be feasible for small stand-alone experimental setups on a laboratory scale. For implementation on a societal scale, long-term reliability and cost-effectiveness studies will be required.

It may be pointed out that modeling and simulation aspects of heat and mass transport in such systems have received less attention. Recovering latent heat to generate vapor will increase Gain Output Ratio (GOR), an approach that can be rewarding in terms of enhancing the Specific Water Productivity (SWP). Latent heat recovery is discussed in the following sections. In addition, salt crystallization, a bottleneck in making such systems reliable on the long time scale for more extended periods and mitigation strategies are discussed in the next section.

2 Fouling Due to Salt Accumulation

Among the parameters that influence specific water productivity, the Gain Output Ratio (GOR) has been identified as a key parameter needing further investigation. However, the practical limitations of implementing latent heat recovery, which critically affects the gain output ratio in these solar-thermal desalination systems, remain a hurdle (Wang et al. 2019; Zhang et al. 2021; Xu et al. 2020a, b). Recent studies show that salt accumulation on the evaporating surface drastically reduces system performance, calling for novel salt mitigation strategies (Li et al. 2020; Xu et al. 2021; Cooper et al. 2018). Therefore, the current state of the art aims to address the following two challenges: (i) improving gain output ratio by efficient latent heat recovery and (ii) achieving a stable evaporation rate for a significantly longer duration (> 30 h). The first is achieved by multi-stage heat recovery (Gao et al. 2023; Deka and Dash 2023) while the second, by designing a suitable salt mitigation strategy (Zhang et al. 2020a, b, 2021).

2.1 Strategies for a Coupled System

The salt mitigation strategies can be classified into two types depending on whether they address only the problem of salt accumulation or provide additional scope to recover the latent heat of condensation. The former case: (i) the vapor-generating surface and the solar radiation absorbing surface are the same, i.e., coupled systems, see Fig. 7, and, the latter case: (ii) systems which have these two processes on separate surfaces, i.e., decoupled systems; see Fig. 9. In case (i), the top cover needs to be transparent which prevent recovering the wasted latent heat of condensation which could have been used multiple times to enhance the water collection rate (Zhang et al. 2021). Thus, for case (i) only the problem of salt mitigation is addressed but not the recovery of latent heat. Accordingly, such systems can be further subdivided in the following way: design of water feed in and material selection to provide least resistance, see Fig. 6 (Zhang et al. 2019; Xu et al. 2019; Ni et al. 2018), superhydrophobic interfaces preventing wetting of the interface but allowing vapor transport (Janus membranes), see Fig. 9a (Zhang et al. 2015; Gao et al. 2019; Xu et al. 2018), designing the evaporator dimension in a way to allow salt accumulation along an isolated location (Yang et al. 2023; Shi et al. 2018; Xu et al. 2020a; Xia et al., 2019; Wu et al. 2020) see Fig. 7b, and using (a) washable fabrics (Kou et al. 2019), (b) self-cleaning mechanisms (Liu et al.2015, 2021), and (c) novel pumping mechanism (Zeng et al. 2019), see Fig. 8. In a typical vapor-generating surface, salt crystallization over time creates a gradient in concentration, because of which the accumulated salt diffuses back to the reservoir naturally in due course of time (Fig. 6b).

This can be explained based on Eq. 2 (Xu et al. 2021), which indicates that materials with high porosity will be able to promote this natural backflow of salt and prevent salt accumulation across day time (Ni et al. 2018).

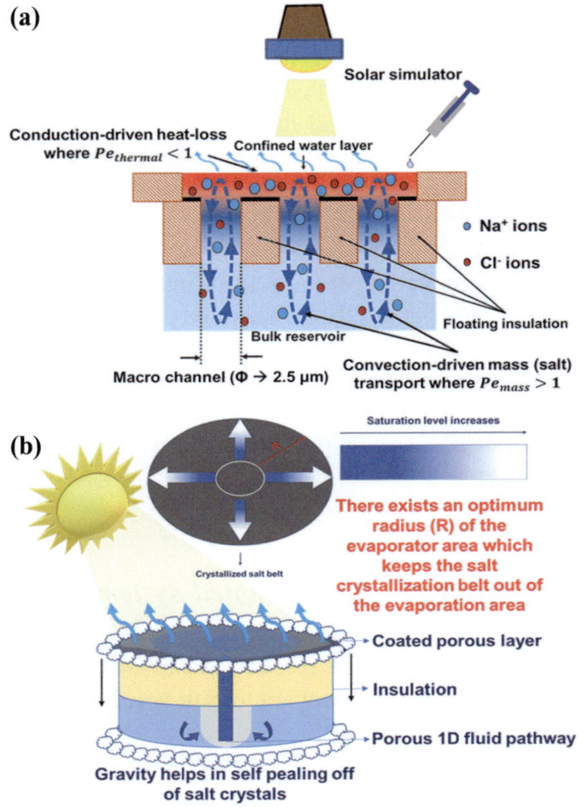

Fig. 7 Schematic showing salt mitigation strategies for coupled systems: **a** confined water layer strategy without wick having convection-driven enhanced salt transport and **b** spatial isolation strategy with salt harvesting having a novel evaporator design which deliberately allows salt accumulation at or any other strategic location

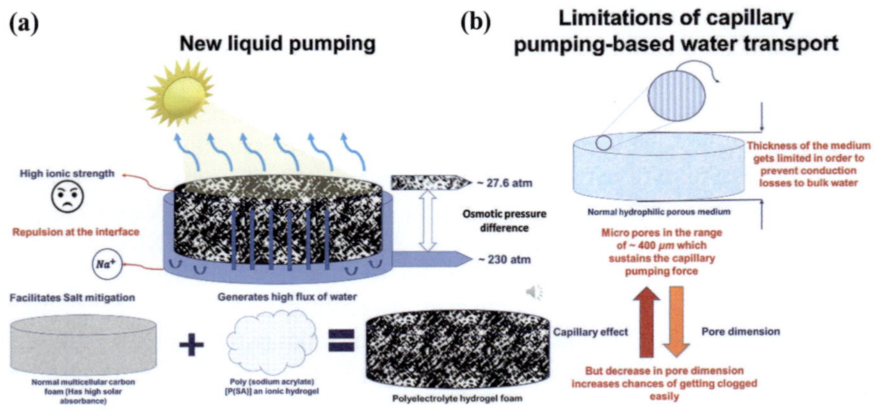

Fig. 8 Novel liquid pumping strategy and its advantages in salt mitigation **a** against the more common capillary **b** pumping often used in interfacial vapor generation systems

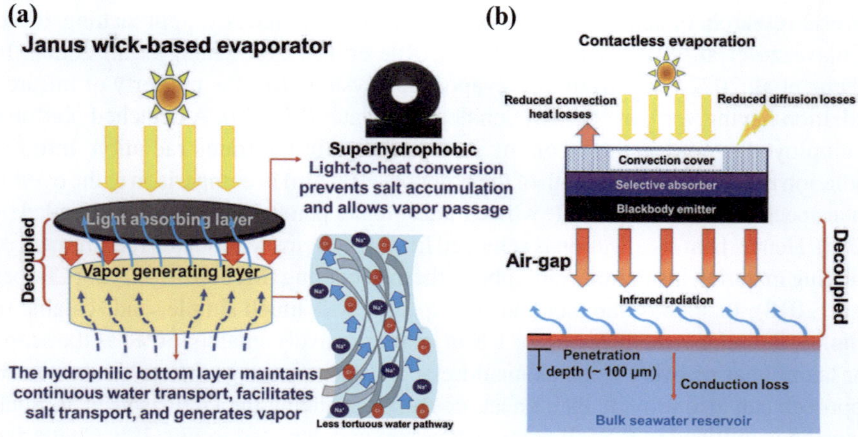

Fig. 9 Schematic showing the decoupling mechanisms adopted for efficient salt mitigation and reliable vapor generation. **a** Janus wick-based evaporator using superhydrophobic light absorbing layer and **b** contactless evaporator using infrared radiation for achieving heat localization

$$J = \frac{\varepsilon D \Delta C}{\tau} \quad (2)$$

Here, J stands for diffusion flux driven by a concentration difference of ΔC, ε is the porosity of the wicking material, and τ is its tortuosity. Zhang et al. (2022) proposed using a confined water layer instead of wick-based strategies, to trigger natural convection leading to salt rejection. This mechanism replaced mass diffusion as shown in Fig. 5a. The authors showed that by selecting an optimal thickness of the confined water layer and water transport passages, one can achieve heat localization as well as natural convection-driven salt rejection.

2.2 *Strategies for Decoupled Systems*

In Coupled systems the evaporator needs to be photo-thermal with proper heat localization to generate vapor with high efficiency and also prevent accumulation of salt. This requires state-of-the-art material innovation relying on complex fabrication methodologies. On the other hand, decoupled systems provide much more flexibility in design and very high efficiencies can be achieved employing commercially well-established materials. The idea of a decoupled system can be classified as devices with upward evaporation and downward evaporation. In downward evaporation systems solar absorption takes place on the front face which transfers the absorbed heat to the back end. Evaporation takes place in a downward direction from a wicking material using the heat from the back face of the absorbing surface; see Fig. 10a. Such systems allow recovering the laten heat of condensation and reuse it to generate more vapor; see Fig. 12. A multi-stage system employing this strategy were reported by several researchers delivering high vapor conversion efficiency, approaching 100% (Chiavazzo et al. 2018; Xu et al. 2020b; Xue et al. 2018; Zhang et al. 2020a, b; Cheng et al. 2022). The front-side evaporation system uses the property of infrared radiation having very less penetration depth in water (Fig. 9b). A detached platform is employed to convert the incoming solar radiation into infrared radiation. Infrared radiation has a penetration depth of the order of ~ 100 μm) in comparison to the overall solar spectrum including visible where much greater penetration depths are needed (~ 40 m). Hence, heat localization is achieved in such systems without the requirement of wicking material, insulation, and a photo-thermal coating (Menon et al. 2020; Cooper et al. 2018). In spite of the fact that such a strategy is much simpler and scalable to employ, the achieved efficiency for 1-Sun is still relatively small (< 50%) as the scope for latent heat recovery remains unaddressed. Thus, a strategy where the front-end approach can also provide latent heat recovery enhancing the GOR is needed. Such a design is outlined in the following discussion and depicted in Fig. 10b. Chatterjee et al. (2023) reported a two-phase closed-loop thermosyphon (TPCT)-based decoupling strategy. The evaporator of the thermosyphon is used to absorb solar irradiation that is transported to the condenser section of the thermosyphon with minimal losses. The vapor generation takes place from the condenser end of the thermosyphon. Thus, decoupling the vapor-generating surface and the solar-absorbing surface is achieved in a front-side configuration. This provides scope of recovering latent heat and opens up the possibility of utilizing low-grade energy in other forms such as industrial waste heat for water purification.

Thermosyphon and heat-pipe integrated solar collectors are commercially mature technology in the domain of solar water heating. This decoupling strategy is viable since the fabrication aspects are greatly simplified.

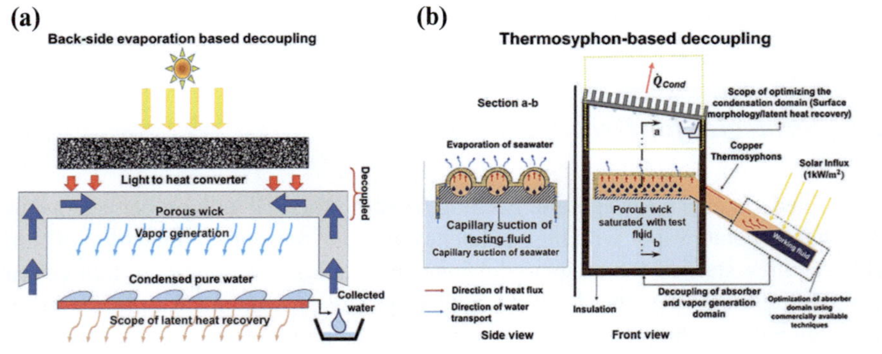

Fig. 10 Schematic showing the decoupling mechanism in **a** backward evaporation and **b** forward evaporation-based systems enabling latent heat recovery in addition to efficient salt mitigation for higher water productivity (Chatterjee et al. 2023, adapted with permission from Elsevier)

3 Enhancement of Condensation Heat Transfer Coefficient

The above discussion was limited to open system configurations, concerned only with reliable vapor generation in an efficient manner. However, to produce potable drinking water, the generated vapor needs to be suitably condensed. In Fig. 1, it was shown that the conventional solar stills relying on a transparent condensing surface have several limitations. Some of the disadvantages of the traditional design are absence of latent heat recovery and losses associated with reflection and scattering from the condensed droplets on the condensing surface. The following section describes methodologies proposed in the literature to address these limitations.

3.1 Single-Stage Water Collection Strategies

Kabeel et al. proposed a novel water collection strategy to address losses associated with residual drops interfering with the incoming solar irradiation. They used a flat still configuration instead of the traditional inclined design. Using capillary wicks to collect the condensed water instead of gravity made the water collection process more efficient. Additionally, the authors proposed a superhydrophilic coating on the condensing surface, Fig. 11a, to form a thin condensate layer instead of a distribution of drops. Film formation assisted the capillary wicks to transport the condensed water from the condensing surface to a collection tray and also reduced losses associated with reflection and scattering. Wang et al. (2021) proposed a condensing surface design involving an inverted honeycomb structure. The proposed configuration ensured that solar absorption and condensing surfaces are on opposite sides, eliminating the extent of optical losses. The inverted honeycomb structure was fabricated using highly thermally conducting copper and coated with hydrophobic nanostructures. This step ensured dropwise condensation while maintaining a much lower degree of subcooling than in a conventional design, where the surface is exposed to incoming solar irradiation. Thus, a higher condensation heat transfer coefficient was possible, resulting in superior water productivity when compared to a traditional design (Fig. 12).

4 Modeling Aspects

4.1 Transient Computational Approach

Interfacial vapor generation employing heat localization to generate vapor efficiently relies primarily on evaporation from wicking materials. The transport process in such porous structures is a function of the geometric and thermal properties of the wicks and the far-field conditions. Yadav et al. (2023) investigated the effect of such key

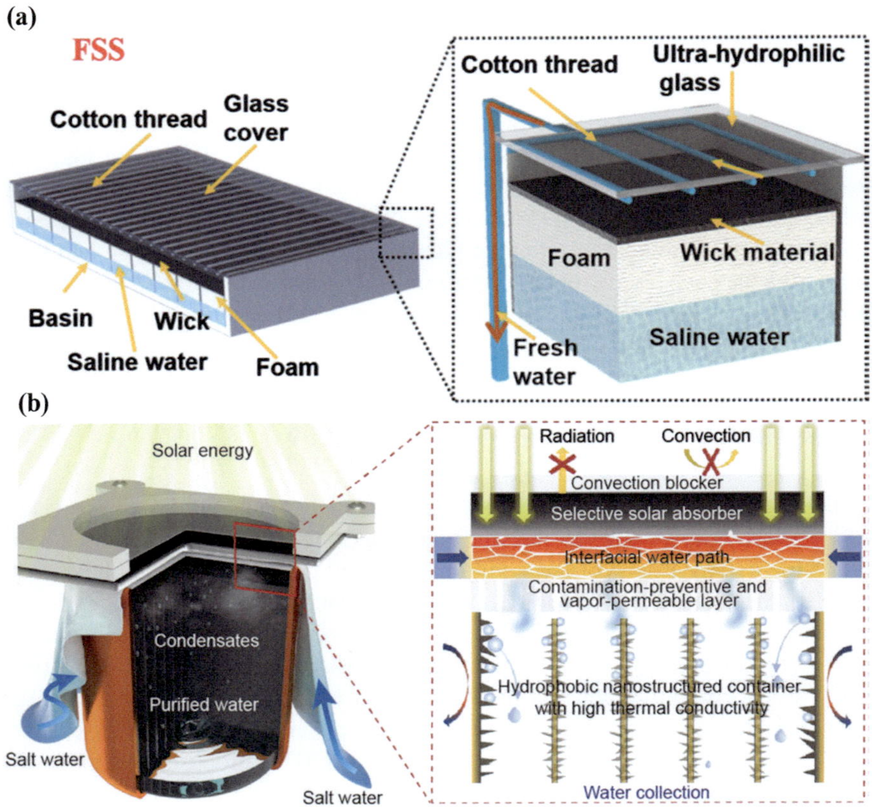

Fig. 11 Schematic showing enhancement in water collection by novel condensation mechanisms: **a** combination of super hydrophilic surface for condensation and porous wicks for water collection and **b** combination of superhydrophobic surface for condensation and high thermally conductive material (Zhao et al. 2024, adapted with permission from Elsevier)

properties on the transport dynamics of both pure and simulated sea water from a single-layer wire mesh at low input heat fluxes (1, 5, and 10 kW/m^2), typical for the intended application. Evaporation mass flux of pure water from different mesh structures (stainless steel and nylon meshes with different pore sizes) were evaluated using non-invasive infrared thermography. The local transport of salt inside the pores of such meshes for evaporation of simulated seawater was visualized using stereo microscopy. A finite element-based computational model was formulated to visualize and quantify the local transient transport dynamics at the pore level. Arbitrary Lagrangian–Eulerian (ALE) formulation was employed to solve the coupled transport equations on a deforming computational grid, using the Fick's law of diffusion at the water-air interface. Clear deviation from the diffusion-based approach was reported for input heat fluxes higher than 1 kW/m^2. Higher evaporation rates were

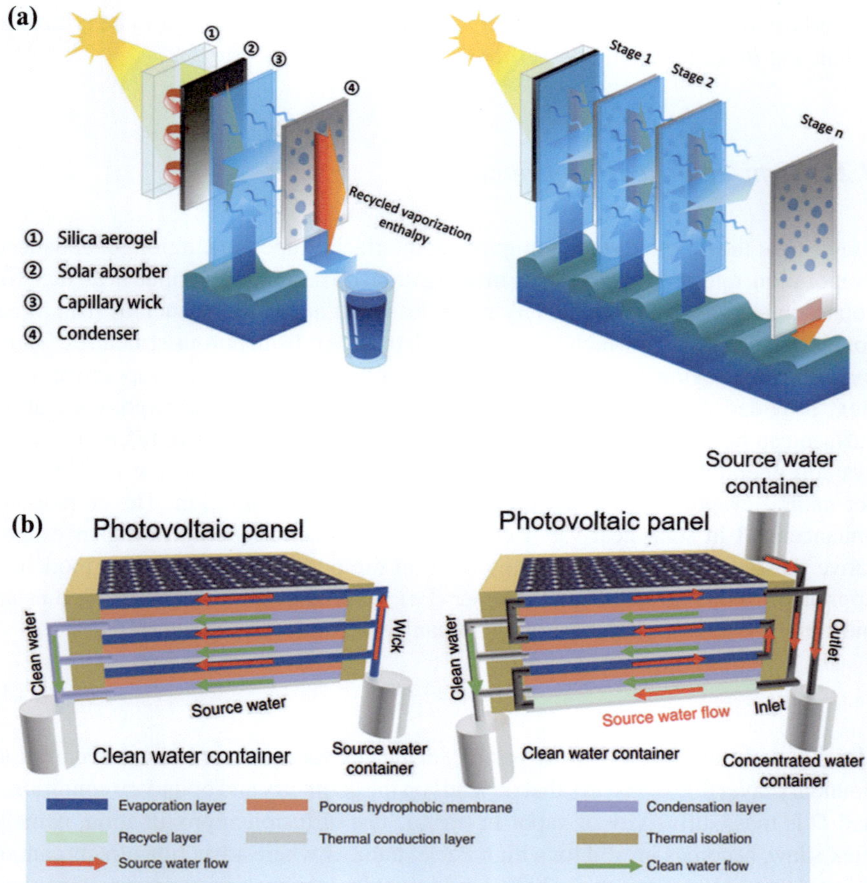

Fig. 12 Schematic showing enhancement in water collection by a novel multi-stagging strategy that utilizes latent heat available during condensation multiple times to generate additional vapor. **a** Use of direct solar energy radiation and **b** solar-PV integrated for potable water production as well as power generation (Zhao et al. 2024, Adapted with permission from Elsevier)

observed in meshes with smaller wire diameter and higher thermal diffusivity. Additionally, the critical role of local thin capillary films along the liquid meniscus and mesh interface was exemplified. The role of thermocapillary convection on evaporation at low input heat fluxes was shown to be minimal using simulation results. The authors also reported the effect of pore size on the nature of salt crystallization and distribution. The growth and distribution of salt concentration gradient within the pores was illustrated using computational results complementing well with the experimental observations. Figure 13a shows the local transport mechanisms within the pores of the meshes employed while 13b shows the growth and distribution of salt concentration gradient within the pores during evaporation gradually leading toward dry out. Local transport effects at the pore level were shown to be critical in enhancing overall evaporation at the system level.

4.2 Semi-empirical Approach

Majority of the reported studies employing interfacial vapor generation for enhanced evaporation rate are open system investigations. The range of input heat flux for solar applications is comparatively much less than applications such as heat pipes for electronics cooling which also deals with transport from porous structures. Thus, modeling employing Ficks's law of diffusion is used to predict the evaporative mass flux, provided the advection effects due to vapor flow over the vapor-generating surface can be ignored (Beyhaghi et al. 2014). This approach is only applicable to systems operating at very low input heat fluxes (typically much below 1 kW/m^2) or for natural evaporation relying only on gradient in concentration. The continuous enhancement in solar-heat-vapor conversion efficiencies leading to higher evaporative mass fluxes pose reasonable doubt on the purely diffusion-based modeling approach. In this regard, Peclet number (Pe), defined in Eq. (3), can be used as an indicator of whether advection of vapor can be neglected.

$$\mathrm{Pe} = \frac{u_{\mathrm{int}}\delta}{D} \tag{3}$$

Here u_{int} is the air–vapor mixture velocity at the interface (m/s), δ is the concentration boundary layer thickness (m), that depends on the geometry and boundary conditions, and D is mass diffusivity of vapor in (m^2/s). The diffusion approximation, namely Fick's law, becomes invalid for a high Peclet number where advection effects cannot be neglected (Vaartstra et al. 2020). Additionally, changes in geometry, size, and operating conditions, where evaporative mass fluxes are of moderate magnitude, need to consider the effect of gravity as well (Peng et al. 2020). The dimensionless quantity that determines the strength of buoyancy effects is Grashof number (Gr), defined in Eq. (4).

$$\mathrm{Gr} = \frac{g\Delta\rho L_c^3}{\rho\nu^2} \tag{4}$$

Here, g is acceleration due to gravity, $\Delta\rho$ is the difference in the air–vapor density between the water surface and the far-field ambient (kg/m^3), L_c is the characteristic length scale (m) of the evaporator, ν is the kinematic viscosity of the air–vapor mixture (m^2/s).

Evaporation operating in a convective mode is classified into two categories: natural convection and forced convection. The evaporative flux is determined according to the dominant mode of convective heat transfer. In this regard, a non-dimensional number is defined in terms of the Richardson number, Eq. (5):

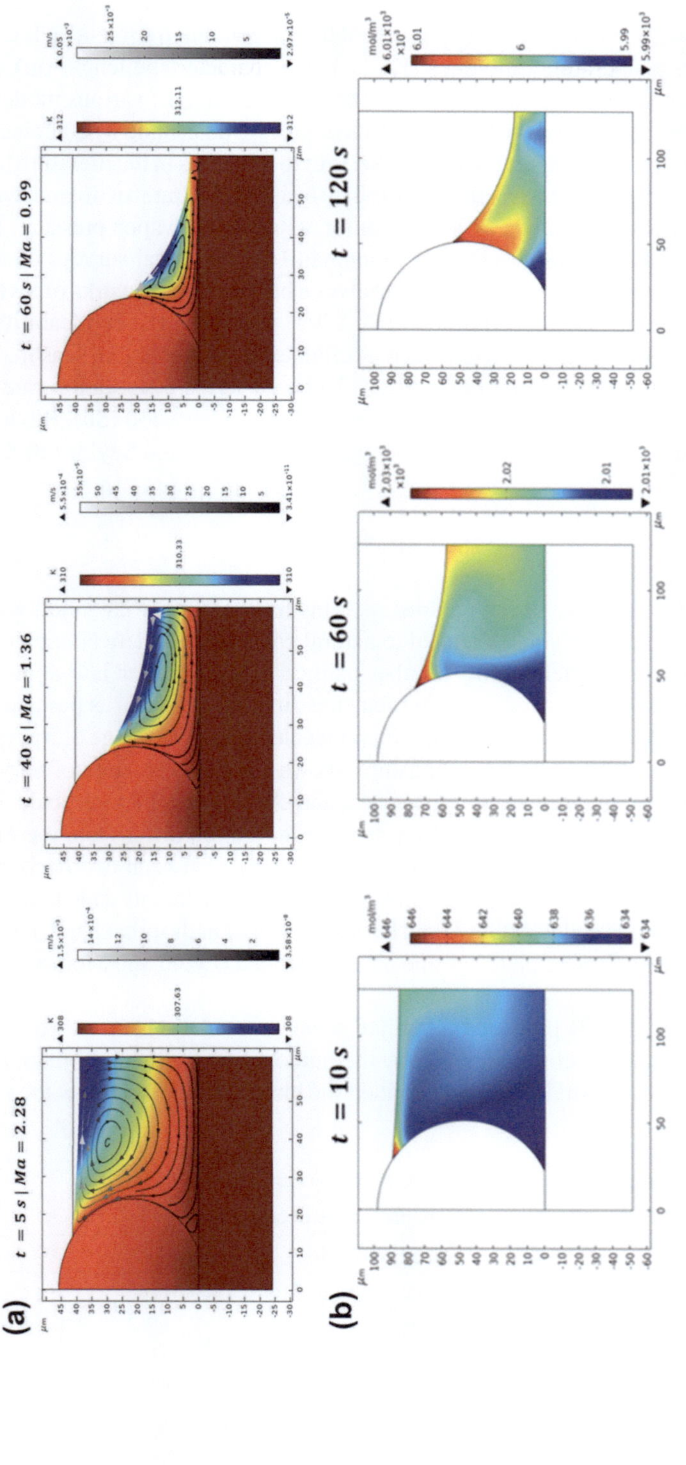

Fig. 13 COMSOL Multiphysics simulation of heat and mass transport along the interface and within the pores of the wick structures used in interfacial vapor generation systems: **a** effect of Marangoni convection on heat transfer and **b** nature of salt transport and evolution of concentration gradient at various time instants (Yadav et al. 2023, adapted with permission from Elsevier)

$$\mathrm{Ri} = \frac{g\beta\Delta T L_c}{U_0^2} = \frac{\mathrm{Gr}}{\mathrm{Re}^2} \tag{5}$$

Here, β is the coefficient of thermal expansion of the gaseous environment (1/K), ΔT is the applicable temperature difference (K), L_c is the characteristic length (m), and U_o is the bulk velocity (m/s). Table 1 presents the classification of various modes of convective heat transfer based on the Richardson number to determine whether inertia (Re) or gravity (Gr) dominates evaporation rate. Previous studies in the literature have parameterized the natural convection-controlled evaporation rate of inland water bodies such as lakes and ponds as a function of wind speed, vapor pressure, and the size of the corresponding water body (Pennman 1948). Critical surveys of such equations developed from field surveys and trials can be found in the works of Sweers (1976), Sparrow et al. (1983), and Smith et al. 1994). However, the applicability of such correlations is reduced by factors such as climate, solar insolation, season, and geographical location (Varju and Poós 2022). To characterize evaporation rate for natural convection, a power law relationship between the Sherwood (Sh), Rayleigh (Ra), and Schmidt (Sc) numbers has been suggested by Bower and Saylor (2009) in the form:

$$\mathrm{Sh} = A\mathrm{Ra}^m\mathrm{Sc}^n \tag{6}$$

where *A*, *m*, and *n* are curve fitting parameters. This form of power law relationship is common in heat transfer studies related to natural convection, where Nusselt (Nu) number is related to Rayleigh (Ra) number using a similar power law approach (Chavanne et al. 2001). There is a certain robustness in this approach, as pointed out by Bower and Saylor (2009), which makes it applicable for a wider range of transport processes. The exponent in such relationships shows similarity in spite of varying boundary conditions, quantity to be transported, and characteristic length scale. The boundary conditions of natural convection-driven evaporation differ from the more common Rayleigh–Bénard setup where heat is transported between two horizontal plates with no-slip boundary conditions and the characteristic length scale being the distance between the two plates (Glove and Dropkin 1959). On the other hand, natural convection from a horizontal plate has one end unbounded (Lyold and Moran 1974).

Goldstein et al. (1973) proposed a similar power law relationship as Eq. 6 for mass transfer by natural convection. Here, the authors considered sublimation of naphthalene from horizontal plates and defined the characteristic length as the ratio

Table 1 Mode of heat transfer classified in terms of the Richardson number

Richardson number (Ri)	Dominant mode of convection
< 0.1	Forced convection
0.1 < Ri < 10	Forced and natural (mixed) convection
Ri > 10	Natural convection

of area to perimeter. Although, mass transfer here is by natural convection, the difference with evaporation phenomena is in the hydrodynamic boundary condition at the interface. For sublimation, it is no-slip, while, for a liquid-vapor interface, a non-zero interfacial velocity will be present.

Though numerous correlations have been proposed in the literature to determine the evaporative flux from the water surface of inland water bodies by natural convection, these correlations are sensitive to the operating conditions and the size of the evaporating surface. Additionally, the empirical constants vary depending on whether the surface temperature of the evaporating surface is higher or lower than the temperature of the far-field ambient. The Rayleigh number (Ra), defined below in Eq. (7), dictates the direction of convective flow:

$$\mathrm{Ra} = \Delta\rho_G \frac{gL_c^3}{\nu_G \rho_G D_{\mathrm{LG}}} \tag{7}$$

Here, $\Delta\rho_G$ is the difference in density of the gaseous environment over the length scale considered (kg/m^3), L_c is the characteristic dimension of the evaporator (m), ν_G is the kinematic viscosity (m^2/s), ρ_G is the density (kg/m^3), and D_{LG} is the binary diffusion coefficient of vapor in the bulk environment (m^2/s). A positive Rayleigh number indicates gas–vapor flow from the evaporating surface toward the far field; a negative Rayleigh number indicates flow from the far-field ambient flow toward the evaporating surface with the possibility of condensation (Sparrow et al. 1983). However, a few published correlations are useful for operating conditions realized in solar-based interfacial evaporation adopted for desalination applications. Kulshrestha et al. (2024) selected four such correlations based on Rayleigh number, characteristic dimensions, surface temperature, and far-field relative temperature and humidity relevant to solar desalination (Poós and Varju 2019). Three correlations were selected for imposed heat flux where Ra > 0 (Rao and Radhakrishnan 1977; Kumar and Arakeri 2015; Bower and Saylor 2009) and one for the dark evaporation case (without imposed heat flux) case, where Ra < 0 (Sparrow et al. 1983). In Fig. 14, the experimentally obtained evaporative heat flux in non-dimensional form is compared against the ones predicted by the chosen correlations. It is shown that the match is quite sensitive to the operating conditions under which the correlations were derived. Moreover, the correlations do not consider possible enhancement in evaporation rate from thin films and from porous materials. The evaporation rates are grossly under-predicted by the chosen correlations at low heat fluxes. Hence, it is necessary to carefully examine available correlations and their applicability to determine evaporative mass flux for solar-based interfacial vapor-generating systems.

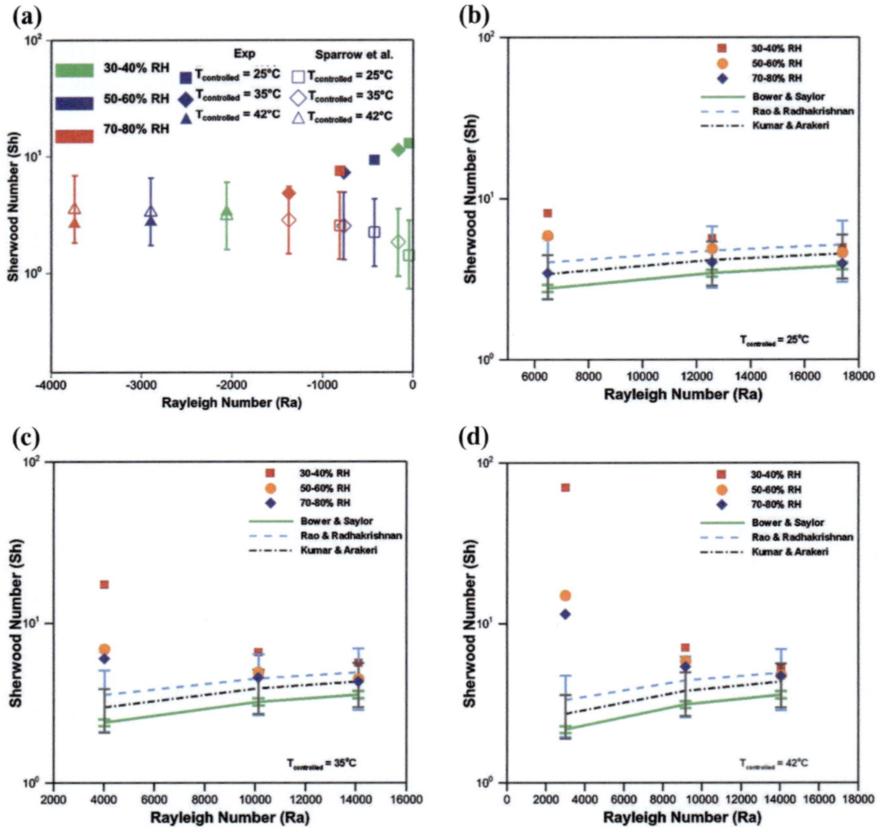

Fig. 14 Comparison of the evaporative mass transfer coefficient obtained experimentally with the ones predicted using existing correlations for various Rayleigh numbers and far-field conditions: **a** buoyancy driven down-flow configuration and **b–d** buoyancy driven up-flow configuration (Kulshrestha et al. 2024, adapted with permission from Springer Nature)

5 Closure

Interfacial vapor generation-based solar-thermal desalination devices are a promising alternative with several advantages in their favor. They can address the critical limitations of traditional solar still technology. Reported water productivity of such systems are multiple times higher than what is available from a conventional solar still. The fact that such technologies are not limited by higher concentrated brine solutions as the working fluid is an added advantage when compared to commercially adopted membrane-based technologies. However, most studies have confined themselves to improving solar absorptivity and solar-to-vapor conversion efficiency. The proposed strategies in terms of new materials are not impactful enough in terms of enhancement in the Specific Water Productivity (SWP). Recent innovations involve addressing fundamental heat and mass transport issues. Efficient salt mitigation strategies to make the systems reliable for long-term operation and large-scale commercial viability are one of the prime areas of interest. Innovative water collection strategies that enhance condensation heat transfer and harness the latent heat release to generate more vapor are limited. Zero liquid discharge (ZLD) is emerging as the new area of interest for desalination, potable water generation, and harnessing rare earth minerals. Mining of valuable minerals from seawater distillate is receiving fresh attention as an alternative pathway. Most of the reported studies use solar energy as the primary energy source, which limits the upper limit of input heat flux. However, it has been shown that incorporating passive devices such as thermosyphons and heat pipes can potentially harness other forms of energy, such as industrial waste heat. The effect on marine ecosystems caused by large-scale brine disposal into the water bodies is a growing concern. Incorporation of such interfacial vapor generation systems with other medium and large-scale desalination plants in a hybrid mode provides the opportunity to address such practical challenges. Hence, researchers will need to invest more in pilot-scale projects rather than producing laboratory-scale experimental setups to prove long-term reliability and commercial viability of new technologies.

References

Adams S, Zhang JZ (2016) Unique optical properties and applications of hollow gold nanospheres (HGNs). Coord Chem Rev 320–321:18–37

Awad FS, Kiriarachchi HD, Abouzeid KM, Özgür Ü, El-Shall MS (2018) Plasmonic graphene polyurethane nanocomposites for efficient solar water desalination. ACS Appl Energy Mater 1(3):976–985

Bower SM, Saylor JR (2009) A study of the Sherwood–Rayleigh relation for water undergoing natural convection-driven evaporation. Int J Heat Mass Transf 52(13-14): 3055–3063. https://doi.org/10.1016/j.ijheatmasstransfer.2009.01.034

Beyhaghi S, Geoffroy S, Prat M, Pillai KM (2014) Wicking and evaporation of liquids in porous wicks: a simple analytical approach to optimization of wick design. AIChE J 60(5):1930–1940

Chavanne X, Chillà F, Chabaud B, Castaing B, Hébral B (2001) Turbulent Rayleigh–Bénard convection in gaseous and liquid He. Phys Fluids 13(5): 1300–1320. https://doi.org/10.1063/1.1355683

Chatterjee D, Kulshrestha T, Khandekar S (2023) Continuous vapor generation for thermal-desalination applications using a thermosyphon based heat localization strategy. Desalination 555:116492

Chen C, Kuang Y, Hu L (2019) Challenges and opportunities for solar evaporation. Joule 3(3):683–718

Cheng S, Li Y, Jin B, Yu Z, Gu R (2022) Designing salt transmission channel of solar-driven multistage desalination device for efficient and stable freshwater production from seawater. Desalination 531:115688

Chiavazzo E, Morciano M, Viglino F, Fasano M, Asinari P (2018) Passive solar high-yield seawater desalination by modular and low-cost distillation. Nat Sustain 1(12):763–772

Cooper TA, Zandavi SH, Ni GW, Tsurimaki Y, Huang Y, Boriskina SV, Chen G (2018) Contactless steam generation and superheating under one sun illumination. Nat Commun 9(1):5086. https://doi.org/10.1038/s41467-018-07494-2

Cui L, Zhang P, Xiao Y, Liang Y, Liang H, Cheng Z, Qu L (2018) High-rate production of clean water based on the combined photo-electro-thermal effect of graphene architecture. Adv Mater 30(22):1706805

Deka N, Dash S (2023) Multistage interfacial thermal desalination system with metallic evaporators. Desalination 556:116576

Evaporation of water from pools

Gao M, Connor PKN, Ho GW (2016) Plasmonic photothermic directed broadband sunlight harnessing for seawater catalysis and desalination. Energy Environ Sci 9(10):3151–3160. https://doi.org/10.1039/c6ee00971a

Gao S, Dong X, Huang J, Dong J, Maggio F. Di, Wang S, Guo F, Zhu T, Chen Z, Lai Y (2019) Bioinspired soot-deposited Janus fabrics for sustainable solar steam generation with salt-rejection. Glob Chall 3(8):1800117

Gao J, Zhang L, You J, Ye Z, Zhong Y, Wang R, Wang EN, Xu Z (2023) Extreme salt-resisting multistage solar distillation with thermohaline convection. Joule 7(10):2274–2290

Goldstein RJ, Sparrow EM, Jones DC (1973) Natural convection mass transfer adjacent to horizontal plates. Int J Heat Mass Transf 16(5): 1025–1035. https://doi.org/10.1016/0017-9310(73)90041-0

Ghasemi H, Ni G, Marconnet AM, Loomis J, Yerci S, Miljkovic N, Chen G (2014) Solar steam generation by heat localization. Nat Commun 5:4449

Kim JU, Lee S, Kang SJ, Kim TI (2018) Materials and design of nanostructured broadband light absorbers for advanced light-to-heat conversion. Nanoscale 10(46):21555–21574

Kou H, Liu Z, Zhu B, Macharia DK, Ahmed S, Wu B, Zhu M, Liu X, Chen Z (2019) Recyclable CNT-coupled cotton fabrics for low-cost and efficient desalination of seawater under sunlight. Desalination 462:29–38

Kulshrestha T, Chatterjee D, Khandekar S (2024) Effect of far-field ambient conditions on interfacial solar vapor generation using a two-phase closed thermosyphon. Exp Comput Multiph Flow 6(3):199–213

Kumar N, Arakeri JH (2015) Natural convection driven evaporation from a water surface. Procedia IUTAM 15:108–115. https://doi.org/10.1016/j.piutam.2015.04.016

Li X, Xu W, Tang M, Zhou L, Zhu B, Zhu S, Zhu J (2016) Graphene oxide-based efficient and scalable solar desalination under one sun with a confined 2D water path. Proc Natl Acad Sci USA 113(49):13953–13958

Li X, Lin R, Ni G, Xu N, Hu X, Zhu B, Lv G, Li J, Zhu S, Zhu J (2018) Three-dimensional artificial transpiration for efficient solar waste-water treatment. Natl Sci Rev 5(1):70–77

Li H, Yan Z, Li Y, Hong W (2020) Latest development in salt removal from solar-driven interfacial saline water evaporators: advanced strategies and challenges. Water Research 177:115770

Liu Y, Chen J, Guo D, Cao M, Jiang L (2015) Floatable, self-cleaning, and carbon-black-based superhydrophobic gauze for the solar evaporation enhancement at the air-water interface. ACS Appl Mater Interfaces 7(24):13645–13652

Liu KK, Jiang Q, Tadepalli S, Raliya R, Biswas P, Naik RR, Singamaneni S (2017) Wood-graphene oxide composite for highly efficient solar steam generation and desalination. ACS Appl Mater Interfaces 9(8):7675–7681

Liu H, Chen C, Wen H, Guo R, Williams NA, Wang B, Chen F, Hu L (2018) Narrow bandgap semiconductor decorated wood membrane for high-efficiency solar-assisted water purification. J Mater Chem A 6(39):18839–18846

Liu Z, Zhong Q, Wu N, Zhou H, Wang L, Zhu L, Jiang N, Zhu B, Chen Z, Zhu M (2021) Vertically symmetrical evaporator based on photo-thermal fabrics for efficient continuous desalination through inversion strategy. Desalination 509:115072

Lloyd JR, Moran WR (1974) Natural convection adjacent to horizontal surface of various planforms. J Heat Transf 96(4): 443–447. https://doi.org/10.1115/1.3450224

Ma X, Jia X, Yao G, Wen D (2022a) Double-sided suspending evaporator with top water supply for concurrent solar evaporation and salt harvesting. ACS Sustain Chem Eng 10(38):12843–12851
Ma X, Jia X, Yao G, Wen D (2022b) Umbrella evaporator for continuous solar vapor generation and salt harvesting from seawater. Cell Rep Phys Sci 3(7):100940
Menon AK, Haechler I, Kaur S, Lubner S, Prasher RS (2020) Enhanced solar evaporation using a photo-thermal umbrella for wastewater management. Nat Sustain 3:144–151
Ni G, Zandavi SH, Javid SM, Boriskina SV, Cooper TA, Chen G (2018) A salt-rejecting floating solar still for low-cost desalination. Energy Environ Sci 11(6):1510–1519
Poós T, Varju E (2019) Review for prediction of evaporation rate at natural convection. Heat Mass Transf 55(6) 1651–1660. https://doi.org/10.1007/s00231-018-02535-4
Peng G, Sharshir SW, Hu Z, Ji R, Ma J, Kabeel AE, Liu H, Zang J, Yang N (2021) A compact flat solar still with high performance. Int J Heat Mass Transf 179: 121657. https://doi.org/10.1016/j.ijheatmasstransfer.2021.121657
Peng G, Deng S, Sharshir SW, Ma D, Kabeel AE, Yang N (2020) High efficient solar evaporation by airing multifunctional textile. Int J Heat Mass Transf 147:118866
Penman HL (1948) Natural evaporation from open water bare soil and grass. Proc R Soc Lond A Math Phys Sci 193(1032): 120–145. https://doi.org/10.1098/rspa.1948.0037
Shi Y, Zhang C, Li R, Zhuo S, Jin Y, Shi L, Hong S, Chang J, Ong C, Wang P (2018) Solar evaporator with controlled salt precipitation for zero liquid discharge desalination. Environ Sci Technol 52(20):11822–11830
Smith CC, Löf G, Jones R (1994) Measurement and analysis of evaporation from an inactive outdoor swimming pool. Solar Energy 53(1): 3–7. https://doi.org/10.1016/S0038-092X(94)90597-5
Sparrow EM, Kratz GK, Schuerger MJ (1983) Evaporation of water from a horizontal surface by natural convection. J Heat Transf 105(3): 469–475. https://doi.org/10.1115/1.3245609
Sweers HE (1976) A nomogram to estimate the heat-exchange coefficient at the air-water interface as a function of wind speed and temperature; a critical survey of some literature. J Hydrol 30(4): 375–401. https://doi.org/10.1016/0022-1694(76)90120-7
Vaartstra G, Zhang L, Lu Z, Díaz-Marín CD, Grossman JC, Wang EN (2020) Capillary-fed, thin film evaporation devices. J Appl Phys 128(13):130901
Wang P (2018) Emerging investigator series: the rise of nano-enabled photo-thermal materials for water evaporation and clean water production by sunlight. Environ Sci Nano 5(5):1078–1089
Wang J, Li Y, Deng L, Wei N, Weng Y, Dong S, Qi D, Qiu J, Chen X, Wu T (2017) High-performance photo-thermal conversion of narrow-bandgap Ti_2O_3 nanoparticles. Adv Mater 29(3):1603730
Wang F, Xu N, Zhao W, Zhou L, Zhu P, Wang X, Zhu B, Zhu J (2021) A high-performing single-stage invert-structured solar water purifier through enhanced absorption and condensation. Joule 5(6):1602–1612
Wang Z, Horseman T, Straub AP, Yip NY, Li D, Elimelech M, Lin S (2019) Pathways and challenges for efficient solar-thermal desalination. Appl Sci Eng
Wilson HM, Rahman ARS, Parab AE, Jha N (2019) Ultra-low-cost cotton based solar evaporation device for seawater desalination and waste water purification to produce drinkable water. Desalination 456:85–96
Wu L, Dong Z, Cai Z, Ganapathy T, Fang NX, Li C, Yu C, Zhang Y, Song Y (2020) Highly efficient three-dimensional solar evaporator for high salinity desalination by localized crystallization. Nat Commun 11(1):521
Xia Y, Hou Q, Jubaer H, Li Y, Kang Y, Yuan S, Liu H, Woo MW, Zhang L, Gao L, Wang H, Zhang X (2019) Spatially isolating salt crystallisation from water evaporation for continuous solar steam generation and salt harvesting. Energy Environ Sci 12(6):1840–1847
Xu W, Hu X, Zhuang S, Wang Y, Li X, Zhou L, Zhu S, Zhu J (2018) Flexible and salt resistant Janus absorbers by electrospinning for stable and efficient solar desalination. Adv Energy Mater 8(14):1702884
Xu Y, Liu D, Xiang H, Ren S, Zhu Z, Liu D, Xu H, Cui F, Wang W (2019) Easily scaled-up photo-thermal membrane with structure-dependent auto-cleaning feature for high-efficient solar desalination. J Membr Sci 586:222–230

Xu J, Wang Z, Chang C, Fu B, Tao P, Song C, Shang W, Deng T (2020a) Solar-driven interfacial desalination for simultaneous freshwater and salt generation. Desalination 484:114423

Xu Z, Zhang L, Zhao L, Li B, Bhatia B, Wang C, Wilke KL, Song Y, Labban O, Lienhard JH, Wang R, Wang EN (2020b) Ultrahigh-efficiency desalination: via a thermally-localized multistage solar still. Energy Environ Sci 13(3):830–839

Xu K, Wang C, Li Z, Wu S, Wang J (2021) Salt mitigation strategies of solar-driven interfacial desalination. Adv Funct Mater 31(8):2007885

Xue G, Chen Q, Lin S, Duan J, Yang P, Liu K, Li J, Zhou J (2018) Highly efficient water harvesting with optimized solar thermal membrane distillation device. Glob Chall 2(5–6):1800001

Varju E, Poós T (2022) New dimensionless correlation for mass transfer at evaporation of open liquid surface in natural convection. Int Commun Heat Mass Transf 136:106–102. https://doi.org/10.1016/j.icheatmasstransfer.2022.106102

Yadav P, Chatterjee D, Khandekar S (2023) Transport phenomena and dynamics of an evaporating water meniscus at low heat fluxes. Int J Heat Mass Transf 214:124368

Yang J, Pang Y, Huang W, Shaw SK, Schiffbauer J, Pillers MA, Mu X, Luo S, Zhang T, Huang Y, Li G, Ptasinska S, Lieberman M, Luo T (2017) Functionalized graphene enables highly efficient solar thermal steam generation. ACS Nano 11(6):5510–5518

Yang Z, Li D, Yang K, Chen L, Wang J, Zhu X, Chen B (2023) Optimized water supply in a solar evaporator for simultaneous freshwater production and salt recycle. Environ Sci Technol 57(35):13047–13055

Zeng J, Wang Q, Shi Y, Liu P, Chen R (2019) Osmotic pumping and salt rejection by polyelectrolyte hydrogel for continuous solar desalination. Adv Energy Mater 9(38):1900552

Zhang Q, Yang H, Xiao X, Wang H, Yan L, Shi Z, Chen Y, Xu W, Wang X (2019) A new self-desalting solar evaporation system based on a vertically oriented porous polyacrylonitrile foam. J Mater Chem A 7(24):14620–14628

Zhang L, Xu Z, Bhatia B, Li B, Zhao L, Wang EN (2020) Modelling and performance analysis of high-efficiency thermally-localized multistage solar stills. Appl Energy 266:114864

Zhang Y, Xiong T, Nandakumar DK, Tan SC (2020) Structure architecting for salt-rejecting solar interfacial desalination to achieve high-performance evaporation with in situ energy generation. Adv Sci 7(9):1903478

Zhang L, Xu Z, Zhao L, Bhatia B, Zhong Y, Gong S, Wang EN (2021) Passive, high-efficiency thermally-localized solar desalination. Energy Environ Sci 14(4):1771–1793

Zhang L, Li X, Zhong Y, Leroy A, Xu Z, Zhao L, Wang EN (2022) Highly efficient and salt rejecting solar evaporation via a wick-free confined water layer. Nat Commun 13(1):849

Zhang L, Tang B, Wu J, Li R, Wang P (2015) Hydrophobic light-to-heat conversion membranes with self-healing ability for interfacial solar heating. Adv Mater 27(33): 4889–4894 https://doi.org/10.1002/adma.v27.33

Zhao Z, Wang C, Wei D, Wang F (2024) Condensation device design represents a critical step for solar-driven water evaporation toward practical applications. Cell Rep Phys Sci 5(2):101794

Zhou L, Tan Y, Wang J, Xu W, Yuan Y, Cai W, Zhu S, Zhu J (2016) 3D self-assembly of aluminium nanoparticles for plasmon-enhanced solar desalination. Nat Photonics 10(6):393–398

Zhu L, Gao M, Peh CKN, Ho GW (2019) Recent progress in solar-driven interfacial water evaporation: advanced designs and applications. Nano Energy 57:507–518

Environmental Aspects, Socio-Economic Implications and Challenges

Ramesh Narayanaswamy

1 Introduction

The earlier chapters provided the reader with adequate background as well as the scientific and engineering principles governing the design and functioning of several devices that make use of water vapour present in the atmosphere to generate liquid water. One should bear in mind that there are several variations of these air-to-water machines available as either technology demonstrator devices or fully operational commercial devices that use a combination of scientific principles to generate water. For example, some devices rely on fog and its deposition on mesh-type surfaces only whereas certain other devices rely on the capture of fog on an adsorbent substrate in combination with, or without the use of an external energy source (e.g., solar energy) to facilitate the extraction of water from the atmosphere. Also available are air-to-water machines that utilise waste heat from captive thermal energy conversion devices, in combination with atmospheric water vapour capture to produce water. Based on a review of available literature, what we found was that every class of device has certain dependencies on the environment, that aids in enhancing its performance, as well as some limitations that hinder its output, and hence there is no single criteria with which to compare performance, or techno-socio-economic implications. Therefore, in what follows we will discuss briefly the environmental and socio-economic implications as applicable to groups of devices the operate on some common principle. Further down, we will outline some of the technological challenges as well.

R. Narayanaswamy (✉)
School of Civil and Mechanical Engineering, Curtin University, Bentley, WA, Australia
e-mail: r.narayanaswamy@curtin.edu.au

R. Narayanaswamy et al. (eds.), *Sustainable Water Harvesting*, Springer Tracts in Mechanical Engineering, https://doi.org/10.1007/978-981-95-3373-2_10

2 Performance Evaluation of AWG Machines

The performance evaluation of air-to-water generation machines is very limited due to the availability of data that is fit enough for an objective comparison. As an equipment or device, the quantity of water generated is known as the yield of the device. The yield from any air to water machine depends on the following factors: (i) local environmental conditions that modify water output; (ii) thermodynamic capability of the device in the sense of how or to what extent can thermodynamics be used to enable functioning of the device; (iii) management (collection and distribution) of the water generated, and (iv) parasitic losses such as those that occur due to evaporation of water droplets from the collection device, or wind-driven loss of droplets. When these machines are deployed in remote locations, consumers often require access to renewable energy sources such as wind or solar energy to distribute water from the point of collection to where it is required to be delivered. Therefore, when evaluating the performance efficiency of these machines, the variations in the employed method of water capture precludes the concept of a universal and unique performance index parameter, with which to compare these machines. As indicated earlier, based on a complete review of literature, although individualised and device-based testing and evaluation of AWG devices exist, there is no universally accepted performance evaluation criteria or design standard, upon which various devices can be compared. Our study has shown that this aspect of air-to-water generation machines is clearly an area of research worth pursuing. This is also applicable for the techno-economic evaluation of such machines. It is imperative that such studies are underway, as the demand for air to water machines is likely to penetrate more parts of the world in the coming future, due to water scarcity. According to the Grand View Research Market Analysis Report (2023), the global market growth for AWGs indicates a potential compounded annual growth rate of about 9.4% for the period 2024–2030, with the Asia Pacific region contributing to the highest share of the revenue (36%). The report also highlights the fact that the worldwide industrial sector accounted for about 74% of the use of AWGs in various forms.

3 Economic Analysis of Air–Water Machines

The economic feasibility or performance essentially is related to the installation cost of the device plus the cost of energy required to operate the unit (pumping and filtration), investment, maintenance costs, and the cost of water delivered to consumers. For machines that are all of one design, but operating at various locations, one could estimate the economic performance based on water output, energy required to run the unit, and factors that affect the output such as ambient temperature, local wind speeds, and relative humidity by considering a yearly average variation for these parameters. The economic analysis presented by researchers in published literature is based upon mathematical relationships that are derived for a specific need, under a set of certain assumptions for the air to water machine in mind. Since air to water machines are available in various designs, operating at several locations around the globe, we conclude that it is impossible to compare the various machines as they are fundamentally and functionally different. For example, Moghimi et al. (2021) provides a comparison of the economics of water yield at ten locations for an atmospheric water generator under variations of environmental and economic conditions; the study compares AWG machines and bottled water for the purpose of economic analysis.

4 Environmental Impact of Air-to-Water Machines

Air-to-water machines is mostly used for water generation in arid to hyper-arid regions of the world. Use of these machines will hinder the free movement of fog as these machines are intended to maximise the capture of condensed water vapour droplets from fog. This means that, over time, the landscape is, to some extent limited in the soil-nutrient production due to fog impact on the topsoil matrix. Fog is known to be a component in the development of earth-surface nutrients in such regions that are devoid of seasonal rainfall and hence contributes to the biodiversity of the region as outlined by Sun et al. (2024). Our search of the published literature on the operation of fog catchers did not provide any information on the impact of fog catchers on the flora and fauna of the local environment. However, it is worth pursuing research in life and environmental sciences, on the effect of these machines on the short-and long-term effects on soil nutrients, as well as the effect of these machines on the free migration of birds. It is also of interest to note here, that the present author's survey of research work on the effect of air-to-water machines provided no information on any impact (favourable or unfavourable) on the quality of climate available in fog environments (Veloso et al. 2024) and localities. This is also a topic of research that could be pursued.

5 Water Quality of Fog-Harvested Water

In some of the earlier chapters, we had outlined the fact that there are many ways to capture water vapour from the atmosphere, such as through fog harvesters, or some form of sorbent-based capture devices. From a general standpoint, fog or dew-based water generation systems are not without contaminants, although these can be minimised through appropriate water quality treatments. Several factors can contribute to contamination. The fog itself primarily may have trace elements of atmospheric pollution, the collection meshes and tanks being exposed to the ambient environment likely causing corrosion, adherence of micro-organisms, accumulation of wind-blown organic and inorganic debris, as well as many forms of contamination that can occur in pumping and piping systems of the distribution network.

Schunk et al. (2018) describes fog water collection and its analysis by comparing water quality of fog water (from Morocco) using German (DIN), European (EN) Standards as well as the World Health Organisation (WHO) guidelines. Their work concluded that the inorganic water quality from fog meshes were found to be acceptable and better than local wells, excepting the first collection of water from fog meshes for several fog events. It must be noted that some of the fog events were also accompanied by rain. The study recommends that although fog water was considered safe for domestic consumption, further testing is required to rule out organic and microbiological contaminants. Kaseke et al. (2018) summarises an excellent study that concludes that both dew and fog-based methods of converting atmospheric water vapour into potable water will not replace conventional drinking and industrial water generation methods, and that more research is required to elaborately establish the true water quality of water obtained from these sources. We are of the view that water quality is also becoming increasing important in both conventional and non-conventional methods of water generation and certainly holds the potential for future research, especially considering atmospheric pollution in various parts of the world.

6 Challenges and Research Gaps

A comprehensive literature review on atmospheric water harvesting indicates that this topic has fascinated scientists from physical and biological sciences, as well as engineers for over the past several years, and continues to dominate the sustainable water technologies area. As outlined in other sections in this book, all aspects of water harvesting from the capture of fog nuclei to economic and water quality analysis provides adequate scope for scientific challenge and research. The significant point to note here is that there are several types of water harvesting devices, and no single standard design is universally considered as operationally efficient. From this point of view itself, research work on water harvesting is interesting and drives scientists and engineers to develop new technologies, by taking advantage of the science underlying solid–liquid–vapour interactions. In what follows, we have identified areas of research importance, when pursued could lead to the development and acceptance of atmospheric water harvesting devices and solar desalination incorporated into domestic and industrial applications. Techno-economic analysis and water quality are two areas of research that commonly runs through all methods for sustainable water harvesting considered in this book and is therefore emphasised as potential common research gaps for reasons of healthy living and well-being for all communities across the globe. Other areas of research specific to some of the promising passive methodologies are summarised below as future research areas for both fog-harvesting-based devices and thermal (solar) desalination methodologies.

6.1 Surface Engineering of Vapour Capture Substrates

Weather and climate changes affect the production and movement of fog, and hence passive atmospheric water harvesting poses challenges in yields during low relative humidity levels of the local environment. Designing fog capture surfaces that are highly receptive to fog nuclei, by creating "engineered surfaces," enable maximisation of fog nuclei capture by resorting to changes in surface wettability. These surfaces are generated by using chemical coatings whereby a thin uniform layer of a coating is adhered to a surface, or by machining structures such as micro- or nano-pillars (of varying cross section) on the surface using advanced fabrication methods involving lasers, lithography, and 3D printing. While chemical coatings seem relatively easier to apply, it has the downside of reliability. Water droplets leaving a substrate has been known to produce wall shear stresses that deteriorates coating to wall surface adhesion and hence advance the peeling of chemical coating sooner than anticipated. Coating peel off is further accelerated as the substrate is exposed to ambient weather conditions all times, causing surface temperature fluctuations and wind effects to adversely affect the weathering of chemical coatings on water capture surfaces. In this regard, mechanically fabricated engineered surfaces using micro-or nano-pillared surfaces have more reliability compared to chemically coated surfaces.

- Droplet deposition on micro- and nano-pillared surfaces continues to attract considerable research attention and will continue to be an important area of research for atmospheric water harvesting devices. Fog meshes that are used for water capture in passive fog harvesting uses meshes of various cross sections such as cylindrical, rectangular, as well as being available in different weave formats. Raschel weave has been commonly found to be employed for water capture, with meshes being highly flexible and of metallic and non-metallic origin. Incorporation of surface engineering of mesh elements in combination with a suitable weave will likely enhance water yields, while also offer the required stability and reliability of the fog meshes under adverse environmental conditions.
- Computational modelling of the deposition of condensed water vapour droplets and its accumulation growth rate is a research topic that needs attention. This modelling should consider the effect of ambient conditions and year-round weather to enable accurate prediction of water yield, to within acceptable accuracy levels. The computational modelling can be used to a priori assess the water yield at locations on the planet, before setting up elaborate water capture devices on large acres of land. This mathematical assessment using modelling will help in judicious selection of land sites while deciding on massive financial investments for setting up large-scale water capture devices.

6.2 *Adsorbent-Based Water Harvesters*

The adsorption-based systems broadly adopt a two-step process to deliver water from atmospheric water vapour. Water vapour is captured on the adsorbent surface at the first step; this adsorbed water vapour is driven off (or phase-transferred) using externally supplied energy in the second step. The whole process is 'green' in the way energy requirements to convert the adsorbed water vapour to liquid water is envisaged; in most cases, this energy is utilised from solar, or waste heat energy (from industry) sources. The system is much more complicated than a passive fog harvesting device, as there are more technological elements in the production of water. The performance of these systems depends on the materials chosen as adsorbents (shape, size, packing density), vapour and heat losses, and more importantly the adsorption kinetics. Adsorbents such as silica gel, zeolite, and metal organic foam (MOF) have been tested for enabling vapour capture with varying levels of effectiveness. In our view, promising research areas in this field include the following:

- Desiccant materials and surfaces, as well as efficient regeneration of water vapour from vapour captured desiccant surfaces
- Use of sustainable materials as vapour capture devices
- Energy consumption requirements including water yield.

6.3 *Dew Capture Using Selective Surfaces and Radiative Cooling*

The enhancement of radiative cooling using special purpose coatings on surfaces or the introduction of special materials as dew capture surfaces is made use of, for dew collection and generation of water in this passive methodology for sustainable water generation. The special coatings can be in the form of pigments that are developed for conditions that suit a certain environmental landscape, keeping in mind fog availability for appreciation of dew water yields. The main drawbacks in this method of water capture are the rate of radiative heat exchange between the surface and the night sky, ambient conditions and the wettability characteristics of the substrates. In our opinion, the essential controllable factor is the wettability of the substrates. Research in materials science engineering holds the potential for enhanced water capture using selective surfaces.

7 Thermal Desalination Using Solar Stills

In this section, we cover the environmental aspects and research challenges for thermal desalination, using solar energy which is also a sustainable or "green" methodology for water generation if embodied in the right way. Solar stills encompass many elements of the science of thermodynamics that affect the conversion of brackish water into potable water. The phase transformation process, wherein vapour generation is created and vapour condensation realised using suitable methods is an area worth pursuing research. There are two aspects that need attention; one is vapour generation, and the other is vapour condensation.

Accordingly, research literature classifies the approach as direct and indirect solar desalination methods. Popular solar desalination methods fall under these two categories: direct solar desalination using traditional solar stills or indirect solar desalination using membrane distillation, flash evaporation, multi-distillation, and so forth.

One of the drawbacks of solar desalination in terms of its effect on the environment is to do with the discharge of brine (or brackish water) into the ocean after it is subjected to solar-induced change of phase. This is considered a serious environmental (bio-diversity) pollution issue. The high density of the brine or brackish water retained in the solar collector when discharged into the ocean is known to affect marine feeding habitats and hence alter migratory patterns of ocean-living creatures, as well as seafood yield from the oceans for human population. For captive solar stills, the discharge of the highly concentrated brine is a waste disposal issue if it is not adjacent to a large waterbody such as oceans. Barring the drainage of the highly-concentrated saline residue, literature appears to evidence no environmental challenges.

Daily productivity of the solar still is dependent on several environmental factors such as ambient temperature, wind speed, solar radiation intensity. An increase in productivity (i.e. the yield of fresh water) has been observed with an increase in the ambient temperature as well as solar radiation intensity, while an increase in wind speed has mixed results reported on the productivity of the solar stills. Solar stills pose no adverse effects on the environment in which it operates. There are several versions of solar stills available in the open literature and hence it is difficult to single out a particular design that provides the required enhancements in the yield irrespective of the ambient conditions.

The productivity of traditional solar stills is very low, considering the heat losses, regardless of the effect of local environmental conditions on the performance of such systems. However, use of advanced materials in the construction of these devices including the modifications on the solar radiation absorbent surface has enabled enhancements in efficiency. There are myriads of techniques being tried out such as the use of a two-phase closed thermosyphon and sustainable materials such as natural fabrics (Chatterjee et al. 2023); graphene (Kim et al. 2018), hydrogels (Kim et al. 2018; Yin et al. 2018), bionic systems (Liu et al. 2017) and nano-photothermal materials (Wang 2018). It is clear that emergent desalination technologies are focussing on new membrane materials and membrane technology, mostly operating in the indirect desalination mode, which also includes the flash evaporation mode. In conclusion, we believe that solar-driven desalination is a very promising growth area addressing the decarbonisation aspects, and hence truly work as a sustainable, water generation method.

The research on the vapour transformation aspect mentioned earlier will provide performance enhancement of the solar still technology, which is essentially the main technological impediment in the widespread use of solar stills in regional and remote localities, especially for captive water capture stations. In terms of the economics of operation of solar desalination units, it is found to be cost-effective after accounting for the initial expenditure incurred in establishing such units. There are several locations on this planet that only provide brackish water, which is not potable; the water is unable to be used even for irrigation purposes, or other domestic uses within communities. Solar stills have a lot of potential in such localities and truly deserves a push for widespread use and commercialisation to address the water requirements of remote and regional communities. Solar stills don't rely on the availability of fog (as in fog harvesting devices) and so these devices can be implemented anywhere there is plenty of solar radiation availability. Of course, rainfall infested localities may not benefit from a solar desalination device. It is in this context that the use of hybrid methods (for example, a combination of fog harvesting and solar desalination) sounds promising, where the use of a combination of technology equipment can take advantage of local climate conditions to enable a year-round facility for sustainable water generation.

References

Atmospheric water generator market size, share & trends analysis report by product (cooling condensation, wet desiccation), by application (industrial, commercial, residential), by region, and segment forecasts, 2024–2030. Report ID: 978-1-68038-480-2

Chatterjee D, Kulshrestha T, Khandekar S (2023) Continuous vapour generation for thermal-desalination applications using a thermosyphon based heat localization strategy. Desalination 555:116492

Kaseke KF, Wang L, Kaseke KF (2018) Fog and dew as potable water resources: maximizing harvesting potential and water quality concerns. Geohealth 2(10):327–332. https://doi.org/10.1029/2018GH000171

Kim K et al (2018) Mesoporous three-dimensional graphene networks for highly efficient solar desalination under 1 sun illumination. ACS Appl Mater Interfaces 10(18):15602–15608

Liu Z et al (2017) High-absorption recyclable photothermal membranes used in a bionic system for high-efficiency solar desalination via enhanced localized heating. J Mater Chem A 5(37):20044–20052

Moghimi F, Ghodussi H, Asiabanpour B (2021) Is atmospheric water generation an economically viable solution? Clean Technol Environ Policy 23:1045–1062

Schunk C, Trautwein P, Hruschka H, Frost E, Dodson L, Derhem A, Bargach J, Menzel A (2018) Testing water yield, efficiency of different meshes and water quality with a novel fog collector for high wind speeds. Aerosol Air Qual Res 18:240–253. https://doi.org/10.4209/aaqr.2016.12.0528

Sun X, Amelung W, Klumpp E, Walk J, Mörchen R, Böhm C, Moradi G, May SM, Tamburini F, Wang Y, Bol R (2024) Fog controls biological cycling of soil phosphorus in the Coastal Cordillera of the Atacama Desert. Glob Chang Biol 30(1):17068. https://doi.org/10.1111/gcb.17068

Veloso JV, Boehm C, Schween JH, Vicencio Veloso J, Böhm C, Schween JH, Löhnert U, Crewell S (2024) A comparative study of the atmospheric water vapor in the Atacama and Namib Desert. Glob Planet Chang 232:104320. https://doi.org/10.1016/j.gloplacha.2023.104320

Wang P (2018) Emerging investigator series: the rise of nano-enabled photothermal materials for water evaporation and clean water production by sunlight. Environ Sci Nano 5(5):1078–1089

Yin X et al (2018) Macroporous double-network hydrogel for high-efficiency solar steam generation under 1 sun illumination. ACS Appl Mater Interfaces 10(13):10998–11007

Conclusions

K. Muralidhar, Malay K. Das, and Ramesh Narayanaswamy

1 Background

Clean water is defined as water within the appropriate temperature range; free of undesirable taste, smell, and color; and free of toxins, cancer-causing agents, pathogenic microorganisms, and other potential health risks. Clean water is essential for the maintenance of public health, quality of life, poverty alleviation, and the growth of the economy. Fresh water on the planet is less than 3% of the total water available, the rest being oceans. In freshwater, 30% is groundwater and 0.3% is surface water in the form of rivers and lakes. Atmospheric water vapor is around 0.001% of the total, but is sustainable enough to provide drinking water for the global population.

Despite this overall availability, water is abundant in the northern hemisphere of the planet and is highly seasonal and non-uniform in the equatorial and tropical regions. The worldwide distress map of potable water is highlighted in Chap. 1. According to the Planning Commission report (NITI Aayog report on *Composite Water Management Index*: 2.0—A Tool for Water Management, 2019), a country such as India is witnessing the worst water crisis in its history, though it has improved in recent years. More than 600 million people face high-to-extreme water stress. About three-quarters of the inhabitants do not have safe and ensured drinking water on their premises. About 84% of rural households do not have access to piped water facilities.

K. Muralidhar (✉) · M. K. Das
Department of Mechanical Engineering, Indian Institute of Technology Kanpur, Kanpur, Uttar Pradesh, India
e-mail: kmurli@iitk.ac.in

M. K. Das
e-mail: mkdas@iitk.ac.in

R. Narayanaswamy
School of Civil and Mechanical Engineering, Curtin University, Bentley, WA, Australia
e-mail: r.narayanaswamy@curtin.edu.au

R. Narayanaswamy et al. (eds.), *Sustainable Water Harvesting*, Springer Tracts in Mechanical Engineering, https://doi.org/10.1007/978-981-95-3373-2_11

Less than 30% of available water can be considered free of contamination. As far as water quality is concerned, India ranks quite low. The situation is alarming and needs to be comprehensively addressed to ensure the long-term developmental goals and aspirations of society.

2 Alternatives

Methods of deriving potable water from oceans, lakes, and rivers exist, for example, those based on reverse osmosis and flash evaporation. These technologies are energy efficient but the purification plants have to be located on-site, i.e., along the coast. Pure water produced has to be stored and transmitted to the end user, and this is often the source of difficulty. In addition, these technologies impose a significant carbon footprint, unless renewable energy is utilized to power the equipment. Thus, despite the availability of mature *centralized* technologies for clean water production, water distribution networks are expensive and incapable of providing fresh water at a reasonable cost, particularly in remote areas to satisfy the needs of the rising human population and increasing demand.

Well-established processes such as multistage flash technology (MSF) and reverse osmosis (RO) require large amounts of electric power to create vacuum conditions and high pressures, respectively. This results in a large carbon footprint for the desalination plant, a substantial amount of slurry, and sensitivity to the price and availability of oil. At present, the globally installed desalination capacities are contributing 76 million tons of CO_2 per year and it is expected to grow to 218 million tons by 2040. In 2019, global CO_2 emission is estimated to grow to 43.2 giga tons per year, 20% higher than the 2013 value of 36.1 giga tons per year. Moreover, owing to their fossil-fuel dependence, conventional desalination techniques are less suitable for decentralized water production.

Water as a commodity is increasingly becoming a scarce resource, motivating the need to identify and develop *decentralized* technologies for alternative sources of water to feed and sustain life. Two approaches, namely, the thermal technique (Chaps. 6, 8, and 9) and fog capture (Chap. 7) are described at length in the present text. The latter is limited by the fact that fog is formed under prescribed thermodynamic conditions of pressure, temperature, and humidity while thermal techniques have greater generality. Both methods should be seen as alternatives to the large-scale clean water production techniques currently in practice. However, the water production rate is quite small, and the focus of research is often to enhance water productivity, in units of liters of pure water produced in unit time and unit area. The previous chapters of this text outline configurations that can be used in this context.

Thermal techniques adopt an evaporation-condensation route to water purification (Chaps. 4 and 5). Thus, the text includes chapters on the thermodynamics of humid air (Chap. 3), fundamentals of evaporation (Chap. 4), and mathematical modeling of dropwise condensation on textured surfaces (Chap. 5).

Water vapor in humid air is now a recognized source of potable water (Chaps. 1 and 2). The method of extraction of water is by condensing it on a specialized surface that encourages the formation of liquid drops instead of a film; such devices are broadly classified as atmospheric water generators (Chaps. 5 and 6). The latent heat released from drops can be rapidly taken away with smaller temperature differentials than what would be required for liquid films. On the other hand, the formation of drops requires the surface to be hydrophobic to the condensing liquid. While most surfaces are naturally hydrophilic which encourages film formation, they can be textured to make them hydrophobic. The simplest approach is to use chemical coatings but on the scale of a practical application, physical patterning might be preferred in light of their mechanical strength and durability.

Impure and wastewater bodies can be treated to produce clean water thermally by a combined evaporation–condensation process. Solar-driven distillation systems are being developed, converting solar energy into heat for local evaporation. These are useful in applications as diverse as desalination of seawater, separation of liquids, thermal management, and wastewater treatment. The research challenge here is to enhance evaporation rates from heated water bodies jointly with maximized condensation rates. Modifications in the evaporator surface including wettability apart from other augmentation techniques become important. Some of these interventions include the use of heat pipes, thermosyphons, multistaging, and latent heat recovery systems combined with thermal energy storage approaches that allow for 24-h operation of the equipment. Newer methods described in the present text (Chaps. 8 and 9) describe the use of woven fabrics and photothermal materials for creating thin films of water that evaporate and the route of interfacial heating as a substitute for heating the entire body of water. The advantage of a woven fabric is the control on pore diameter and wettability, permitting water transport in one direction simultaneously with salt diffusion in the other. Overall, a low-cost water purification apparatus can be conceived, driven by low-grade heating at the evaporator end and a high-performance condenser rejecting heat to the ambient, effectively producing copious amounts of potable water.

Earth's atmosphere is the storehouse of abundant amounts of water present in the form of water vapor and droplets and can be explored for fresh water. The harvesting of atmospheric water vapor under extreme conditions is practiced by plants and insects and is a source of inspiration for design. It has led to the emergence of engineered surfaces for droplet capture and condensation in steam condenser design and evaporation in cooling of high heat flux electronic components. A profound weakness of the water collection approach from the ambient is the small rate of water produced, calling for strategies to accelerate the entire process.

Natural fog represents another untapped source of potable water that has not been harnessed effectively, especially in arid climatic zones. If we scrutinize nature, we find that there are several examples where plants and animals use textural and chemical features on their body surfaces to harvest and effectively utilize this precious resource. With such nature-inspired techniques, several research groups in the world are trying to optimize artificial fibers and surfaces for effective fog harvesting. Woven meshes of fibers are quite useful for this purpose if proper design of their wettability, length scales of the weave and meshes, and density of fabric material is achieved. Literature suggests that appropriate tuning of the wetting characteristics of the surfaces, reducing the wire radii, and optimizing the wire spacing will lead to more efficient fog collection. While some preliminary design pointers are available, the knowledge base is quite limited. Fog harvesting is not technically a thermal technique but its efficiency relies on several fluid dynamic processes and is of interest.

3 Thermal Techniques

Given the wastefulness of converting electricity to thermal energy, sustainability and the development of clean energy sources have emerged as significant challenges and the need of the hour. The utilization of renewable energy sources such as solar (and possibly, waste heat) are crucial in achieving these goals. Despite the widespread availability of solar energy, it remains an underutilized renewable resource. In recent years, the utilization of two-phase heat transfer in harvesting solar energy has taken center stage in various application areas. One approach adopted in an atmospheric water generator is where a Peltier element cools a hydrophobic copper surface below the dew point inducing dropwise condensation of moisture. The Peltier element is driven by a solar photovoltaic system, making the entire system dependent only on a renewable energy platform.

Solar-driven water evaporation systems have been developed, which involve the conversion of solar energy into heat for localized evaporation. These systems are useful for diverse applications such as desalination of seawater, energy generation, separation of liquids, pollution detection, seawater mining, and wastewater treatment. Various attempts have been made to increase the evaporation rates of heat and mass fluxes in these systems. These are closely connected with the characteristics of the evaporator surface including wettability that impact the solar-driven evaporation performance. Two-phase heat transfer is also perceived as one of the optimal solutions for dissipating high heat fluxes, as in the cooling of electronic equipment. While technological advances require heat exchangers to be compact with increased capacity for heat transfer rate per unit area, two-phase heat transfer provides a realizable approach to achieving this objective. Liquid-vapor phase change processes involve high heat flux and negligible change in temperature arising from the absorption or release of latent heat. In low-temperature systems that work away from the boiling point, latent heat is large in comparison to the sensible heat, allowing for the design of a heat exchange system with a smaller surface area.

In solar energy applications of water purification and desalination, evaporation should be accompanied by condensation wherein pure water is recovered as distillate and latent heat is rejected closer to ambient conditions. To this extent, understanding the evaporation of impure water at interfaces, in thin films and contact lines, filmwise and dropwise condensation, and droplet dynamics are of prime importance. Impurities may be in the form of dissolved salts or particulate suspensions. In a water droplet, evaporation rates including those at the contact line and condensation rates are substantially high and can contribute to a higher water production rate. Hence, research is increasingly focussed on these subjects, aiming at a full understanding of interfacial and phase change phenomena.

The phase change behavior of liquid films and droplets of salt water on a textured surface is complex, depending on the relative magnitude of various forces at appropriately small lengths and time scales. However, evaporation of an aqueous solution in the form of a liquid film from a hydrophilic surface would be advantageous as will condensation in the form of droplets over a hydrophobic surface. The interplay among capillary, viscous, gravity, and inertia forces gives rise to a complex multiscale interaction of the liquid medium with the solid surface, depending on its wettability. In one approach to controlling heat fluxes, surface engineering tools are in a state of development so that drops of various shapes and sizes can be realized. In another, a liquid film is allowed to spread within the pore space of a suitably manufactured porous substrate. The dynamics of the spreading process are governed by the interfacial properties of the liquid as well as the morphology and wettability of the bounding surface. Liquid films greatly increase the evaporation area and generate long three-phase contact lines at which evaporative mass fluxes are enhanced. Jointly, a patterned superhydrophobic surface is essential to sustain a dropwise condensation cycle.

A topic of fundamental research is quantifying evaporation rates of impure water or aqueous solutions in carefully engineered porous media where the liquid is present as a thin film with long contact lines. Jointly, condensation rates over a suitably textured surface will be determined in the presence of noncondensable gases such as air. The two steps can be integrated into a water purification system that will absorb solar energy, evaporate water in a porous structure of controlled porosity, condense the vapor separately, and reject heat to the environment. Additional attention is to be paid to removing the salt that has crystallized along the flow path. With carefully selected pore geometry on the evaporation side and the texture of the condensing surface, prototypes with enhancement in the rate of clean water production from impure water can be developed.

4 Solar Distillation

Clean water production and water purification are known to be energy-intensive processes. In addition, the concern for a clean environment has become vocal, and green technologies are increasingly being adopted. Thus, a shift from high efficiency electrical and pressure-driven approaches such as reverse osmosis is seen in favour of thermal technologies that can nevertheless, accommodate concentrated brine and function with unskilled labour. A body of literature has emerged, dealing with renewables such as solar energy with the clean water objective from wastewater and seawater. Prominently, low-temperature systems, specifically those in which water does not boil are attractive since they negate the need for specialized corrosion-resistant materials. Such systems carry out an evaporation-condensation cycle with solar energy as the source and the normal ambient as the sink. Since the rates of evaporation and condensation tend to be slow, water productivity is low. The goal of the research is then to develop enhancement techniques that will make water production a commercially viable technology.

Solar still is an old technology used to distill water with dissolved and suspended impurities. It has received tremendous attention over the years; several research groups have carried out extensive studies in order to improve the productivity of solar stills. Despite these efforts, there has been only a small improvement in the productivity and efficiency of stills. Often, the added complexities brought about by the basic design have shown that they are not economically viable solutions. Several unconventional designs like triangular stills, tubular stills, hemispherical, and multiple slopes have been tried without achieving a clear advantage. Additional usage of phase-change material to manage the variability of solar energy, copper and aluminum oxide nanoparticles to improve the heat transfer coefficient, reflectors, and vibratory harmonic effect have shown an increase in output but with an increase in complexity and cost. New approaches include re-using the latent heat release that has shown promise. Nevertheless, the production output of such units has typically hovered around 6–8 l/m^2-per (24 h) day.

Parametric investigations point out that solar intensity, wind speed, and environmental temperature have significant effects on the production rate; however, these cannot be controlled, being meteorological parameters. Others include inclination angle, insulation thickness, and water depth. The inclination angle can be optimized according to the latitude angle of the location. The minimum water depth to avoid drying out is preferred and an increase in insolation increases the output. In the limit, the water interface with air alone may be heated, intensifying the overall evaporation step.

The incorporation of heat pipes with evacuated or vacuum-type solar collectors has been investigated for solar still configurations on a few occasions to heat the water in bulk. Heat pipes have been used in combination with phase-change material and thermoelectric devices to power pump-like devices but the improvement in performance relative to the additional cost has not been satisfactory.

Another significant issue that has gained attention is brine disposal management. Disposal of concentrated brine from the distillation plants along the sea coast has been the obvious choice but can have a significant effect on the neighboring marine life. Chemicals like copper and chlorine are added at various stages in order to prevent bacterial growth and corrosion ultimately accumulating in the food chain. The desalination plants operating on brackish water far inland have significant issues in managing their waste. The focus should now be on developing a zero liquid discharge (ZLD) process based on renewable energy sources. The mining of concentrated brine for precious metals is research in progress, whose contribution to the financial viability of the solar distillation technologies is a distinct possibility. The ZLD concept is one where the entire industrial and municipal wastewater is reused after recycling, solid waste is confined and no liquid is discharged into clean water.

Decentralized pure water production is important for regions that have neither the infrastructure nor the economic resources to run MSF and RO plants that are sufficiently distant from large-scale production facilities to make pipeline distribution prohibitive. Solar desalination has thus emerged to be a better alternative to address the issue of small-scale portable drinking water supply for decentralized, off-grid, and arid regions. Solar-powered RO using photovoltaic technology is the most mature and widely used desalination system having over 65% of market share, as far as renewable energy use is concerned. This is closely followed by electrodialysis using PV technology. Thermal technologies circumvent the efficiency limitations imposed by the second law of thermodynamics but require careful material selection and further research to establish cost-effective processes.

5 Surface Texturing

Specialized surfaces play an important role in accelerating phase change phenomena. Textured surfaces are required to be hydrophobic to encourage droplet formation during condensation and assist drainage. Surface texturing is often achieved by uniform chemical coatings in a laboratory but scaling up the process is a challenge. In addition, chemical coatings are known to wear out due to the repeated motion of liquid droplets over them. An alternative is to create physically textured surfaces, for example, a distribution of micron-sized pillars that create a pattern. Droplet interaction and evaporation as well as condensation of water vapor over such patterned surfaces are of interest. An important goal of research is to improve the predictability of interfacial phenomena for both evaporation and condensation via carefully conducted experiments juxtaposed against mesoscale simulation.

Hydrophilic porous media increase the surface area of evaporation and create thin films and contact lines of high mass flux. Hydrophobic surfaces and mesh-like coils sustain dropwise condensation in which a high drainage rate of potable water is sustained. Related interfacial and phase change phenomena require careful analysis.

While commonly available surfaces tend to be hydrophilic, hydrophobic surfaces relative to water may be made on a laboratory scale using chemical texturing as well as physical texturing. The preparation of super-hydrophilic and super-hydrophobic surfaces requires special attention. Substrates such as glass and PDMS are easier to texture but cannot be used in heat transfer applications involving condensation and evaporation where high thermal conductivity of the substrate material is essential. Copper may be chemically textured resulting in protrusions in the form of microflowers that display superhydrophobicity and low contact angle hysteresis.

For physical texturing, an air-cooled CO_2 ytterbium fiber laser micromachining system has been used by various authors producing up to 1200 dots per inch engraving. A comparable electric discharge machine can be used for texturing, for example, with a CNC wire-cut electro-discharge (EDM) system. Laser texturing is generally found suitable for lower aspect ratio pillars and the wire-EDM method produces higher aspect ratio pillars on copper. Surface wettability can now be tuned by adjusting the power, translational speed, and pulsing frequency of the laser.

In summary, evaporation, condensation, and fog capture require surface wettability that is tuned to the process in progress and related design objectives. Accordingly, hydrophilic, hydrophobic, and superhydrophobic surfaces may all be required to enhance water productivity as a whole.

6 Fog Harvesting

Fog, for the present discussion, is a collection of suspended water droplets in air, that was once seen as a parasitic loss of visibility, and fog collection as water, an improvement to fog dissipation. With time, it is now seen as a source of pure water, provided fog can form naturally and repeatedly, under the available thermodynamic conditions. On a field scale, fog carried by air is collected by one or more layers of cloth mesh placed well above the ground. The fog collector is placed perpendicular to the direction of the prevailing wind so that water particles coalesce by the impact of fine water droplets on the polypropylene mesh netting filament. As more and more of these particles collect on the mesh, they slide down and get collected. Since a thermal technique essentially provides a water distillate, water from fog is less pristine and may require additional processing before consumption.

Water capture by meshes from fog carried by the main flow is a simple technology to set up, though the mechanics of droplet capture, coalescence, and drainage are multifactorial, requiring both laboratory and theoretical investigations. Experimental studies are often carried out in a fog tunnel where the moisture content and drop size can be systematically varied. Woven meshes of fibers are quite effective for this purpose since their wettability, length scale of the weave, and density of fabric material are controllable. An optimized fog capture process can be implemented on a field scale, but the literature shows that such studies are presently in progress.

7 Human Health

The techniques under discussion for extracting pure water from moist air are equally applicable in an unrelated but important domain. In a health-related context, it is now known that Covid-like viruses are transmitted in the atmosphere in the form of aerosols and water droplets, particularly indoors and in air-conditioned venues such as classrooms, conference halls and shopping malls. Hence, a fundamental study on remediation will include examining physical phenomena occurring on the scale of micron-scale water drops. These phenomena include evaporation, condensation, droplet interaction with physically and chemically textured surfaces, deposition on mesh-like surfaces, droplet coalescence, thermal effects arising from surface as well as atmospheric heating, and possible drainage. In view of the small size of the droplets, interfacial phenomena become prominent and influence thermodynamic phase behavior and transport. The subject under discussion will lead to the design of interventionist strategies that help remove the virus from the air, at the cost of transferring latent heat to a cooler surface, thus improving air quality in interior spaces.

8 Open Questions

Research on decentralized methods of pure water collection is very much in progress. The overarching goal is to improve water productivity on a liters-per m^2-per-(24 h) day basis, 10–12 units being the minimum acceptable. The second consideration is continuous water production round the day, while consumption of actual grid electricity should remain minimal. Enhancement techniques include promoting dropwise condensation and drainage, latent heat recovery by multi-staging, interfacial heating, and increasing three-phase contact lines in a woven fabric. Decoupling evaporation and condensation steps will help maximize their respective performances. Fog harvesting devices have to be designed to suit the location of fog formation, specifically the drop size distribution and airspeed. It is necessary to highlight here the scaling consideration, namely, the need for pilot-scale studies over and above those on the laboratory scale so that peak performance seen in a small apparatus is reproduced in a larger prototype. Viewed as a phase-change heat exchanger, the distillation setup can be evaluated for the destruction of exergy. Despite the technology being developed for remote areas, several components such as hydrophobic surfaces, wicks, and fabric require proprietary technology that is cost intensive. Given conflicting demands that arise from these intermediate steps, the overall process will have to be optimized for performance as well as cost.

Some of the fundamental research questions that emerge in the field of sustainable water harvesting are listed below:

1. Droplet shapes, stability, and dynamics during interaction with wires and mesh screens.
2. Interaction of water drops with electrically charged mesh screens.
3. Interaction of droplets in a turbulent flow field with and without mesh effects.
4. Drop shapes, spreading, movement, and instability over physically and chemically textured surfaces.
5. Evaporation of three-phase contact lines in surface textured porous media.
6. Effect of capillarity on evaporation from pore spaces.
7. Fundamentals of interfacial phenomena and heat transfer during evaporation and condensation in clear media, membranes, and water-saturated fabric.
8. Importance of surface texturing and pore space design in maximizing evaporation and condensation rates with minimal temperature penalty.
9. Hybrid approaches combining membranes on a medium scale with thermal desalination of the residual brine.

10. Manufacturing options for specialized surfaces, wicks, and fabric.
11. System integration studies in which conflicting factors preclude joint optimization of component and device performance.

Discussion in the previous chapters will serve as a useful starting point for simulations and experiments, and for pursuing the research directions outlined above.

Bibliography

Fosso-Kankeu E, Alili AA, Mittal H, Mamba B (2023) Atmospheric water harvesting: development and challenges, water science and technology library book 122. Springer Nature, New York

Michaelides EE (2021) Energy, the environment, and sustainability. CRC Press (Taylor and Francis), UK

Mukherjee A (2022) Riverine systems: understanding the hydrological, hydrosocial and hydro-heritage dynamics. Springer Nature, New York

Pietz DA, Zeisler-Vralsted D (2021) Water and human societies: historical and contemporary perspectives. Springer Nature, New York

Rosso R (2024) Five easy pieces on water: essentials of water science explained by an engineering scholar. Springer Nature, New York

MIX
Papier aus verantwortungsvollen Quellen
Paper from responsible sources
FSC® C105338
FSC
www.fsc.org

If you have any concerns about our products,
you can contact us on
ProductSafety@springernature.com

In case Publisher is established outside the EU,
the EU authorized representative is:
Springer Nature Customer Service Center GmbH
Europaplatz 3, 69115 Heidelberg, Germany

Printed by Libri Plureos GmbH
in Hamburg, Germany